WINE AND THE VINE

The products of the grape vine are amongst the most diverse of any agricultural crop. This is not only the result of differences in geology and climate; it also reflects the labour of countless generations of vine growers and wine makers, each set in their own distinctive human context. It is this interaction of people and environments, creating specific cultural identities, that lies at the heart of any understanding of the emergence and spread of viticulture and wine production.

Wine and the Vine provides an introduction to the historical geography of viticulture and the wine trade from prehistory to the present. Throughout, the rich symbolic and cultural significance of wine is related to its evolution as a commercial product. The book discusses both the numerous symbolic roles assigned to wine and the vine by people of different religions and also the internationalisation of wine production and marketing. Particular themes which form a focus for analysis include the role of the Roman Empire in influencing the spread of viticulture; the importance of political factors in determining the contours of the medieval wine trade; the use of wines and vines as social symbols throughout history; the global spread of viticulture under colonialism and imperialism; and the role of transnational corporations in the modern wine industry.

Unlike previous books, *Wine and the Vine* sets the development of viticulture and the wine trade within a particular theoretical framework, considering them as one expression of changing interactions between social, economic, political and ideological structures.

Tim Unwin is a Reader in Geography at Royal Holloway, University of London.

WINE AND THE VINE

An Historical Geography of Viticulture and the Wine Trade

Tim Unwin

Routledge
Taylor & Francis Group

LONDON AND NEW YORK

First published 1991
Paperback edition first published in 1996
by Routledge
2 Park Square, Milton Park, Abingdon,
Oxon, OX14 4RN

Simultaneously published in the USA and Canada
by Routledge
270 Madison Ave, New York NY 10016

Routledge is an imprint of the Taylor & Francis Group

Transferred to Digital Printing 2004

British Library Cataloguing in Publication Data
A catalogue record for this book is available from the British Library

Library of Congress Cataloguing in Publication Data
A catalogue record for this book has been requested

ISBN 0–415–14416–7

For those who have shared their wisdom and their wine with me

CONTENTS

Place-names: a note	x
List of figures	xi
List of tables	xiii
Preface to the paperback edition	xiv
Preface	xxiii
Acknowledgements	xxv

1 THEMES IN THE HISTORICAL GEOGRAPHY OF VITICULTURE | 1
The historical geography of viticulture | 3
Individuals and structures: a theoretical perspective | 5
Symbols and society | 8
Production and exchange | 10
Cultural landscapes of the vine | 14

2 VITICULTURE AND VINIFICATION | 27
Ampelography and the vine | 29
Vinification: chemistry and the human milieu | 46
Elevage: the ageing and maturing of wine | 54
Wine traditions: ancient and modern | 57

3 THE ORIGINS OF VITICULTURE: SYMBOLS AND MYSTERIES | 58
Sources and evidence | 61
The origins of viticulture and vinification | 63
The prehistoric symbolism of wine and the vine | 77
Wine and the vine: symbols of fertility and luxury | 91

4 WINE IN THE GRAECO-ROMAN ECONOMY | 94
Viticulture and the wine trade in the Greek economy | 95
Viticulture and vinification in the Roman world | 101

vii

CONTENTS

The spread of viticulture in the Roman world   110
The Roman wine trade   119
The Greek and Roman contribution to viticulture   131

5 VITICULTURE AND WINE IN THE EARLY MIDDLE
AGES   134
Christianity and a divided Empire   135
Wine and the vine in Christian symbolism   139
Viticultural continuity and survival   144
The Islamic conquests and their effects on viticulture   150
The extent of viticulture c. AD 1000   155
Wine in a Christian world   165

6 MEDIEVAL VITICULTURE AND THE WINE TRADE   166
Viticulture and wine production 1000-1450   167
The market for wine   175
The medieval wine trade   180
Political alliance and the medieval wine trade   192
The fourteenth century: crisis and change   202

7 WINE IN THE AGE OF DISCOVERY   205
The creation of a new world, 1500-1750   206
Discovery and conquest: the vine in Latin America   214
European viticulture in the fifteenth and sixteenth centuries   221
The structural reorganisation of viticulture and the wine trade   231

8 CAPITAL IN THE SPHERE OF PRODUCTION   233
Alcohol and alcoholism in seventeenth century Europe   234
Production in new environments   242
The creation of new wines   252
The seventeenth century: new wines and new attitudes   266

9 CRISES AND EXPANSION: THE RESTRUCTURING OF
VITICULTURE IN THE NINETEENTH CENTURY   270
Quality or quantity: the polarisation of demand   271
The nineteenth century crises in European viticulture   283
*The spread of viticulture to the new lands of Australia,
New Zealand and California*   296
The crises and restructuring of global viticulture   306

CONTENTS

10 THE DOMINANCE OF CAPITAL IN THE TWENTIETH
   CENTURY: DEMARCATION, SECTORAL INTEGRATION
   AND THE CREATION OF FASHION   310
   Scandal and demarcation: the delimitation of privilege   311
   From wine merchant to global corporation: the restructuring
   of the retail wine trade in Britain   325
   Capital, technology and fashion   343
   The future of capital   354

11 CONCLUSION   357

   Appendix: historic wine measures   364
   Bibliography   367
   Index   394

PLACE-NAMES: A NOTE

The spelling of place-names presents very serious problems, since not only are places spelled differently in different languages, but the spelling of such names has also changed through time. In general, the names currently in use in the country where a place is located have been preferred, except where there is a very widely used English language alternative, and then only for a capital city. Thus Rome, rather than Roma has been preferred, but Marseille and Lyon have been left in their French spelling. Belgian names, with either French or Flemish alternatives, have been rendered in English, as with Ghent and Bruges. Coastlines and national boundaries, where shown are those existing at date of publication.

LIST OF FIGURES

1	Cartoon by Blachon	2
2	Cartoon by Barbe	11
3	Pommard and Beaune from the Côte de Beaune	16
4	Pinhão at the junction of the rivers Douro and Pinhão	20
5	Robert Mondavi winery, Oakville, Napa Valley	23
6	The global distribution of viticulture	35
7	The composition of a grape	36
8	The life cycle of *Phylloxera vitifoliae*	41
9	The traditional method of treading grapes: the vintage at Quinta da Foz, Pinhão, northern Portugal	51
10	The construction of a modern winery, with stainless steel fermentation tanks, Cave des Hautes Côtes, Beaune	52
11	The origins and spread of viticulture in south-west Asia and the eastern Mediterranean	65
12	Relief from royal palace at Nineveh illustrating Ashurbanipal and his queen drinking under a bower of vines	67
13	Vintage scene depicted on the walls of the tomb of Khaemwese (Thebes No. 261) c.1450 BC	70
14	Vintage scenes from Dynastic Egypt	72
15	Banquet scene from the tomb of Nebamun (Thebes) c.1450 BC	74
16	Carved image of a Hittite god at Ivriz	76
17	Statue of Bacchus with a personification of the vine (AD 150–200)	88
18	Roman statue of Priapus (first–second century AD)	92
19	Red Figure Kylix (480 BC) depicting wine drinking	100
20	The spread of viticulture in Gaul	114
21	Roman wine amphorae	121
22	The distribution of Dressel 1 amphorae in Europe and the western Mediterranean	122
23	A Roman bar at Ercolano (Herculaneum) (AD 79)	125
24	Marble relief illustrating the transport of wine by ox caft	130
25	Vintage scene from the roof of the Church of Santa Costanza, Rome	143

26 Vine cultivation in an illustration from the Khamseh of
    Nizami (1595)   156
27 The distribution of vineyards along the rivers Rhine and
    Mosel before 1050   163
28 Illumination for the month of September, showing the vintage
    at the Château de Saumur   170
29 Woodcut of Virgil and Maecenas, showing different types of
    vine training   172
30 Treading the vintage from the early fourteenth century manuscript
    of Queen Mary's Psalter   173
31 Vintage scene from a Flemish Book of Hours of the Blessed
    Virgin (c.1500)   174
32 Main European wine trade routes c.1250   183
33 The global spread of viticulture, 1500–1800   219
34 Early sixteenth-century miniature from a Bréviaire illustrating
    the arrival of wine at Bruges   228
35 The village of Hautvillers   260
36 Port lodges at Vila Nova de Gaia   267
37 The spread of phylloxera in France   288
38 The vintage in California: wood engraving after a drawing
    by Paul Frenzeny   305
39 Advertisement for the wines of Côtes du Roussillon and Côtes
    du Roussillon Villages   351

LIST OF TABLES

1 Subgenera and main species of the genus *Vitis*   30
2 Main vine varieties used for wine production, classified by species 32
3 Main vine pests and diseases   38
4 Wine exports from Gascon ports during the fourteenth century   199
5 Wine exports from Bordeaux during the fifteenth century   201
6 The 1855 classification of the wines of the Gironde   281
7 Alcohol consumption in the United Kingdom, 1930-88   336
8 Changes in the consumption of wine, spirits and beer, 1976-86   343

PREFACE TO THE PAPERBACK EDITION

In the five years that have passed since the first publication of *Wine and the Vine*, the global wine industry has seen considerable change. Levels of wine consumption have continued to fall in the old producing countries of Europe, while the importance of New World producers such as Chile, Australia and South Africa has continued to rise in the market place. Technological changes in vine cultivation and wine making have also played a significant role in transforming the wines that we drink, with international wine makers now travelling the globe and using their technological expertise to produce new wines, often specially designed to suit the tastes of specific north European markets.

These changes have been matched by the publication of an increasing amount of research on the history of wine, and of major new texts on viticulture and oenology. In the historical sphere, there have been detailed regional case studies such as that of Piemonte edited by Rinaldo Comba (1992, 1994), new accounts of specific countries and periods including Loubère's (1990) *The Wine Revolution in France*, and more obscure yet beautifully illustrated works such as the collection of papers by the geographers Alain Huetz de Lemps, Jean-Robert Pitte, Xavier de Planhol and Philippe Roudié (1990) entitled *Les Vins de l'Impossible*. Likewise, Pierre Spahni (1995) has written an invaluable survey of the contemporary international wine trade, providing not only a broad overview of recent changes, but also detailed analysis of specific import markets and exporter profiles. On the viticultural and oenological side, major new texts include Boulton, Singleton, Bisson and Kunkee's (1996) comprehensive *Principles and Practices of Winemaking*; Ron Jackson's (1994) ambitious *Wine Science*, Ough's (1992) *Winemaking Basics*, John Gladstones' (1992) somewhat controversial *Viticulture and Environment*, and the second volume of Coombe and Dry's (1992) excellent *Viticulture*. As well as these, 1994 saw the publication of Jancis Robinson's edited *The Oxford Companion to Wine*, which provides a wealth of readily accessible information on all aspects of wine, from its origins and spread to the complexities of its contemporary production and legislation. While the main arguments of the first edition of *Wine and the Vine* remain little affected by such research,

xiv

the publication of this paperback edition provides an opportunity to add further detail and clarification with respect to certain of the themes explored, and also to draw attention to some of the more pertinent recent literature. For ease of publication, this is undertaken in this extended Preface, with cross references where appropriate to the pages of the original text.

Chapter 1 provides a broad framework within which to view the historical geography of viticulture and the wine trade. Iain Stevenson (1991: 213), in reviewing the original edition, argued with respect to the section on individuals and structures that 'This diversion seems to serve little purpose save perhaps the negative one of scaring off casual readers', and he recommends that the general reader skips pages 5–8 so that they can go on to 'find much of accessible interest in the book'! This is a sentiment expressed elsewhere in reviews of the book that appeared in the popular press, and it might well therefore be an appropriate strategy for those wishing simply to pick up a few historical anecdotes about wine. However, as Stevenson also notes, *Wine and the Vine* is not just a descriptive account of themes in the history of wine (see Johnson, 1989). Rather it seeks to do two specific things: first, it uses an analysis of wine to explore the complex interplay between culture and the physical environment throughout history; and second, it seeks to situate this within a particular formulation of the way in which societies function. Both of these concerns lie at the heart of contemporary geographical practice (Unwin, 1992a), and an understanding of the broad framework provided in Chapter 1, with its focus on the interaction between social, economic, political and ideological structures, is essential for a full appreciation of the subsequent chapters (see also Unwin, 1994a).

Chapter 2 gives a basic summary of some of the key aspects of viticulture and oenology necessary to comprehend the physical and chemical reasons for developments in vine growing and wine making in the past; it is by no means meant to be a comprehensive guide to contemporary wine making (for major new texts see Coombe and Dry, 1988, 1992; Jackson, 1994; Boulton *et al.*, 1996). Recent viticultural and oenological developments in the New World, and particularly in Australia, have meant that some aspects of this Chapter do not now reflect the best of current practice. In particular, four additional points could well be emphasised; first, the increasing importance of canopy management, which is now widely used to alter the exposure of leaves and fruit to the sun (p. 33) (see Smart and Robinson, 1991); second, the significant potential of genetic engineering techniques in vine breeding which are being used to introduce particular characteristics into vine stocks (see Mullins, Bouquet and Williams, 1992); third, the growing trend towards organic and even biodynamic viticulture (Dutel, 1990; Rousseau, 1992); and fourth, the greatly increased knowledge of chemical and biochemical processes involved in wine making which means that modern wine makers now have an enormous array of different options available to them in designing their wines (for scientific background see for example Jackson,

1994; for the consumer context see Millson and Hawes, 1994).

Certain details of wine-making processes in Chapter 2 could also have been expanded and clarified further. Thus in the discussion of fermentation (pp. 45, 47, 346), the distinction between inoculated and spontaneous fermentations could have been made clearer. Traditionally, fermentations were spontaneous, but in recent years wine makers particularly in the New World have increasingly used selected cultured yeasts to inoculate the must so that the fermentation proceeds in a predictable fashion. The use of active dry cultured yeasts is merely one of the more recent technological developments to enable such yeasts to be delivered in an appropriate form to the winery. Another point to note about yeasts is that it is now widely recognised that the key wine yeast *Saccharomyces cerevisiae* is most commonly found in cellars and on winery equipment rather than in the vineyard (Martini, 1993). With reference to methods of preventing bacterial spoilage and oxidation (pp. 54, 346), it should be emphasised that legislation in most wine producing areas limits the range of chemical additives permitted. Most countries only allow sulphur dioxide and sorbic acid, and serious wine makers are increasingly trying to reduce the amounts used to an absolute minimum, preferring instead to concentrate on improved sanitation and other techniques to optimise the stability of wines. The discussion of *terroir* in Chapter 2 could also have been extended to include mention of a crucial distinction between Old World and New World wine making. In the New World, many wine makers draw their grape supplies from numerous different locations, seeking to optimise the fruit characteristics that they want for their wines. In such circumstances, the notion of *terroir* is meaningless, and most of the key features of the wine will be determined in the processes of wine making and maturation. However, in the Old World, where grape growing and wine making is much more integrated, and individual wines are often made from specific plots of vines, then it is possible to identify particular characteristics in wines that derive from the environments in which the grapes are grown (Bessis, Leneuf and Fournioux, 1994; Riou, Morlat and Asselin, 1995). Indeed, there are strong grounds for arguing that further advances in the quality of New World wines may well result from increased attention being paid to the identification of the optimum growing conditions for particular vine varieties (Elliott-Fisk and Noble, 1992).

Turning to Chapter 3, recent advances in the analytical techniques used by archaeologists have provided further evidence concerning the origins of viticulture and wine making. In particular, the use of transmission and diffuse-reflectance FT-IR spectrometry has enabled a red stain on jars from the Godin Tepe site in western Iran, dating from the Late Uruk period (*c.* 3500–2900 BC), to be identified as containing a carboxylic acid that is most probably tartaric acid, and thus indicative of a grape product (Michel, McGovern and Badler, 1993). The shape of the jars, with a narrow mouth and elongated neck, suggests that they were filled with a liquid, and this is thus the earliest good evidence for wine making anywhere in the world. However, the gradual process of grape selection, by which fruits best suited for wine making were brought into cultivation, and then the acquisition of the skills that would consistently turn them into a flavoursome beverage, are likely to have preceded this date by some considerable time (p. 63). The subsequent genetic evolution of different grape varieties, and their regional distribution in antiquity still remain subjects of considerable debate. Although numerous mentions of different grape varieties are cited by classical writers such as Columella and Pliny, it is extremely difficult and hazardous to try to equate these with modern varieties, and it is hoped that future DNA (deoxyribonuleic acid) based research will shed light on the genetic origins of different modern varieties of *Vitis vinifera* (Thomas, Cain and Scott, 1994). With reference to the Greek and Roman world, further archaeological research and textual analysis is increasing our depth of knowledge about classical viticulture and wine making. For an overview of ancient Greek agriculture, Isager and Skydsgaard's (1992) recent book provides a good introduction, and examples of detailed research on Roman landscapes and wine estates can be found in the edited volumes by Barker and Lloyd (1991) and Frenzel (1992). Rathbone's (1991) comprehensive analysis of rural society in 3rd century AD Egypt also provides fascinating insights into the practices of viticulture and wine making and their significance to the Egyptian rural economy at this time.

For the early medieval period, my tentative critique of the role of Christianity in the maintenance of viticulture and wine making (pp. 144–8) has been extended in Unwin (1992b), which draws attention to the very low overall consumption of wine in religious ceremonies at the end of the first millennium AD. An analysis of the evidence for Saxon and early Norman viticulture in England has also been published in Unwin (1990), wherein it is argued that the evidence for wine making in England prior to the Norman conquest is much less strong than is often argued. Much research on medieval wine continues to be undertaken in Europe (see for example Comba, 1990), but little of this has been translated into English (although see Schenk's 1992 study of Franconia), and there remains a need for further broad syntheses of this expanding literature.

With reference to Chapter 7, Nitz (1993) has edited a major collection of works by geographers on the early-modern world-system, which includes several mentions of the role of the wine industry during this period. Moreover, the section of this chapter on the development of viticulture in Latin America (pp. 216–20) provided only a partial summary of the spread of wine making in this region, and a much more comprehensive account, drawing on a range of original Spanish sources, can be found in Dickenson and Unwin (1992). This not only explores the botanical evidence, but also examines the effects of alcohol on traditional social and ideological customs of the indigenous populations of the area (Gruzinski, 1979). Detailed historical accounts by Brown (1986) on Arequipa, and Cushner (1980) on Peru should also be referred to.

New and revised books on the major wine producing regions of the world discussed in Chapter 8 appear with great regularity (see for example Liddell and Price, 1992; Peppercorn, 1991), and most include some comment on a region's history. For the history of port wine (pp. 262–6), though, the definitive account is that provided by Conceição Andrade Martins (1990), which documents in great detail the development of the industry from its origins to the end of the 1980s. This highlights the periodic cycles of crisis and expansion associated with the emergence of capitalist relations of production in the wine industry. Briggs' (1994) historical account of Haut-Brion gives considerable detail of the role of individual families such as the Pontacs in the development of the New French Clarets, and adds useful depth to supplement the summary description of Bordeaux on pages 256–9.

Chapter 9 focuses on the physical crises of European viticulture and the expansion of the wine industry in the New World in the nineteenth century. Both are subjects which have received considerable recent attention. Pouget's (1990) examination of the fight against phylloxera in France is an excellent survey of the various methods used to try to eliminate phylloxera, and there is a definite need for further such analyses for other European countries. The key factor influencing the sudden emergence of phylloxera in Europe appears to have been that, before the late 1850s, trade in American vines was in the form of cuttings and seeds, but thereafter rooted plants were also sold, and it was these that carried overwintering eggs. The widespread destruction of vines by phylloxera on the west coast of North America in the last few years is once again costing vast amounts as vineyards are having to be re-planted on new rootstock; as Morton (1993: 9) comments 'problems of the past still plague growers today. In the field of rootstocks, there has been little progress, especially in the U.S. where past lessons are being re-learned at great expense.'

Much detailed research on the history of the Australian wine industry in the nineteenth century has recently appeared, with Bell's (1993, 1994) studies of the South Australian wine industry being particularly useful. Norrie (1990) also provides a survey of the vineyards of Sydney, and Aeuckens et al. (1988) have published an account of early wine-makers in the Barossa. The more recent development of Australia's wine industry in the twentieth century has been summarised by Halliday (1995).

Finally, Chapter 10 sketches some of the main features of the wine industry in the twentieth century, focusing on three themes in particular: vineyard demarcation, sectoral integration, and the creation of fashion. In all three areas there have been considerable developments over the last five years. Debates over the use of vineyard demarcation systems, particularly in the New World, have continued unabated. In America, the introduction of legislation in the late-1970s led to the creation of a number of Approved

Viticultural Areas during the 1980s, but in Australia attempts to create controlled appellations, as with Margaret River in Western Australia and Mudgee in New South Wales, have so far been largely unsuccessful (Stevenson, 1988). In Europe, Italy (pp. 322, 324) introduced a new wine denomination code in 1992, supplementing the DOCG and DOC regulations with a new *Indicazione Geografica Tipica* (Fregoni, 1992). Spahni's (1995) analysis of the international wine trade expands in great detail on the bare summary of sectoral integration touched on in Chapter 10. In the retail trade, the increasing power of the supermarkets (p. 341) across Europe is a theme which should be emphasised; even in France, it is estimated that 42 per cent of total national wine sales were sold in supermarkets in 1990 (Gille, 1992; Spahni, 1995). Another key feature of the international wine trade over the last decade has been the strength of new wine producing areas in the market place; the boom in Australian wine exports to Europe is now, for example, being challenged by Chile, and, with the collapse of apartheid, South Africa is once again a wine exporting force to be reckoned with. Turning to the section on fashion (pp. 347–53), much more recent research has been undertaken on the role of advertising in the wine industry (for a review see Unwin, 1992c). Moreover, the rising power of the anti-alcohol lobby is reflected in the highly controversial Evin Law in France in 1992, which severely restricts the advertising of all alcohol products, including wine (Casamayor, 1993).

To these themes should be added three others which are becoming increasingly significant towards the end of the twentieth century. First, there is the observation that while European countries are seeking to reduce the area of their vineyards in the face of falling wine consumption, many countries in the southern hemisphere, notably Australia, are embarking on massive new planting schemes. The early 1990s have seen considerable debate in the European Commission and Parliament concerning the optimum strategy to be adopted to reduce wine production (for a review see Unwin, 1994b), and as yet no definite solution has been adopted. Second, the last five years have seen increasing debate over the health effects of wine consumption. The anti-alcohol lobby has become increasingly powerful in Europe and North America, and remains a source of concern for the alcohol beverage industry. However, the health implications of moderate wine drinking remain hotly debated (Delin and Lee, 1992), and recent Danish research (Grønbæk et al., 1995) even suggests that, for those who drink up to five glasses of wine a day, the risk of dying steadily decreases with an increasing intake of wine. Finally, some mention should also be made of the possible future influence on the wine industry of research currently being undertaken in viticultural and oenological laboratories throughout the world. Key research underway, for example, includes the work being directed by Pat Williams at The Australian Wine Research Institute on the factors influencing the flavour of grapes and the development of techniques that might help grape growers and wine makers exercise greater control over wine flavours. A major technological change of a very different nature has been the advance in computer networking, and the very rapid expansion in wine related information that is now available, for example, on the World Wide Web (for an introduction to the sites currently available, see Hawkins, 1996). *Wine and the Vine* puts these contemporary changes into a historical perspective, and explores the social, economic, political and ideological context within which wine has become part of the culture of those who drink it.

While *Wine and the Vine* owes much to colleagues and friends mentioned in the Preface to the first edition, my subsequent research has been closely influenced by many other people, and it is appropriate here to acknowledge my thanks to them. Terry Lee has not only been a marvellous source of inspiration, but he also commented at length on the first edition of *Wine and the Vine*, and was instrumental in arranging for me to spend time at The Australian Wine Research Institute in 1994. Many people in Australia made that an unforgettable experience, and I am particularly grateful to Peter Leske, Creina Stockley, Paul Henshke, Rae Blair, Janet Currie and Eveline Bartowsky for all that they did for me during that visit. Funding from the Sir Robert Menzies Centre for Australian Studies is also gratefully acknowledged. I have continued to learn much from colleagues in Britain, and would particularly like to thank Jasper Morris, Hazel Murphy, John Dickenson and Gerry Fowles, for all that they have taught me about the wine industry, the geography of wine, and wine chemistry. Finally, a great debt of gratitude is owed to all of those wine makers, too many to name individually, who have shared their wine and their wisdom with me over the last two decades. This book owes an enormous amount to them.

REFERENCES

Aeuckens, A., Bishop, G., Bell, G., McDougall, K. and Young, G. (1988) *Vineyard of the Empire: Early Barossa Vignerons 1842–1939*, Adelaide: Australian Industrial Publishers.

Barker, G. and Lloyd, J. (eds) (1991) *Roman Landscapes: Archaeological Survey in the Mediterranean Region*, London: British School at Rome.

Bell, G. (1993) 'The South Australian Wine Industry 1858–1876', *Journal of Wine Research* 4(3), 147–63.

Bell, G. (1994) 'The London market for Australian wines 1851–1901: a South Australian perspective', *Journal of Wine Research* 5(1), 19–40.

Bessis, R., Leneuf, N. and Fournioux, J.-C. (1994) 'Les bases de la typicité des vins: le cépage et le terroir', *Pour la Science* 203, 48–55.

Boulton, R.B., Singleton, V.L., Bisson, L.F. and Kunkee, R.E. (1996) *Principles and Practices of Winemaking*, New York: Chapman and Hall.

Briggs, A. (1994) *Haut-Brion*, London: Faber and Faber.

Brown, K. (1986) *Bourbons and Brandy: Imperial Reform in Eighteenth Century Arequipa*, Albuquerque: University of New Mexico Press.

Casamayor, P. (1993) 'Il giornalismo francese e la legge Evin', *Il Consenso, Trimestrale di Analisi Sensoriale e Bolletino del Seminario Permanente Luigi Veronelli* 7(3), 33–41.

Comba, R. (ed.) (1990) *Vigne e Vini nel Piemonte Medievale*, Alba and Cuneo: Famija Albèisa, Edizioni l'Arciere.

Comba, R. (ed.) (1992) *Vigne e Vini nel Piemonte Moderno*, Alba and Cuneo: Famija Albèisa, Edizioni l'Arciere, 2 vols.

Comba, R. (ed.) (1994) *Vigne e Vini nel Piemonte Antico*, Alba and Cuneo: Famija Albèisa, Società per gli Studi Storici, Archaeologici ed Artistici della Provincia di Cuneo, Edizioni l'Arciere.

Coombe, B.G. and Dry, P.R. (eds) (1988) *Viticulture, Volume 1: Resources in Australia*, Adelaide: Australian Industrial Printers.

Coombe, B.G. and Dry, P.R. (1992) *Viticulture, Volume 2: Practices*, Adelaide: Winetitles.

Cushner, N.P. (1980) *Lords of the Land: Sugar, Wine and Jesuit Estates of Colonial Peru, 1600–1767*, Albany: State University of New York Press.

Delin, C.R. and Lee, T.H. (1992) 'Psychological concomitants of the moderate consumption of alcohol', *Journal of Wine Research* 3(1), 5–24.

Dickenson, J. and Unwin, T. (1992) *Viticulture in Colonial Latin America: Essays on Alcohol, the Vine and Wine in Spanish America and Brazil*, Liverpool: University of Liverpool Institute of Latin American Studies.

Dutel, G.H. (1990) 'The viticultural and oenological aspects of organic wine production', *Journal of Wine Research* 1(3), 225–30.

Elliott-Fisk, D.L. and Noble, A. (1992) 'Environments in Napa Valley, California and their influence on Cabernet Sauvignon wine flavors', in H. De Blij (ed.) *Viticulture in Geographical Perspective*, Miami: Miami Geographical Society, 45–72.

Fregoni, M. (1992) 'Changes to Italy's wine legislation: the *Nuova Disciplina delle Denominazioni d'Origine dei Vini*', *Journal of Wine Research* 3(2), 123–36.

Frenzel, B. (ed.) (1992) *Evaluation of Land Surfaces Cleared from the Forests in the Mediterranean Region during the Time of the Roman Empire*, Strasbourg: European Science Foundation.

Gille, P. (1992) *Le Rôle de la Grande Distribution dans la Filière Vinicole Française – Essai d'Analyse*, Montpellier: INRA.

Gladstones, J. (1992) *Viticulture and Environment: a Study of the Effects of Environment on Grapegrowing and Wine Qualities, with Emphasis on Present and Future Areas for Growing Winegrapes in Australia*, Adelaide: Winetitles.

Grønbæk, M., Deis, A., Sørenson, T.I.A., Becker, U., Schnohr, P. and Jensen, G. (1995) 'Mortality associated with moderate intakes of wine, beer, or spirits', *British Medical Journal* 310, 1165–9.

Gruzinski, S. (1979) 'La mère dévorante; alcoolisme, sexualité et déculturation chez les Mexicas (1500–1550)', *Cahiers des Amériques Latines* 20, 5–36.

Halliday, J. (1995) *A History of the Australian Wine Industry, 1949–1994*, Adelaide: Winetitles.

Hawkins, A. (1996) 'Communicating via computer: reading the label of WWW', *Journal of Wine Research* 7(1) (in press).

Huetz de Lemps, A., Pitte, J.-R., de Planhol, X. and Roudié, P. (1990) *Les Vins de l'Impossible*, Grenoble: Editions Glénat.

Isager, S. and Skydsgaard, J.E. (1992) *Ancient Greek Agriculture: an Introduction*, London: Routledge.

Jackson, R.S. (1994) *Wine Science: Principles and Applications*, San Diego: Academic Press.

Johnson, H. (1989) *The Story of Wine*, London: Mitchell Beazley.

Liddell, A. and Price, J. (1992) *Port Wine Quintas of the Douro*, London: Sotheby's Publications.

Loubère, L.A. (1990) *The Wine Revolution in France*, Princeton: Princeton University Press.

Martini, A. (1993) 'Origin and domestication of the wine yeast *Saccharomyces cerevisiae*'. *Journal of Wine Research* 4(3), 165–76.

Martins, C.A. (1990) *Memória do Vinho do Porto*, Lisboa: Istituto de Ciências Sociais, Universidade de Lisboa.

Michel, R.H., McGovern, P.E. and Badler, V.R. (1993) 'The first wine and beer: chemical detection of ancient fermented beverages', *Analytical Chemistry* 65(8), 408A–13A.

Millson, P. and Hawes, C. (eds) (1994) *Horizon: Designer Wines*, London: BBC.

Morton, L.T. (1993) 'History of viticultural "Armageddon": European phylloxera crisis', *Practical Winery & Vineyard* 14(3), 9–11.

Mullins, M.G., Bouquet, A. and Williams, L. (1992) *Biology of the Grapevine*, Cambridge: Cambridge University Press.

Nitz, H.-J. (ed.) (1993) *The Early-Modern World-System in Geographical Perspective*, Stuttgart: Franz Steiner Verlag.

Norrie, P. (1990) *Vineyards of Sydney: Cradle of the Australian Wine Industry*, Sydney: Horwitz Grahame.

Ough, C.S. (1992) *Winemaking Basics*, Birmingham NY: Food Products Press.

Peppercorn, D. (1991) *Bordeaux*, London: Faber and Faber, 2nd ed.

Pouget, R. (1990) *Histoire de la Lutte Contre le Phylloxéra de la Vigne en France*, Paris: INRA.

Rathbone, D. (1991) *Economic Rationalism and Rural Society in Third-Century A.D. Egypt*, Cambridge: Cambridge University Press.

Riou, C., Morlat, R. and Asselin, C. (1995) 'Une approche intégrée des terroirs viticoles: discussions sur les critères de caractérisation accessibles', *Bulletin de l'O.I.V.* 60, 93–106.

Robinson, J. (ed.) (1994) *The Oxford Companion to Wine*, Oxford: Oxford University Press.

Rousseau, J. (1992) 'Wines from organic farming', *Journal of Wine Research* 3(2), 105–21.

Schenk, W. (1992) 'Viticulture in Franconia along the River Main: human and natural influences since AD 700', *Journal of Wine Research* 3(3), 185–204.

Smart, R.E. and Robinson, M. (1991) *Sunlight into Wine: a Handbook for Winegrape Canopy Management*, Adelaide: Winetitles.

Spahni, P. (1995) *The International Wine Trade*, Cambridge: Woodhead Publishing.

Stevenson, I. (1991) Review of *Wine and the Vine: an Historical Geography of Viticulture and the Wine Trade*, *Journal of Wine Research* 2(3), 213–14.

Stevenson, T. (1988) *Sotheby's World Wine Encyclopedia*, London: Dorling Kindersley.

Thomas, M.R., Cain, P. and Scott, N. (1994) 'DNA typing of grapevines. A methodology and database for describing cultivars and evaluating genetic relationships', *Plant Molecular Biology Reporter*, 25, 939–49.

Unwin, T. (1990) 'Saxon and early Norman viticulture in England', *Journal of Wine Research*, 1(1), 61–75.

Unwin, T. (1992a) *The Place of Geography*, London: Longman.

Unwin, T. (1992b) 'Continuity in early medieval viticulture: secular or ecclesiastical influences?', in H. De Blij (ed.) *Viticulture in Geographical Perspective*, Miami: Miami Geographical Society, 7–22.

Unwin, T. (1992c) 'Images of alcohol: perceptions and the influence of advertising', *Journal of Wine Research* 3(3), 205–34.

Unwin, T. (ed.) (1994a) *Atlas of World Development*, Chichester: Wiley.

Unwin, T. (1994b) 'European wine sector policy and the UK wine industry', *Journal of Wine Research* 5(2), 135–46.

PREFACE

This book has been written as an introduction to the changes that have taken place in viticulture and the wine trade since prehistoric times. It is also, more specifically, an historical geography of viticulture, consciously seeking to blend together an understanding of processes of social and economic change in particular places and at particular times. As such, it builds on a tradition represented by the French historical geographer, Roger Dion, and developed by such scholars as Marc Bloch and Fernand Braudel. I owe much to their style of writing, and their grasp of the complexities of such changes.

In any undertaking such as this, that crosses traditional disciplinary boundaries of scholarship, many debts are incurred. My first encounters with landscapes of the vine, and with the fruits of its cultivation, were as an eager participant on family holidays, and for these as well as for their continuing encouragement I am deeply indebted to my parents. My academic interest in viticulture began in 1974, and it is with pleasure that I recall the journey of exploration through the vineyards of Europe undertaken with John Candler in the summer of that year. Since then, countless wine makers and people within the wine trade have offered me something of their knowledge, and it is with particular gratitude that I record the help and assistance offered to me in this respect by John and Janet Burnett, Louis Bouzereau-Bachelet, Maria Cálem, Jean-François Cholet, and Blandine and François Rocault. Numerous colleagues throughout the world have offered useful guidance, and have pointed me in directions that I would not otherwise have explored, and among them I am particularly appreciative of advice given by Stephanie Besse, Peter Credland, Felix Driver, Julian Jeffs, James Lapsley, Fergus Kelly, Edmond Maudière, Alberto Melleli, Anngret Simms, Kenneth Stevenson, Bruce Tiffney and Paul Tweddle. The following have very kindly read various parts of the manuscript and have contributed greatly to my own understanding of the subject through their perceptive comments and wise suggestions: David Baverstock, Dick Bowes, Jasper Morris, Hazel Murphy, Hans-Jürgen Nitz, Jancis Robinson, and Iain Stevenson. For financial support, I am most grateful to the British Academy for a grant to enable me to visit California. Many thanks are also due to Ron Halfhide and Justin Jacyno for their cartographic work, to Axel Borg at the Shields Library of the University of California, Davis, for his considerable bibliographical assistance, to Valerie Allport for obtaining numerous items for me through the inter-library loan service, to the staff of other libraries and museums who have provided me with relevant material, and to the students who have participated in my course on the historical geography of viticulture at Royal Holloway and Bedford New College, University of London for their enthusiasm and encouragement.

At Routledge, I am particularly grateful to Alan Jarvis and Alison Walters who provided support where necessary, and have been very patient with the shortcomings of the manuscript.

Above all, though, I must thank my wife Pam and daughter Jenny. Both have had to put up with my frequent absences necessitated by the writing of this book, and both have accompanied me around numerous vineyards, wine cellars and museums in different parts of the world. Moreover, with her flair for identifying the central elements of any argument, Pam has also read the whole manuscript, and I am deeply indebted to her for her numerous comments and suggestions, which have unquestionably improved the text.

Tim Unwin
Englefield Green

ACKNOWLEDGEMENTS

The author and publishers wish to thank the following who have kindly given permission for the use of copyright material:
R. Blachon for Figure 1; Barbe for Figure 2; Figures 12, 13, 15, 17, 18, 19 and 24 are reproduced by courtesy of the Trustees of the British Museum; The Trustees of the British Library for Figures 26, 30 and 31; Photographie Giraudon for Figure 28; the Bayersiche Staatsbibliothek of Munich for Figure 34; Dover Publications Inc. for Figures 29 and 38; Hirmer Verlag for Figure 16; the Groupement Interprofessionnel des Côtes du Roussillon for Figure 39; J.M. Dent and Sons for extracts from *The Rubáiyát of Omar Khayyám*; extracts from *The Canterbury Tales, Piers the Ploughman, Li Po* and *Tu Fu*, the *Ruba'iyat of Omar Khayyam* and the *Epic of Gilgamesh* are reproduced by permission of Penguin Books Ltd.; and extracts from *New Songs from a Jade Terrace* are reproduced by permission of Unwin Hyman Ltd. Extracts from the Bible have been published with permission of Hodder and Stoughton, and the author is particularly grateful to Dick Douglas for his helpful advice with respect to their inclusion.

Although every effort has been made to trace the owners of copyright material we would like to take this opportunity to apologise to any copyright holders whose rights may have been unwittingly infringed.

1

THEMES IN THE HISTORICAL GEOGRAPHY OF VITICULTURE

Les thèmes de la vigne et du vin, quand j'y arrivai, me frappèrent par la beauté de l'illustration qu'ils trouvaient en France et par l'ampleur de leur résonance historique. [The themes of the vine and wine, when I came to them, struck me by their beauty of illustration in France, and by the depth of their historic resonance.]

(Dion, 1959: vii)

There are few agricultural crops whose products are as subtly diverse as those of the grape vine, *Vitis vinifera*. Moreover, this diversity is reflected at a range of scales, from the global variations between wines in different continents, to the local differences between adjacent vineyards which can entitle the wine from one one to be called a *Grand Cru* while that of its neighbour remains a simple *Appellation Communale*. This diversity is not only the outcome of differences in geology and climate, but it is also the result of the labour of countless generations of vine growers and wine makers, each set in their own distinctive human context. It is this historical interaction between people and the environment, creating a specific cultural identity, that lies at the heart of any understanding of the emergence and spread of viticulture and wine production. The essence of this distinctive interaction has been well captured in Figure 1 by the French cartoonist Blachon (Humoristes Associés, 1980), who depicts the roots of a wine bottle spreading out into the soil beneath a grass covered field. In the background, set among rolling hills and under a cloudless sky, lies a small village, the home of the *vignerons*, and also the repository of the cultural heritage that gives the wine its own special identity. Dominating the skyline is the spire of the church, emphasising the important role played by religion, not only in the daily lives of the inhabitants of the village, but also in the development and maintenance of a viticultural tradition.

However, there is much more to an understanding of the emergence and spread of viticulture than merely its expression as a product of the interaction of people in a particular environment. Viticulture and wine

Figure 1
Source: Humoristes Associés (1980)

have played fundamental economic, social, political and ideological roles
in different parts of the world throughout history. Thus, in an economic
context, the use of wine as an exchange currency by the Romans in Gaul
is well attested (Tchernia, 1983), the significance of the wine trade with
northern Europe for the inhabitants of Bordeaux in the medieval period
has been widely examined (Carus-Wilson, 1954; James, 1971), and the
role of trans-national capital in the development of the wine industry
during the twentieth century is beginning to attract scholarly attention
(Briggs, 1985). Likewise, the social significance of viticulture and wine
consumption has also varied considerably in the past. The distinction
between viticulture as part of a peasant subsistence economy in much of
southern Europe in the medieval period at a time when wine
consumption in the north was becoming one of the social attributes of
the urban rich has thus been explored at some length by Dion (1959), and
Hyams (1987) has written extensively on the social history of the wine
vine. Moreover, wine and the vine have had important ideological
significance and symbolism, from their efflorescence in Bacchic ritual to
their use in the Christian Eucharist. To these factors must also be added

the political importance of viticulture and the wine trade, expressed not only in the famous edict of Domitian in the first century AD (Tchernia, 1986), but also in the effects of the Hundred Years War on the Anglo-Gascon wine trade (James, 1971) and the influence of the 1703 Methuen Treaty on Portugal's wine production and trade relationships (Francis, 1966).

The cultural landscape of viticulture can thus be seen to be an expression of transformations and interactions in the economic, social, political and ideological structures of a particular people at a specific place. Against such a framework, this book seeks to do two things. At one level it explores the emergence and spread of viticulture and wine consumption throughout history. In so doing, it also aims to develop a greater understanding of the underlying structural changes which have generated these reconstructed images of viticulture and wine in the past.

THE HISTORICAL GEOGRAPHY OF VITICULTURE

Given the importance of wine and the vine in European society it is surprising how little research has been undertaken on the historical geography of viticulture and the wine trade (Dickenson, 1990). As Dickenson and Salt (1982: 159–60) comment, the geography of wine

> may be studied from a variety of perspectives and encompasses the influence of the physical environment, historical diffusion of the vine and viticulture, economic geographies of cultivation and marketing, political influences on trade and production, and cultural perceptions of landscapes, products and people

but despite this apparently wide appeal, viticulture and wine have largely been ignored by geographers.

Most academic literature on wine and viticulture not surprisingly emanates from the countries which have long traditions of wine production, most notably France and Germany. More recently, though, a considerable volume of material has been published in America and Australia, reflecting the extensive research being undertaken in departments and institutes of oenology and viticulture, such as those at the University of California, Davis, and at Roseworthy College in South Australia. The bulk of this literature, however, is concerned with contemporary practices of viticulture and vinification, and the historical development of the wine industry receives but a passing mention in introductory sections or chapters of major texts (Amerine and Joslyn, 1970; Coombe and Dry, 1988). Where geographers have contributed to the literature, it has mainly been in the context of research on subjects such as slope processes in vineyards (Richter, 1980), or on general descriptions of vineyard regions (de Blij, 1985b; Baxevanis, 1987).

3

The classic study by a geographer on the historical development of viticulture is undoubtedly Roger Dion's *Histoire de la Vigne et du Vin en France des Origines au XIXe Siècle*, published in 1959, reprinted in 1977 and still without an English translation. This scholarly work explores the environmental and locational factors underlying the emergence of viticulture in different parts of France, and then traces the development of wine production from its Greek origins, through the medieval period up to the advent of champagne in the eighteenth century. Since the early 1960s, however, a considerable number of other French regional studies of viticulture have been undertaken, most notably by Pijassou (1980) and Roudié (1988), but the majority have concentrated primarily on analyses of the post-medieval period, and as Dickenson and Salt (1982) note, many are surprisingly insubstantial in nature. This French research activity culminated in a colloquium on the historical geography of viticulture held in Bordeaux in 1977, the proceedings of which were published in two volumes edited by Huetz de Lemps (1978). In his bibliographic conclusion to these volumes Huetz de Lemps lists 701 books and articles by French writers, of which 470 are concerned specifically with France. Most are local or regional studies of viticulture in the period after the seventeenth century, and few address wider issues concerning the marketing, distribution or consumption of wine. Not surprisingly, given their historical importance, the main areas of France represented in his survey are the Bordeaux region, Languedoc and Burgundy.

Within the English language, the most comprehensive geographical book on wine and viticulture is de Blij's (1983) *Wine: a Geographic Appreciation*, which discusses the subject under the broad headings of historical geography, physical geography, political geography, economic geography, and cultural landscapes, before providing a series of regional geographies of viticulture in different parts of the world. However, few individual themes are tackled in depth, and the historical development of viticulture and the wine trade is examined in a mere forty pages. A very differerent geographical approach to the subject is found in Stanislawski's (1970) exploration of the cultural landscape of the vine in Portugal. Here, he argues that the vine serves as a clue to understanding the character of Portugal's regions:

> Its cultivation is immemorially important, not only as a source of livelihood, but also as part of the balance of culture which is concerned both with existence and with more subtle satisfactions as well. More than a producer of a satisfying and nutritive beverage, its contribution to the landscape is aesthetically pervasive and economically indicative of local character. It creates landscapes that are unique and bear social and economic implications.
>
> (Stanislawski, 1970: 4–5).

Apart from these two important studies, English-speaking geographers have until recently contributed little to an understanding of the historical practice of viticulture, and in their survey of the literature of wine Dickenson and Salt (1982) note that less than twenty papers on viticulture were published in the major English language geographical journals between 1945 and 1982. Since then, *The Viticultural Geography Newsletter*, first published in 1986, has attempted to bring together Anglo-American geographers within the field, and further publications (de Blij, 1985b; Baxevanis, 1987; Peters, 1989; Dickenson, 1990) have emerged, but there still remains a dearth of material on the historical geography of viticulture.

Given the substantial work of historians and archaeologists on the history of viticulture and the wine trade (Bassermann-Jordan, 1907; Billiard, 1913; Marescalchi and Dalmasso, 1931-7; Tchernia, 1986; Lachiver, 1988; *Van Rank tot Drank*, 1990) it is essential to define a theoretical framework within which to situate the present essentially geographical enquiry. De Blij (1983), for example, has interpreted the historical geography of viticulture very broadly as being the historical background to his geographical appreciation of wine, involving an analysis of the origins of viticulture, its expansion under the Romans, the post-Roman survival of viticulture and its subsequent global diffusion. However, he makes little attempt to integrate this empirical material into an overall theoretical framework. Likewise, the papers in Huetz de Lemps' (1978) edited volumes on the *Géographie Historique des Vignobles* are presented without any overview or introduction concerning the broader issues involved, and the individual papers stand largely as empirical examples of viticulture, vinification and the cultural landscape of the vine in past places.

INDIVIDUALS AND STRUCTURES: A THEORETICAL PERSPECTIVE

Since the 1970s increasing numbers of geographers have attempted to identify and analyse the way in which individual action takes place within a framework of structural constraint (Gregory, 1978). More specifically, they have been concerned with the extent and manner in which social and economic structures, which by definition are seen as determining human behaviour, are themselves changed or influenced by human action. Duncan (1985: 187), for example, has neatly captured the dilemma involved with this undertaking, noting that 'traditional approaches have either erred on the side of overstressing the power of structures, as in structural Marxism, or conversely have granted near autonomy to the individual, as in neoclassical economic and behavioural geography'.

5

One solution to this dilemma has been proposed by the sociologist Giddens, with his theory of structuration, according to which

> all social action consists of social practices, situated in time-space, and organised in a skilled and knowledgeable fashion by human agents. But such knowledgeability is always 'bounded' by unacknowledged conditions of action on the one side, and unintended consequences of action on the other.
>
> (Giddens, 1981: 19)

For Giddens (1981: 19), 'the structural properties of social systems are simultaneously the *medium and outcome of social acts*'. An important contribution of Gidden's theory of structuration has been the attention it has paid to questions of ideology, which have generally been ignored in much geographical literature in this field. As Macintyre and Tribe (1975: 4) point out, the traditional 'doctrine of historical materialism has in the past posed many problems for Marxists', largely because of the tendency to reduce the materialist conception simply to its economic 'base' or 'infrastructure'. In such a crude conception, political and ideological action in the 'superstructure' become fully subordinated to the economic 'base'. In other words, politics and ideology tend simply to be seen as being determined by the particular economic structure prevailing at any given time. Althusser (1969; Althusser and Balibar, 1970), attempted to overcome this shortcoming by identifying three instances of social activity in any social formation: economic practice, political practice and ideological practice. He also drew a distinction between dominating and determining instances. In Macintyre and Tribe's (1975: 8) words,

> According to Althusser, Marx showed that the economic *or* the political *or* the ideological instance could be dominant in a particular social formation. The role of the economic base, says Althusser, is to determine *which instance is dominant*, by which is meant the instance in which social conflicts are formulated.

In his effort to address the basic question of human action within a structuralist framework, Giddens has also focused particular attention on the question of ideology. Thus Duncan (1985: 175), has argued that 'Central to this structuration view is the concept of ideology which plays a key mediating role between various actors and the social system of which they are both products and producers'. One way in which ideology finds its physical expression is through the production and use of symbols, and among such symbols wine and the vine have always been of considerable importance in European society. Central to the present analysis, therefore, is an examination of the way in which this symbolism of the vine has been reproduced and transformed.

Fundamental to the above arguments is the existence of a distinction between economic, social, political and ideological structures or instances in any given 'society'. The ways in which economic and social characteristics are maintained involve both power relationships and also the means by which these are legitimated. It is here that ideology and politics can be seen as playing such important roles. Political structures can be interpreted as controlling the power relationships within social formations, and ideological structures as providing their legitimation. Habermas (1976, 1978), in his development of critical theory, thus stresses the need to develop a critique of ideology, in order to understand the way in which late-capitalist society is reproduced. In Geuss's (1981: 2–3) words, 'the very heart of the critical theory of society is its criticism of ideology. Their ideology is what prevents the agents in the society from correctly perceiving their true situation and real interests'. Hence, critical theory 'makes its own that concern for the rational organization of human activity which it is its task to illuminate and legitimate' (Horkheimer, 1976: 220). Such arguments, developed in the context of late-capitalism, suggest that the prevailing ideological structure serves to legitimate and maintain social and economic relations, to the benefit of those in power. They can, however, also be used as a framework for examining the role of ideology in pre-capitalist societies, and an analysis of the ideological significance of wine provides one avenue of insight into such an undertaking.

Societies are transformed by collective human action, and yet the structure of society itself also influences the extent to which individuals are able to effect that transformation. Out of this interaction between individuals and structures, material products are formed through action in the form of labour. Thus at many different places in the past, vineyards have been carved out of hillsides, images of gods have been shaped from inanimate wood and rock, illustrations of rituals and festivals have been painted on the walls of tombs, and people have become inebriated through the consumption of fermented grape juice. Understanding both the past and the present is achieved through the interpretation of these material products in the context of theoretical propositions. This book therefore seeks to explore the way in which economic, social, political and ideological structures have found their expression through human action in the cultivation of vines and the production of wine. In so doing, it argues that no one structure should be seen as either dominant or determinant. Rather than assuming the determinant nature of the economic base, as does Althusser, it instead seeks to explore the complex *interactions* between structures and individuals as they have varied in space-time. It is this variation that has led to the complex and changing use of wine throughout history.

It is rarely possible, however, simply to interpret a material product of society as being solely in the domain of, for example, the economy or the ideology of a group of people. The particular significance of wine is that its economic, social, political and ideological importance have continuously varied through space and time. Thus, through an analysis of the role of viticulture and the wine trade, it is possible to gain a broad understanding of the structural transformation of society from a number of different perspectives. Against such a background, though, it is possible to identify two broad themes that have played a significant role throughout the history of viticulture and wine production. These are the role of wine and the vine as symbols of both social and ideological importance, and the ability of the owners and exploiters of vineyards to derive economic profit from a crop over and above its value in purely nutritional terms. It is to a brief theoretical consideration of these relationships between symbols and society, and between production and exchange, that attention now turns.

SYMBOLS AND SOCIETY

The vine and wine are among the most important symbols of societies that have emerged around the shores of the Mediterranean (Goodenough, 1956; Stanislawski, 1975). However, their precise role in ideological and social terms is still not well understood. Considerable anthropological work has been undertaken on the meaning of symbols (Douglas, 1975; Lévi-Strauss, 1966), but it is only in recent years that historians (Bourdieu, 1977; Darnton, 1986) and some geographers (Cosgrove, 1984; Cosgrove and Daniels, 1988) have begun to explore the meaning of symbols and ritual in any depth. In the case of geography, most of this work has been concerned with the symbolism of landscape, and with its understanding as 'an ideologically-charged and very complex cultural product' (Cosgrove, 1984: 11). The cultural landscapes of viticulture fit well into such an interpretation, but the symbolism of wine and the vine goes far deeper than simply its overt expression in different types of viticultural landscape.

In helping us to make sense of what we do not understand, symbols play an important role in the maintenance of social structure. However, as Darnton (1986: 219) stresses, 'symbols convey multiple meanings and . . . meaning is construed in different ways by different people'. In understanding the symbolism of wine and the vine it is essential to comprehend the metaphorical relations between the symbol and the symbolised. Put another way, it is important to grasp the different shades of meaning that people have given to attributes of the vine and to wine at different periods in the past. As Darnton (1986: 223) again points out

People can express thought by manipulating things instead of abstractions . . . Such gestures convey metaphorical relations. They show that one thing has an affinity with another by virtue of its color, or its shape, or their common position in relation to still other things.

The use of the vine as a symbol of fertility can, for example, be understood in part by the observation that each spring new growth develops rapidly from the apparently 'dead' wood of an old vine. Once this metaphorical relation is established, other products of the vine, including grapes and the wine made from them, can then in turn be incorporated into the wider symbolic representation of fertility. Such symbolism, with its direct expression of ideological concepts associated with death and rebirth, can be seen as having emerged as part of a particular religious consciousness developed in the early civilisations of south-west Asia and the eastern Mediterranean.

Moreover, the use of the vine and wine in Jewish and subsequently Christian symbolism has also played an important role in the spread of viticulture in the post-Roman era. There is considerable debate, though, concerning the relative importance of economic and ideological factors in the maintenance and expansion of viticulture in the medieval period. This in part results from another aspect of the role of symbols in society, that of power and domination. As Bourdieu (1977: 411) has argued 'Le pouvoir symbolique, pouvoir subordonné, est une forme transformée, c'est-à-dire méconnaissable, transfigurée et légitimée, des autres formes de pouvoir'. [Symbolic power, a subordinated power, is a transformed, that is to say unrecognisable, transfigured and legitimated, form of the other types of power.] Symbols can, and often are, used by those with economic or political power to retain their position of control and dominance. Symbolic power can therefore be interpreted as merely another way in which the powerful attempt to retain their control. In the context of medieval viticulture, once the necessity of wine for the Christian Eucharist had been established, it then became possible for the owners of vineyards to generate profit from the sale of such wine. While the use of wine for religious purposes only consumed a small fraction of the total wine produced, its symbolic significance assured it an important market. Moreover, as wine became incorporated as an essential element in the symbolic ritual of the burgesses of northern Europe (Dion, 1959), it became incumbent on the wine producers and merchants to enforce and sustain its symbolic importance if they were to continue to generate profits from its production and sale.

By the fourteenth century wine had become an essential part of the culture of the ruling classes of northern and western Europe, just as it had of the Greeks and Romans who centuries earlier colonised most of the Mediterranean. As the Romans had taken the practice of viticulture with them in their conquests, so too did the European powers who invaded and came to dominate much of the Americas, Africa and Australasia. Another important theme in the historical geography of viticulture is therefore the way in which cultural traits become embedded within certain societies, and how these traits are then taken with people as part of their cultural assemblage when they settle in new lands. The development of viticulture in North America, Chile, Australia and South Africa can in part be understood by the desire of the first Europeans to arrive in these regions to recreate their own cultural identities in new places. However, it was also fundamentally connected with the creation of new kinds of economy and society associated with the expansion of the colonial and imperial interests of the European metropolitan powers.

The symbolic role of wine and the vine is not merely a feature of the past. One has only to look at the advertisements for wine to be found in newspapers and magazines today to appreciate something of the social symbolism of wine in twentieth century capitalist society. This symbolism, though, is more than just the expression of a particular middle class, or bourgeois, lifestyle, designed to encourage consumers to purchase a particular brand of alcohol. The seductive imagery of many wine advertisements, particularly in France, and the latent eroticism of Barbe's cartoon (Humoristes Associés, 1980) depicted in Figure 2, serve to highlight that the symbolism of wine and the vine lies very deep in the cultural psyche of much of European society even today. Figure 2 in particular returns us to the symbolism associated with Dionysiac ritual more than 2,000 years ago, and is a reminder of the fundamental importance of wine and the vine as symbols of fertility, which, though repressed, still retain their social and ideological significance in the twentieth century. The significance of wine in French society today is also amply illustrated by a display in the wine museum at Tours, the caption for which reads as follows: 'Au rythme des heures, sur les degrés des âges s'égrènent tous les rites de passage familiaux de la course à la vie, du berceau à la tombe. Chacun d'eux est sacralisé par des libations'. [To the rhythm of a clock, all the family rites of passage in the race of life, from the cradle to the grave, are marked off all at the correct time. Each of them is celebrated with libations.] Above this caption, each stage of life, from infancy, through adolescence, youth, manhood, adulthood, maturity, to old age, is illustrated with a relevant picture and a pertinent vinous celebration. Infancy (*enfance*) is thus depicted by a first communion celebration, manhood (*âge viril*) by a wedding feast, adulthood (*âge adulte*) by fishing and hunting, and maturity (*âge mûr*) by mothers' day.

PRODUCTION AND EXCHANGE

In addition to their social and ideological significance, two great economic rhythms have underlain the history of viticulture and wine: the

Figure 2
Source: Humoristes Associés (1980)

processes of production and exchange. The methods of vine cultivation, the labour relations involved, and the methods of vinification must all be understood as parts of the production process, while the distribution of wine, its marketing and trade, and its use as an investment are all concerned with exchange. As economic structures have changed, so too have the production and exchange of grapes and wine.

Methods of viticulture and vinification have until recently remained remarkably unchanged over the centuries, and indeed it is still possible in parts of Portugal, Greece and Italy to find wine made in much the same way as it must have been almost 2,000 years ago. It is essential to distinguish between two different systems within which wine has been produced, the one a subsistence polyculture economy, where vines have been cultivated and wine made as one part of a household's wider domestic economy, and the other a market oriented monoculture of vines producing wine for an external demand. Central to an understanding of the development of viticulture and the wine trade have been the ways in which the distribution of these two types of economy have changed in both space and time. Melleli and Perari (1978), for example, have

11

illustrated how the traditional mixed cropping system of Umbria was replaced in the 1960s and 1970s by vineyard monoculture, but this is also a process which has occurred throughout history and it therefore provides a key conceptual underpinning to many of the arguments developed in this book.

A second, and related, aspect of the production process of viticulture that warrants attention is the labour system involved in both the cultivation of vines and in the making of wine from their grapes. Considerable information concerning these matters survives from the Roman period in the writings of, for example Columella (1941, 1954, 1955), Varro (1912) and Cato (1933), and its interpretation has been widely debated (Carandini, 1983; Duncan-Jones, 1974; and Tchernia, 1986). The use of slaves, and how their cost should be accounted for, are thus central to an examination of the role of wine in the Roman economy. The emergence of a revitalised wine industry during the medieval period, and the rise of commercially oriented vineyards, was based on quite a different labour system, associated with the incorporation of viticulture into the feudal economy of central and northern Europe. More recently, within the capitalist mode of production, as the price of labour has increased, there has been a shift towards mechanisation in all aspects of wine production. This is also related to the need to create new products, better suited to a wider market, but the efforts of large commercial vineyard owners to improve labour productivity, through the use of tractors, mechanical grape pickers, and modern fermentation and bottling systems, all need to be interpreted against the wider development of advanced capitalism. As Mandel (1976: 35) has argued, 'Machines are capital's main weapon for subordinating labour to capital in the course of production'.

Wine's importance as an item of trade, particularly in classical antiquity and in the medieval period, makes it of especial interest for an understanding of the emergence of wider exchange mechanisms under different social and economic structures. Key issues which need to be addressed in this context include the effects of political alliances on patterns of economic interaction, the development of credit systems and other financial institutions, and the transformation of certain commodities into capital investments in their own right. These three themes are explored in various different places within the book, with attention focusing in particular on the relationships between political and economic interests in determining political alliances in the context of the fourteenth and fifteenth century Anglo-Gascon wine trade (James, 1971), and also with respect to the 1703 Methuen Treaty between England and Portugal (Francis, 1966). Both of these examples illustrate the extent to which economic considerations underlay the development of political alliances, and also the manner in which political conflicts expressed as warfare have influenced commercial relationships. The history of the Anglo-Gascon wine trade is a classic instance of the way in which warfare seriously disrupted the trade between two countries, but in the surviving popularity of claret in England it also illustrates the way in which underlying economic considerations can overcome temporary political disruptions. The Methuen Treaty, again an outcome of war on the European stage, provides an excellent example of the methods whereby English merchants established factories overseas in the seventeenth and eighteenth centuries, and which, in this instance, led to the development of the considerable trade in port wine between England and Portugal.

These developments are closely related to the emergence of what Wallerstein (1974; 1980) has called the modern world system, and to the establishment by European powers of colonial interests in Africa, America and Asia. Critical to any understanding of the development of modern capitalism is an analysis of the development of new credit and banking systems in the sixteenth and seventeenth centuries (Braudel, 1982). Among other things these enabled trade to take place much more easily over longer distances, and thus facilitated the wider expansion of colonial interests. In examining the development of viticulture in areas far from its Mediterranean core, it is therefore important to address the complex set of issues concerning the nature of colonialism and imperialism (Baumgart, 1982). Although the immediate factors giving rise to the development of viticulture in North America, Southern Africa and Australia were set against the overall context of the increasing global integration of capital, the actual processes involved were in fact very different, ranging from the attempts of the British government to produce its own wine in America rather than importing it from France, to the 'consuming passion' of the Scotsman James Busby for viticulture which led him to establish experimental vineyards in Australia and New Zealand in the second third of the nineteenth century (Ramsden, 1940: 362).

The history of vintage wines reflects a third element of exchange, that involving the circulation of capital. There is good evidence that the Romans, and possibly also the Greeks and Egyptians, had considered wines from certain years to be of particularly good quality and worthy of being kept. The lack of such vintage wines in the medieval period was the result of the inability to store wine for any length of time following the replacement of sealed amphorae by wooden barrels at the end of the second century AD (Tchernia, 1986), and their reappearance had to await the development in the seventeenth century of glass bottles stoppered with cork. The critical features of vintage wines for the present discussion, though, are that they require greater capital investment, and that this is undertaken in the expectation of increased profits. This was as true of the Roman period as it has been of the nineteenth and twentieth centuries.

For wines that are not destined to be sold within a year of their vintage, not only will the return on capital investment take more than a year to be realised, but it is also necessary to have enough barrels and space for storage to keep more than a year's vintage at a time. This requires yet further capital investment. Vintage wines can therefore only be produced by those who have sufficient capital reserves and who have little need for immediate access to money. Such investment of capital is made in the expectation of the realisation of profits over and above those obtainable from the sale of the wine within a year of its production. So far this discussion has been solely concerned with the profit to be accrued by the producer who stores vintage wines in the expectation of acquiring greater profits, but in recent years wine has also attracted attention as a capital investment in itself. Fine quality wine is now frequently bought strictly as an investment, with the buyer often preferring to sell it in the future at a considerable profit, rather than actually taking enjoyment in its consumption. Wine has thus entered the realms of finance capital and the auction houses.

These then are some of the main features of economic production and exchange that will be examined in the specific context of viticulture and the wine trade. Together with the social and ideological factors summarised earlier in this chapter they provide an overall framework within which to construct an historical geography of viticulture. One of the most immediate expressions of these structural tensions, of traditions of cultural production, and of 'irrepressible human experiences cutting across history and geography' (Cosgrove, 1984: 65) lies in the landscape itself. As Sauer's writings (Leighly, 1963) so clearly illustrated, the cultural landscape provides an important key to an understanding of past people and places. This chapter therefore concludes with brief descriptions of three cultural landscapes chosen to illustrate contrasting sets of processes and periods of evolution. The first is that of the Côte d'Or in Burgundy, exemplifying the creation of a great medieval vineyard of quality. The second is that of the Upper Douro valley in Portugal, which developed as a new wine exporting region in the eighteenth century primarily as a result of Portugal's political and economic ties with Britain. The third is that of the Napa Valley in California, which, largely through the application of a range of recent technological innovations, has emerged in the late twentieth century as one of the great wine producing areas of the world.

CULTURAL LANDSCAPES OF THE VINE

The Côte d'Or, Burgundy

The hills of the Côte d'Or in Burgundy produce some of the finest red and dry white wines of the world. The Côte itself is a major fault scarp, dividing the Middle and Upper Jurassic limestone plateau from the Saône rift valley to the east, with the highest quality vineyards being situated on the brown calcitic or limestone soils of the gentle south-east facing slopes (Pomerol, 1989). This is a landscape dominated by vineyards as far as the eye can see, with small villages nestled at the foot of the vine-clad slopes. It is the land of Pinot Noir and Chardonnay, of old monasteries, and of great merchant houses.

Figure 3, illustrates the view looking down over the *Premier Cru* vineyard of Les Rugiens, just above the village of Pommard, with the old walled town of Beaune, the home of most of the merchant houses, the *négociants*, to be seen several kilometres away out on the plain to the north-east. The vines here are almost all of the Pinot Noir variety, destined to produce the noted red wines of the Côte, whereas just to the south, in Meursault and Puligny-Montrachet, the white Chardonnay takes over as the dominant variety. Out on the plain, the brown soils of the marls, sandstones and gravels produce lower quality wines, but as the slopes rise to the west one encounters first the *Appellation Communale* vineyards and then the great *Premier* and *Grand Cru* ones. Generally speaking, the vineyards become more stony, and have better drainage further up the slope, providing better conditions for the production of great wines (Gadille, 1967).

The Côte de Beaune, and its northern neighbour, the Côte de Nuits lie at the heart of the modern area of Burgundian wine production, which also includes vineyards in Chablis, the Côte Challonais, the Mâconnais, Beaujolais and the Hautes Côtes. These, however, represent but a fraction of those that existed before the advent of phylloxera in the second half of the nineteenth century. In the eighteenth century the vineyards of Burgundy covered vast expanses of the valleys of the Yonne and the Armançon, as well as the regions of Auxerrois, Tonnerois and Avallonais. Thus, in the words of Nicolas de Lamare in his *Traité de la Police*, published between 1705 and 1738 the term 'vins de Bourgogne' referred to all those wines produced 'au-dessus du pont de Sens' (Richard, 1978). This great expanse of vineyards had been established over the centuries since the Roman foundations, which Thevenot (1952) fixes between AD 50 and 150, and Dion (1952a, 1952b) locates in the third century AD. Surprisingly, for vineyards not situated directly on a major navigable river, Burgundy flourished, with Richard (1978) suggesting that its success lay in the prestige nature of its wines which were produced in an ideal physical environment. By the thirteenth century the vineyards of Beaune were among the most prestigious in the whole of France, with many being owned by religious orders. Both the abbeys of Cluny and Citeaux thus played an important part in the development of Burgundian viticulture, and the Cistercians in particular achieved a lasting influence on the Côte de Nuits, through the establishment of the

Figure 3 Pommard and Beaune from the Côte de Beaune
Source: author 12th October 1985

famous walled vineyard of Clos de Vougeot.

Following the ravages of the plague in the mid-fourteenth century, and the resurgence of the powerful Duchy of Burgundy under Philip the Bold (1364-1404), a remarkable, and much quoted, piece of legislation was enacted. In an ordinance on 31st July 1395 Philip the Bold denounced the introduction 'd'un trés mauvais et desloyal plant nommé gaamez' [of a very bad and disloyal plant called gamay] (Lavalle, 1855: 37–9; Berlow, 1982; Richard, 1978). He argued that merchants no longer purchased the wines of Burgundy because of the introduction of the Gamay grape and the practice of adding hot water to the wine to make it appear sweeter, a practice which once the effects wore off actually made the wine 'tout puanz' [quite foul]. He also denounced the use of organic fertilisers, claiming that they made the wines unfit for human consumption. All Gamay vines were to be cut down within a month and uprooted by the following Easter, and organic fertilisers were to be banned from future use, with offences against both laws being subject to heavy fines. The population of Dijon and the surrounding region reacted vociferously against this legislation, and after much argument the Duke eventually jaoled the mayor of Dijon and appointed his own governor. It seems, though, that the Gamay and manure were not the real culprits of the crisis. Instead, it is likely that the devastations of the plague, which first erupted in 1348, leading to a dramatic population decline, meant that there was less labour available for the cultivation and maintenance of the vineyards. Productivity thus declined. The Gamay produces at least twice the yield of wine as does the more 'noble' Pinot Noir, and being more adaptable it was an obvious alternative to which the hard pressed *vignerons* could turn. The use of the Gamay and organic fertilisers may well therefore have been realistic ways in which the wine producers were attempting to alleviate the effects of declining productivity. Rather than leading to an economic resurgence, the Duke's legislation created an even greater crisis. In Berlow's (1982: 437) words, 'Productivity declined even more. Speculation in wine sales dropped to an all-time low and the population at large, in both town and countryside, was impoverished'. She concludes that

> The results went far beyond the Duke's expectations. With the destruction of the entrepreneurial class came the decline of the wealth of Burgundy. The resources of the Duchy had once been sufficient to propel its dukes to national power. Now, Burgundy would take a back seat in a state which, while using its name, was centred more and more in the alien territories to the north.
>
> (Berlow, 1982: 438)

The vineyards of Burgundy nevertheless survived, and by the eighteenth century their popularity had revived. Many of them were still owned by the Church and the nobility, and the early 1700s saw the emergence of several of the great *négociant* houses of wine brokers and shippers, with Champy Père et Fils being founded in 1720 and Bouchard Père et Fils in 1731. These *négociants* bought up wine from the producers for sale throughout France and northern Europe, and Arlott and Fielden (1976: 24) have argued that 'The rise of the *négociant* meant the first systematic prospecting of markets both in France and abroad'. However, the end of the eighteenth century saw a dramatic change in the character of the Burgundian wine trade, and with it the emergence of one of its dominant characteristics today. The revolution of 1789 led to the suppression of the religious orders and the flight or death of many of the nobility. As a result, their vineyards were sold off and broken up, with partible inheritance leading to their further subdivision over the next century, and the multiple ownership of vineyards that is found in the region today. The devastations of phylloxera in the late nineteenth century, the introduction of *Appellation d'Origine Contrôlée* legislation during the 1930s, and a dramatic rise in good quality wine production for export, have all in their turn added to the cultural landscape of the vine in Burgundy, making it among the greatest of the world's wine producers.

Even today, underlying the apparent uniformity of the vineyards illustrated in Figure 3, there is therefore still a tremendous diversity of ownership and production practices. A single named vineyard might be in the hands of a dozen or more owners, with some leasing out their land in return for a share of the wine or the profits. The vines themselves may be of very different ages, with some parts of a vineyard having recently been replanted, and others declining in productivity but still improving in quality. Many old traditions survive, but technological improvements have recently begun to make themselves felt, with the introduction of mechanical grape picking machines, and closely controlled fermentation equipment now becoming widespread. But near the northern limits of successful viticulture, Brugundy suffers more than most areas from the unreliability of the weather, and the lack of sunshine sometimes causes serious problems for the ripening of the grapes.

The Douro valley, Northern Portugal

In contrast to the relatively cool weather of Burgundy, the Upper Douro valley in northern Portugal regularly records summer temperatures above 40°C, with temperatures becoming markedly hotter as one travels further east. Here, on the steep valley sides the main problem is one of increasingly high labour costs, and the great difficulty of mechanisation. However, the development of a monocultural landscape of the vine in this region is a relatively new phenomenon, that dates mainly from the eighteenth century (Croft, 1787; Roseira, 1973; Robertson, 1982; Bradford, 1983). Figure 4 illustrates the heart of this grape growing region, with the little town of Pinhão clinging to the valley sides, where the river Pinhão joins the Douro. Nearby lie the *quintas,* or estates, of the main foreign port shippers, such as Croft, Taylor and Symington, and Quinta da Foz, belonging to the Portuguese company A.A. Cálem & Filho Lda can be seen in the middle distance just to the right of the railway bridge. The hot dry summers make it an ideal location for the production of grapes, and the vertically bedded schists enable the roots of the vines to extend deeply down in search of water. However, most of the grapes are grown by thousands of small farmers, each owning a few hectares of land, and the large shipping companies only grow a relatively small percentage of their grape requirements.

Traditionally, the narrow terraced rows of vines were separated by countless vertical stone walls, arduously constructed and annually repaired by hand. In the winter and spring, soil and rock that had been brought downslope as a result of erosion were often carried back up the slope to renew the terraces on the backs of peasant labourers. In the autumn, grape pickers from all over the north of Portugal would descend into the valley for the vintage. While women and children did the picking, the men would carry the bunches of grapes up and down the steep slopes to the quintas in heavy baskets strapped to their shoulders. Then, after a long day's work in the vineyards, they would spend the early hours of the night, treading the grapes in large granite tanks known as *lagares.* In the spring, the new wine would be taken down the fast flowing river, through the rapids, to the coast where it was matured in the lodges at Vila Nova de Gaia. Now, the river has been dammed, and its ferocity tamed, but the harshness of the landscape remains. The traditional treading of grapes has also in most cases given way to the use of modern steel or concrete fermentation tanks, but one or two estates still prefer to treat some of their wines in the time-honoured way.

Viticulture was already established in Portugal in Roman times, and vines are still to be found throughout much of the country to this day, cultivated in a variety of traditional ways (Allen, 1963; Gonçalves, 1984; Read, 1982). In the north-west, in the lush irrigated green landscape of the Minho region, they were often grown up trees, and further south, in the basin of the river Dão, they were trained precariously on small wooden fencelike structures. For Stanislawski (1970), these promiscuous landscapes of the vine are true *Landscapes of Bacchus.* The development of the vineyards of the Douro valley, however, awaited the changing configuration of international politics that occurred at the beginning of the eighteenth century. From their first political alliance in the Treaty of Windsor in 1386, England and Portugal had established close economic links, with English merchants developing among other interests a

Figure 4 Pinhão at the junction of the rivers Douro and Pinhão
Source: author 12th July 1981

profitable export trade in wine from Viana do Castelo in the north-west of the country. Much of this wine, though, was of low quality and came from the immediate hinterland of the Minho. The European wars of the late seventeenth and early eighteenth century had led to the British government placing restrictions on trade with France, and in 1703 the preferential treatment given by Britain to Portuguese wine as a result of the Methuen Treaty paved the way for a considerable expansion of the wine trade between the two countries. British wine merchants went further afield in their search for better quality wine, and large areas of the Upper Douro valley were converted into the steeply terraced vineyards that dominate the landscape today. However, many of the wines were of poor quality, resulting in part from widespread adulteration, and in order to preserve them variable amounts of brandy were often added to the casks before they were shipped to northern Europe. Concern over the declining quality of the wines in the middle of the eighteenth century led the Portuguese government to establish a monopoly over the trade, and one result of this was that the Upper Douro valley became the first officially demarcated wine region in the world.

The popularity of port wine in Britain, and its profitability for the British shippers, during the eighteenth and nineteenth centuries led to a further expansion of the vineyards at the expense of other crops, and by 1850 vines had come to dominate the landscapes of the Upper Douro. Vineyards extended out onto slopes that had previously been covered with woodland or brush, and a monoculture of vines gradually replaced the traditional agricultural system in the few places where it had managed to survive. However, the arrival of phylloxera at the end of the 1860s led to devastation in the vineyards, and caused widespread emigration. Many small farmers left the region, and those that could not afford to replant their vineyards with new grafted vines turned to new crops, such as olives. The decline in labour supply also meant that farmers could not maintain their terrace systems, and many vineyards were abandoned. This crisis was exacerbated by political unrest in Portugal at the end of the nineteenth century, and then by the economic slump of the 1930s, both of which caused further emigration. Following the Second World War the port trade revived, and as demand rose the industry rapidly returned to prosperity. During this period, many of the smaller family owned port shippers merged to form larger companies, and with the increasing vertical integration that has taken place in the industry a number have been taken over by large global corporations. Since the 1970s, and particularly during the late 1980s following Portugal's accession to the European Community, increased labour costs have caused serious problems for the port shippers. Where possible, some mechanisation has taken place, with new terraces suitable for mini-tractors being carved out of the steep schist slopes. Recent financial support from the World Bank has also led to the reconstruction and replanting of several parts of the demarcated area that had previously been abandoned. Figure 4 thus illustrates a great diversity of terrace types, some abandoned, some traditionally steep walled, and others designed for tractors curving around the slopes. However, very few of these slopes are suitable for the large grape picking machines that have replaced the seasonal bands of labourers in many other parts of the world, and the industry is likely to face increasingly serious problems as labour becomes harder and harder to find at the time of the vintage.

The Napa Valley, California

In contrast to the long history of viticulture in the Côte d'Or, and the steeply terraced slopes of the Douro, the vineyards of California are very largely the product of the last fifty years. In the seventeenth century Franciscan monks had brought the Mission vine to New Mexico, and in the eighteenth century most of their missions in California were making wine. By the beginning of the nineteenth century, following the secular-isation of the missions by the Mexican government, a small number of other vineyards had emerged (Teiser and Harroun, 1983). However, the major impetus to this incipient wine industry came in the 1850s when a number of immigrants began experimenting with European varieties of vine in the Sonoma valley (Schoenman, 1979). Leggett (1941) thus records how the number of wineries in California grew from eleven in 1860 to 139 in 1870. However, this nineteenth century expansion of viticulture was cut short by the collapse of the gold rush, the arrival of phylloxera, and then in 1920 the passing of the Prohibition law (Hutchinson, 1969). It was only with the repeal of Prohibition in 1933 that the Californian wine industry began to emerge in its present form (Blue, 1988).

The leading area of Californian viticulture is widely considered to be the Napa Valley (Figure 5), situated about 75 km north-east of San Francisco (Olken et al. 1982; Blue, 1988; Laube, 1988). Here, on the soils of the flat valley floor, only some 5 or 6 km wide, are to be encountered widely spaced rows of Cabernet Sauvignon vines and some of the most modern wine making facilities to be found anywhere in the world. In the spring, huge fans turn the still air to prevent frosts, and because of the very dry conditions drip irrigation is widely practised to help establish young vines. The dramatic expansion of quality wine production here, though, is a very recent phenomenon, with there only being some twenty-five wineries present in the valley in 1960. By 1975 the number had risen to some forty-five, and by 1990 to over 200. Following the end of the Second World War in 1945, there was little demand for wine in the United States of America, and what demand there was tended to be for poor quality wines, produced mainly from the extensive vineyards

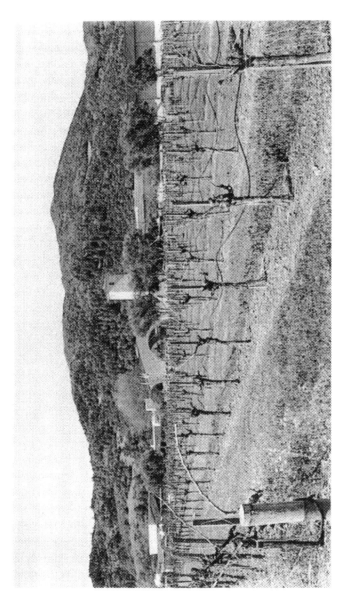

Figure 5 Robert Mondavi winery, Oakville, Napa Valley
Source: author 13th March 1990

planted with Thompson Seedless grapes in the San Joaquin Valley. Nevertheless, during the 1920s and 1930s much research had been done by Winkler, and later Amerine on the best physical environments in which to grow vines for wine in California (Amerine and Winkler, 1944). This had shown that the Napa Valley combined within its short length of about 50 km a good range of different soil and climate conditions suitable for quality wine production, with its southern parts being influenced by the mellowing effects of the San Francisco Bay, and its northern end having much more extreme climatic conditions. Moreover, a small number of wine producers had for several decades shown that it was possible to make fine wines in the valley. Most notably among these were those at the winery established by Charles Krug at St Helena in 1861 and reopened by an Italian immigrant Cesare Mondavi in 1943, at Beaulieu Vineyards under the inspirational leadership of André Tchelistcheff, and at the Inglenook Winery at Rutherford established in 1879 by the Finnish sea captain Gustave Niebaum (Parker, 1979).

It was not, though, until the 1960s, when a number of other social and economic conditions became favourable, that the Californian wine industry, and particularly that of the Napa Valley, really began to develop. The changes that occurred reflected a complex interaction between demand and supply, in which it is not possible to see any single cause as having been dominant. Moulton (1975: 8) has thus suggested that 'Some of the more important factors have been growth in consumer income, changes in wine prices, expanding population, increased advertising, shifting channels of distribution and changes in life style'. Four main groups of factors seem to have been particularly important. First, social changes were taking place in California which encouraged people to turn to wine as a preferred alcoholic beverage. Increasing affluence resulting from the economic growth of the 1960s meant that greater numbers of people could afford to drink quality wines, and indeed they increasingly began to seek after them as a symbol of social status and prosperity. This was enhanced by improved travel opportunities, which enabled many Americans to visit Europe and taste European wines for the first time, thus leading to a general increase in the demand for imported wine. Secondly, however, a number of technical changes generated a situation where it became increasingly easy to make good wines in California. In particular, the introduction of cool fermentation equipment, modern hygienic wine making and bottling facilities, and the use of dry forms of cultured yeast, all improved the quality of Californian wines. This happened at the same time as increasing numbers of graduate students were being trained in oenology and viticulture at the University of California, Davis, (Winkler, 1973), where the driving mentality was towards the eradication of poor quality wines. Thirdly, tax concessions existed which made it advantageous for companies to invest capital in vineyards and wineries. Although profit margins were not high (Moulton, 1984), the industry appeared to have considerable growth potential, and this was therefore a period when a number of companies began to invest in the sector. Typical of the purchases that took place at this time were the acquisition of Beaulieu and Inglenook by Heublein in 1969, and that of Beringer by Nestlé in 1971 (Sorenson and Beringer, no date). Moreover, these conditions also provided the opportunity for the successful development of wineries by enterprising individuals such as Robert Mondavi who, having broken away from his family connections at Charles Krug, established a winery at Oakville in 1966 (Figure 5) (Ray, 1984). However, there was a fourth, far less tangible, element to the expansion of quality wine production at this period, and this was the development of an aesthetic interest in both the production and the consumption of wine. The establishment of wineries in rural areas of California touched an element of romance in the minds of many people, and reflected aspects of what Lapsley (1990, pers. comm.) has called an 'arcadian myth'. The wineries that flourished were also those directed and led by people with vision and energy, the classic example once again being that provided by the notable success of Robert Mondavi.

Central to any understanding of the Californian wine industry, though, is also the recognition that an important distinction exists between grape growing and wine making. This partly results from the American experiences of prohibition, when wine making declined although grape production for table grapes continued. This has led to a complex industrial structure ranging from wineries which have no vineyards of their own, through wineries producing wines largely from their own grapes, to vine growers who make no wine at all. The relative balance of power between the different groups varies depending on the market for wine, and it closely reflects the cyclical pattern of grape prices, which for example peaked in 1973, fell to a nadir in 1976, rose again to record levels in 1981, and then declined again in the mid-1980s (Moulton, 1984). At times of grape shortage and high prices, vine growers have benefited, whereas at times of surplus the wineries which have mainly bought grapes have had the upper hand.

During the 1980s there was a general tendency for the larger wineries to increase their vineyard holdings. Laube (1988) has thus noted that between 1978 and 1988 the amount of land owned by wineries in the Napa Valley doubled, with the percentage of grapes grown by independent farmers decreasing from around 70 per cent to about 35 per cent. Indeed, by 1988, seven main companies, Beaulieu, Beringer, Christian Brothers, Domaine Chandon, Charles Krug, Robert Mondavi and Sterling, owned or controlled one-third of the grape crop, and produced two-fifths of the wine made in the Napa Valley. Land prices increased over tenfold between the early 1970s and the late 1980s, and this, according to

Laube (1988) has made it extremely difficult for many of the smaller wineries to compete with the large producers, often owned by global corporations, which are well financed and able to pay the higher prices prevailing for both land and grapes.

Physical factors in the production of wine

These examples, taken from widely contrasting parts of the viticultural spectrum, have illustrated not only the varied range of economic, social, political and ideological factors that have influenced the emergence of viticulture in the past, but also the great importance of physical factors in determining the nature of the wine industry at any given time and place. The influence of different grape varieties, the emergence of new methods of vinification, the importance of geology and soil type, and the role of phylloxera can all be seen to have played a significant role in creating the viticultural landscapes of northern Portugal, Burgundy and California. Any understanding of the historical geography of viticulture must therefore begin with a basic comprehension of the physical processes involved in cultivating vines and in making wine. The climatic constraints on the cultivation of vines, and the precise nature of the chemical processes involved in vinification, are critical to any appreciation of the emergence and development of viticulture and the wine trade. It is therefore to these physical processes that the next chapter now turns.

2

VITICULTURE AND VINIFICATION

No viticultural subject has generated such diverse opinions as the relation of the vine to its environment and to the composition and quality of the resulting musts and wines . . . The interrelation of climate, geography, and the variety of grape determines the potential for fruit and wine quality. Man, of course, can influence it by pruning, training, irrigation, fertilization, cultivation, time of harvest, sugaring, control of fermentation, and aging. These factors vary from year to year, from region to region, and from variety to variety. But most enologists agree that climate has the greatest influence on wine quality . . . Over and above climatic variables, there may be specific geographical conditions that influence grape maturity and wine quality.

(Amerine and Wagner, 1984: 97, 118)

True wine is the fermented juice of grapes, although the term is frequently applied less correctly to the products of other fruits. In Lichine's (1981: 38) words, 'properly wine comes from grapes and nothing else – and it is the natural product of grapes which have been gathered, carted to the wine shed, pressed, and left in vats until the grape sugar has fermented into alcohol'. Three separate stages are involved in making wine: *viticulture*, the cultivation of grapes; *vinification*, the process whereby these grapes are turned into wine through the fermentation of grape sugar into alcohol; and then the *maturation* and development of wine into a product that is ready to be consumed or sold. In order to understand the changes that have taken place in viticulture and the wine trade it is essential to grasp the central characteristics of these processes, and this chapter therefore provides a basic overview of each in turn.

The physical properties of the vine, and the biochemistry and microbiology of wine making have played a fundamental role in

27

determining both the broad distribution of viticulture throughout the world, and also the specific character of the wine trade at different periods during the past. The basic physiological requirements of the vine have been a major influence not only on the location of vineyards, but also on the spread of vine pests and diseases. Thus, while broad climatic factors have influenced the global distribution of viticulture, local microclimates and soil conditions have in the past largely determined the quality and reputation of a wine. Moreover, the introduction of pests and diseases from North America, against which European vines had no natural resistance, was the main factor responsible for the widespread destruction of European vineyards during the nineteenth century. The indigenous North American vine species, which had developed in association with such pests, had achieved some natural resistance to them, and thus while being the cause of the problem they also offered a solution, through the grafting of European vines onto American rootstocks. Climate and soil, through their influence on the varieties of vine cultivated and the types of wine produced, have also played an important role in influencing the character of the wine trade throughout history. Thus, the increasing preference of north European people for sweet wine during the later medieval period led to a growth in the wine trade between Mediterranean countries which produced such wine and the towns of northern Europe where they were consumed. Similar shifts in fashion have also influenced the wine trade in more recent times, leading to differential benefits for some wine producing regions at the expense of others. The biochemical and microbiological processes causing wines to deteriorate were also significant in determining the great rhythmic patterns of medieval trade. Thus, because wines kept in barrels generally lasted only a year before becoming unpalatable, it was essential for the towns of northern Europe which did not produce wine to ensure a regular annual supply. This was furnished through the activities of merchants who visited wine producing regions soon after the vintage and acquired wines for shipment to the main consuming areas. Once the nature of fermentation and the activities of bacteria had been explained during the nineteenth century, a whole new era of technological change dawned in the wine industry, reaching its culmination in recent decades with the creation of new types of wine designed to suit, or create, entirely new markets. While wine making today is a highly technical science, the same basic chemical and physical properties have always influenced the creation of wines, and an understanding of these conditions provides the essential grounding upon which an understanding of their historical geography can be developed (Winkler, 1962; Amerine and Joslyn, 1970; Winkler et al., 1974; Kozma, 1975; Weaver, 1976; Kunkee and Goswell, 1977; INRA, 1978; Amerine et al., 1980; Coombe and Dry, 1988).

AMPELOGRAPHY AND THE VINE

Ampelography: the science of grape-bearing vines

The origins of the grape vine remain uncertain, although its great age is testified by fossilised vine leaves and seeds in Palaeocene and Eocene deposits (Winkler, 1962; Galet, 1979; Mai, 1987). Approximately 24,000 varieties of vine have been named, of which it is probable that only about 5,000 are genuinely different varieties (Dry and Gregory, 1988). Of these only some 150 are used at all widely, and only nine varieties are considered to produce 'classic' wines (Viala and Vermorel, 1901-10; Robinson, 1986). All vines, though, belong to the genus *Vitis* which has been classified within the family variously known as Vitaceae, Vitidaceae and Ampelidaceae (Willis, 1973), within the botanical order Rhamnales (Cronquist, 1981), which also includes the families Rhamnaceae and Leeaceae. The genus *Vitis* is recognised as including two sub-genera: *Euvitis*, the true grape, and *Muscadinia*, whose fruit is properly called the Muscadine (Table 1). Winkler (1962: 15) has distinguished the physical differences between these two sub-genera as follows,

> The shoots of the species of *Euvitis* – true grapes – have bark that is longitudinally striate-fibrose, shedding at maturity; pith interrupted in nodes by a diaphragm; forked tendrils, with mostly elongated flower clusters; berries adhering to the stems at maturity; and seeds pyriform, with long or short beak. In contrast, the shoots of the species of *Muscadinia* have tight non-shedding bark, with prominent lenticels; nodes without diaphragm; simple tendrils; short, small clusters; berries that detach one by one as they mature; and seeds oblong, without beak.

While *Euvitis* species have been found in Tertiary deposits in both Eurasia and North America, *Muscadinia* has only ever been found in fossil material in North America (Tiffney, pers. comm. 1990), suggesting that this main division of the genus probably took place before the onset of the Quaternary era some two million years ago. In the late Tertiary period it appears that climatic changes led to a division of the genus *Vitis* into a number of separate species, each becoming adapted to a particular environmental niche (Tiffney, 1979). Tiffney and Barghoorn (1976: 188) thus suggest that 'many presently separate, similar species, such as *Vitis coignetiae* of Asia and *V. labrusca* of North America, are derived from the allopatric speciation of a single, perhaps variable, ancestor whose distribution was fragmented by the climatic deterioration of the late Tertiary'. The subsequent regular cycles of cold glacials and warm interglacials during the Quaternary (Lamb, 1982; Lowe and Walker, 1984) further refined the distribution of these species, so that by the time the grape vine was first used to make wine the distributions noted in

Table 1
Subgenera and main species of the genus *Vitis*

VITIS		
EUVITIS		MUSCADINIA
Eurasian species	North American species	North American species
amurensis	**aestivalis**	munsoniana
armata	argentifolia	popenoei
betulifolia	arizonica	**rotundifolia**
coignetiae	**berlandieri**	
Davidii	baileyana	
embergeri	californica	
ficifolia	candicans	
flexuosa	**Champini**	
lanata	cinerea	
Pagnucii	cordifolia	
pedicellata	Doaniana	
pentagona	gigas	
Piasezkii	Girdiana	
reticulata	helleri	
Romanetia	illex	
Retordi	indica (cariboea)	
Thunbergii	**labrusca**	
vinifera	**Lincecumii**	
Wilsonae	**Longii**	
	monticola	
	novae-angliae	
	palmata	
	riparia	
	rufotomentosa	
	rupestris	
	Shuttleworthii	
	Smalliani	
	Simpsoni	
	sola	
	Treleasei	
	vulpina	

Note: Species in bold are those cultivated most commonly for wine making, either in their own right or as rootstock.
Source: derived from Winkler (1962), de Blij (1983) and Robinson (1986).

Table 1 were well established. Two further aspects of the spatial distribution of these species can also be noted: none were present in the southern hemisphere, and the vast majority of them were to be found on the eastern sides of the great northern continental land masses (Amerine and Wagner, 1984).

One of these species, *Vitis vinifera*, a native of Eurasia, is responsible for most of the wine now made. A key feature of all species in the *Vitis* genus is that they hybridise easily, and *Vitis vinifera* in particular is highly heterozygous, readily mutating under different conditions. This has given rise to the great diversity of vine varieties that have so far been identified, and it has also led to considerable problems in classifying the various prehistoric varieties of vine. Wild forms of *Vitis vinifera* are often referred to as *silvestris*, and cultivated varieties as *sativa* (Amerine and Wagner, 1984), but Núñez and Walker (1989: 214-15) caution against this practice. Whatever classification is adopted, it seems that wild *Vitis vinifera* vines were once very widespread throughout Europe, from Scandinavia to north Africa, and that individual varieties gradually emerged in different places, reflecting the varied climatic conditions encountered by the vine (Amerine and Wagner, 1984). The early forms of wild *Vitis vinifera* were dioecious, consisting of both male and female plants, but eventually hermaphroditic forms appeared. Negrul (1938) has argued on the basis of the morphological and biological characteristics of vines found in different ecological habitats, that three main groups, or proles, of *Vitis vinifera silvestris* emerged: *orientalis*, growing in the Caucasus, *pontica* found around the Black Sea, and *occidentalis* encountered in more western parts of Europe. However, this classification is now regarded with some scepticism (Antcliff, 1988; Núñez and Walker, 1989), and it seems safer simply to suggest that, amid the great diversity of wild varieties, some vines with larger and sweeter berries were selected by early farmers, and that this practice eventually led to the appearance of domestic varieties, which have been designated by some authorities as *Vitis vinifera sativa*.

Table 2 lists the varieties of *Vitis vinifera* that are of most importance for wine making. Within each variety a number of different clones have also been identified and propagated, each with different degrees of such characteristics as resistance to disease, and yield. This propensity of vines to adaptation has meant that in recent years a number of hybrids and crosses have been bred, designed to produce certain characteristics that are found in both parent vines. The best known of the crosses between different varieties of *Vitis vinifera* is probably the Müller-Thurgau which was bred in 1882, either as a Riesling × Silvaner cross or possibly as a cross between two Riesling clones. However, there have been a number of other highly successful crosses developed in recent years, such as the Kerner, a Trollinger × Riesling cross launched in 1969 and now the fourth most popular grape in Germany, and two Riesling × Silvaner crosses, the Scheurebe developed in 1919 and the Ehrenfelser dating from 1929. The

Table 2
Main vine varieties used for wine production, classified by species

SPECIES	VARIETY/CULTIVAR	
	Red	*White*
Vitis vinifera	**Cabernet Sauvignon**	**Chardonnay**
	Merlot	**Chenin Blanc**
	Pinot Noir	**Riesling**
	Syrah/Shiraz	**Sémillon**
	Barbera	**Sauvignon Blanc**
	Cabernet Franc	Aligoté
	Carignan	Gewürztraminer
	Cinsaut	Pinot Blanc
	Gamay	Malvasia
	Grenache	Müller-Thurgau
	Nebbiolo	Muscadet
	Pinot Gris	Muscat
	Sangiovese	Palomino
	Tempranillo	Silvaner
	Zinfandel	Trebbiano
		Viognier
		Welschriesling
Vitis labrusca	Concord	
	Catawba	
Vitis rotundifolia		Scuppernong

Note: Species in bold are those classified by Robinson (1986) as 'classic', whereas other *Vitis vinifera* species are those classified as 'major'. Concord, Catawba and Scuppernong fit neither category, but are shown as illustrations of varieties from other species.

Source: derived from de Blij (1983,), Robinson (1986) and Winkler (1962).

main features that German vine breeders, working at the northern limits of viticulture, have tried to achieve include late-budding which reduces the likelihood of frost damage, high yield, good balance between acidity and sugar, and high quality. Thus the Kerner has a reliably higher yield than the Riesling, it is late-budding, and it also has a distinctive flavour which is generally considered to be of high quality. Numerous experiments have also been made with hybrids, which are crossings between different vine species, and although many 'are very resistant to disease, it is generally accepted that they do not produce the very best wines' (Sutcliffe, 1981: 14).

Vine and grape physiology

While it is the grapes that provide the fruit for wine, some knowledge concerning the structure of the vine plant itself is important if the environmental factors governing its growth are to be fully understood. *Vitis vinifera* is hermaphroditic and can thus be propagated vegetatively as well as sexually. Most vines grown from seed are inferior to their parent vines in terms of vigour, productivity and fruit quality, and therefore the propagation of vines for use in vineyards is usually undertaken by cuttings, although layers, buds and grafts are also used (Winkler, 1962).

The root systems of vines are both spreading and descending, with roots often obtaining a depth of four metres, and in some cases reaching a depth of up to fifteen metres. However, certain conditions are unfavourable for the roots of vines, most notably high water tables, which limit aeration, and the presence of an impervious layer in the soil preventing the penetration of roots. Above the ground, the shoot system consists of the trunk, arms, shoots and leaves of the vine. Leaf and fruit buds develop at nodes regularly located along the shoots, but the number of fruit clusters and the relative balance between leaf and fruit buds differs between varieties. The leaves are the main source of food production for the plant, converting light energy into chemical energy through photosynthesis, and producing sugar from which other products are then synthesised.

The methods by which vines are trained and pruned determine the quantity and quality of the crop produced. The basic systems of training are designed to achieve the best possible shape for a vine under particular climatic and environmental conditions. Thus, if frosts are a danger, vines are usually trained high, whereas when heat from the soil is wanted to ripen the grapes they are usually trained low. Pruning is used to achieve a balance between crop yield and vine growth. In recent years, as mechanisation has increased in vineyards a number of new styles of vine training have been introduced, but there are three basic methods: *cordon*, *gobelet* and *espalier*. In *cordon* systems one or two arms are usually allowed to grow from a short trunk parallel to the ground, and from these a number of shoots and spurs are trained along wires or supports. The classic example of a *cordon* system is the Guyot, which is found in many of the great wine making areas of the world, as in Burgundy (Figure 3). When vines are trained according to the *gobelet* method, typical of Beaujolais, several arms are allowed to grow vertically above a short trunk, and these are then tied together to form the shape of a goblet. *Espalier* systems of training, such as the Lenz Moser, the Geneva Double Curtain or the Karl Merz, usually have a long trunk and several canes trained high on a wire trellis or other such support.

The vine has a very distinctive annual cycle, remaining dormant until the mean daily temperature reaches approximately 10°C. As the temperature rises in the spring, shoots elongate rapidly for about eight weeks. At this stage the flower clusters come into bloom, usually when the temperature has reached a mean daily average of 20°C. Following pollination and fertilization, the small berries set and then grow, at first quite rapidly and then more slowly until a stage, known in French as *véraison* (ripening) is reached. At this point berry growth again accelerates, sugar levels increase, acidity decreases, and the colour develops in black grapes. It is generally accepted that on average grapes are ripe and ready for picking approximately one hundred days after flowering, but this figure depends much on the balance between sugar and acidity required for a particular wine. Physiologically, the time of ripening depends mainly on the variety of vine and on the number of days in the growing season when the mean daily temperature rises above 10°C (Winkler, 1938). Over the winter the vine remains dormant, with the accumulated sugar in the plant offering protection against frost (Winkler, 1962). Vines can thus survive low temperatures in their dormant period, down to about −18°C for short periods. However, in the spring, frosts of only −4°C can kill young leaves, and a temperature of −2°C will damage flower clusters (Jackson and Schuster, 1981).

For its best development the above physiological requirements mean that *Vitis vinifera* thrives most successfully in areas with long fairly hot summers and cool winters. When grown in tropical areas the vine is evergreeen, and the lack of a dormant period generally gives rise to very small crops of poor quality grapes. It is therefore widely accepted (Amerine and Joslyn, 1970; de Blij, 1983; Winkler, 1962) that the best areas for viticulture lie between the 10°C and 20°C annual isotherms, equating approximately to the warm temperate zones between latitudes 30° and 50° north and south (Figure 6). Where grapes are produced successfully nearer the equator, as in parts of Bolivia and Tanzania, it is usually because they are grown at higher altitudes, which are somewhat cooler and thus compensate for the higher temperatures to be found at sea level at such latitudes.

While these climatic conditions largely determine the parts of the world where it is possible to grow vines, it is the character and composition of the grapes that influences the nature of the wine produced. Figure 7 provides a schematic illustration of the composition of a typical grape, indicating the sources of the different products of significance in wine making. Throughout the growing season the proportions of the different parts of a grape vary as it ripens from the exterior to the interior. When fully ripe, grapes consist of 70-80 per cent water, which together with sugar, acids and pectins make up most of the pulp (Amerine *et al.*, 1980). The main sugars are glucose and fructose, which provide the basis for the alcohol as well as some of the flavour. Pectins, which are the derivatives of polygalacturonic acid responsible for the texture of the fruit, account for between 0.02 to 0.6 per cent of ripe grapes, and the principal acids found are tartaric and malic, with a small

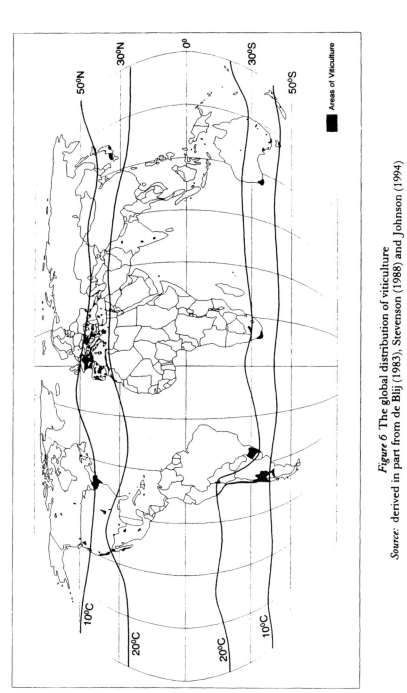

Figure 6 The global distribution of viticulture
Source: derived in part from de Blij (1983), Stevenson (1988) and Johnson (1994)

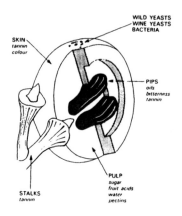

Figure 7 The composition of a grape
Source: derived in part from Johnson (1971)

amount of citric, oxalic and phosphoric acid. While the ripeness of a grape is generally determined by the acid/sugar balance, it must be emphasised that 'The ripe stage is not absolute, nor does it represent the end product in the changes that are proceeding in the berries . . . A grape high in acid and low in sugar may be required for one purpose, and the reverse for another purpose' (Winkler, 1962: 119). Generally speaking pHs of just below 3.0 are found in north European musts, whereas Californian musts at maturity tend to be a little less acid varying from about 3.0 to 3.9 (Amerine *et al.*, 1980: 97).

The skins, which account for between 5 and 12 per cent of the total weight of a ripe grape, provide the source of colour as well as much of the tannin for the wine. As grapes ripen, the chlorophyll which provides their initial green colouring fades, and other colours become more evident. While much is still to be learnt about the pigments, known as anthocyanins, which give grapes their colour, it seems that the main red pigment of *Vitis vinifera* is the monoglucoside of malvidin, known as oenidin (Ribéreau-Gayon and Sudraud, 1957). Tannins occur in all parts of a young grape, but during ripening those in the pulp are hydrolised, so that in a ripe grape most of the tannin is found in the skins, stalks and pips. While tannin also influences the flavour of wine, its most important role is in its maturation, since it stabilises the colour, improves the body, and assists in the process of fining the wine, by which unwanted products are deposited out. Also found on grape skins are the yeasts and bacteria, so important in the process of fermentation. The pips or seeds are the final main component of a grape, constituting less than 10 per cent of the fruit, and containing bitter tannins, oils and resinous material generally found to be undesirable in a wine. Small amounts of other constituents are present in grapes, including a range of vitamins, enzymes, esters, and soil derived minerals, which to varying degrees contribute to the flavour, aroma and colour of the eventual wine (For a full list of the composition of musts and wines see Amerine *et al.*, 1980: 111-12; and Peynaud, 1987: 50-1).

Vine pests and diseases

Vines are affected by a wide range of pests and diseases, as well as the ravages caused by inclement weather, particularly frost and hail (Table 3). Many of the more disastrous pests and diseases were common to North America but unknown in Europe until the nineteenth century. Of these, oïdium and *Phylloxera vitifoliae*, have historically been of most importance. The North American origins of the pests and diseases also provided the source for many cures, since *Euvitis* species indigenous to North America have proved to be much more resistant to the pests and diseases that laid waste the European vineyards of *Vitis vinifera*. Fungus parasites cause the greatest number of vine and grape diseases, but the use of fungicides and good vineyard maintenance limits their effects. In general, the warmer and wetter the climate, the greater will be the difficulty in controlling both fungi and virus diseases. Viruses carried by insects and in the soil, such as Pierce's disease and fan leaf, are the most difficult to control, and the only real cure is to use powerful insecticides in the case of the former and to grub up the vines and fumigate the soil in the latter. Comprehensive accounts of the nature and treatment of the pests and diseases noted in Table 3 are provided in most works on viticulture (see particularly Winkler *et al.*, 1974) and thus only the most important on a global scale are discussed here.

Oïdium, or powdery mildew, attacks all green parts of a vine, producing a fine, powdery fungus. The spores (conidia) are usually spread by the wind, and if left untreated will kill the vine. Unlike many other fungi, it is favoured by cool weather, and develops best in the shade. Until the nineteenth century oïdium was unknown in Europe. It was introduced from North America in the mid-nineteenth century and caused widespread damage until it was discovered that dusting the vines with sulphur gives excellent protection (Ordish, 1987). While *Vitis vinifera* vines are particularly susceptible to oïdium, most American species show some resistance. Downy mildew, otherwise known as Peronospera, is also North American in origin and was probably first introduced to Europe on American vines brought across the Atlantic to combat phylloxera during the 1870s. It needs high temperatures and humidity to develop successfully, and produces an oily stain on the leaves

Table 3
Main vine pests and diseases

FUNGUS PARASITES

Fungi affecting both vine and grapes
Oidium tuckerii (Powdery mildew) caused by *Uncinula necator*
Peronospera (Downy mildew) caused by *Plasmopara viticola* (Berl. & Toni)
Anthracnose caused by *Gloeosporium ampelophagum* (Pass)
Black rot caused by *Guignardia bidwelli* (Ellis)
Black measles caused by *Fomes igniarius* (L. ex Fr.)
Dead-arm caused by *Phomopsis viticola*
Eutypiose

Fungi affecting grapes
Botrytis (Grey rot/pourriture grise) caused by *Botrytis cinerea* (Pers.)
Blue mould caused by *Penicillium* sp.
Black mould caused by *Aspergillus niger* (Van Tiegh)
Rhizopus rot caused by *Rhizopus nigricans* (Ehrenb. ex Fr.)

VIRUS DISEASES
Fan leaf (Court-noué)
Pierce's disease
Yellow mosaic
Leaf roll
Yellow vein
Asteroid mosaic
Corky bark

PESTS
Phylloxera vitifoliae
Grape leaf hopper (*Erythroneura elegantula* Osborn)
Red spider mite (*Tetranychus pacificus* McG.)
Erinose (*Eriophyes vitis* Pgst.)
Grape berry moth (*Polychrosis viteana Clem.*)
Eudemis moth (*Polychrosis botrana* Schiff.)
Cochylis (*Clysia ambiguella* Hübner)
Nematodes
Birds
Wasps
Grasshoppers
Rabbits
Deer

Sources: Winkler (1962), Lichine (1981), MAFF (1980), Jackson and Schuster (1981)

before developing into white patches of mildew on their undersides. The main form of treatment is through regular sprayings of plants with copper sulphate, in the past as Bordeaux mixture, or more recently in the form of systemic sprays along with organic fungicides. Most American vines are resistant to all but the most severe outbreaks of downy mildew, but all *Vitis vinifera* varieties are highly susceptible to the disease. A further recent fungus found to attack vines is Eutypiose, which was first identified in French vineyards in 1977. By 1989, one-third of the vineyards of the Cognac *appellation* and one-fifth of the Cabernet Sauvignon vines between 15 and 25 years old in the Bordeaux region were reported to be affected (Anon., 1989). Currently Eutypiose is proving resistant to chemical fungicides, and since vines take up to six years to reveal its symptoms, the search for a satisfactory cure is likely to be lengthy.

The two most significant virus diseases affecting vines are fan leaf and, in California, Pierce's disease. Fan leaf leads to leaf deformity, yellow colouring of leaves, early maturing canes, enlarged nodes, premature dropping of flowers, and little or no grape crop. It is spread by *Xiphinema index* Thorne & Allen nematodes and also by propagating wood from infected vines. The only cure is to grub up all infected vines, and to leave the soil fallow until the roots of the removed vines have rotted. Many *Vitis vinifera* vines are tolerant of fan leaf when grown on their own roots, but they are highly susceptible to the virus when grafted onto American rootstocks. Fan leaf has therefore developed in Europe in the wake of the use of American rootstock following the outbreak of phylloxera during the late nineteenth century. Pierce's disease, on the other hand, is apparently restricted to California, where it gives rise to a yellowing of leaves followed by their apparent scalding in late summer, the slow growth of shoots in the following season, and the eventual death of the vine. The disease seems to have originated from the Gulf coastal plain region of the United States, where indigenous vines are found to have resistance to the disease (Hewitt, 1958). Since it is spread by insects the only cure is to pull up the infected vines and subsequently to use insecticides to protect new plantings.

The phylloxera aphid *Phylloxera vitifoliae*, also known as *Dactylasphaera vitifoliae* (Shimer), *Daktulosphaira vitifoliae* (Fitch) and *Phylloxera vastatrix* (Planchon), was responsible in the nineteenth century for the most catastrophic disaster to affect European viticulture (Winkler, 1962: 467). While its effects can now be countered, when it first arrived in Europe it led to widespread destruction of vineyards, and considerable social and economic disruption. Unlike many other aphids, its only host is the vine. The life cycle of the phylloxera aphid varies according to the species and variety of vine with which it is associated, and also on the nature of the physical environment in which it is found (Williams and Shambaugh, 1988). Its native habitat is the United States of America to the east of the Rocky Mountains. Here its life cycle has the following stages (Figure 8) (Davidson and Nougaret, 1921; Ordish, 1987; Winkler, 1962):

Stage 1: From an overwintering egg found on a vine, a female fundatrix larva emerges to find a leaf, where it produces a gall by feeding on the sap. On developing into an adult, this female lays eggs which in turn develop into more female larvae. These find new leaves, make galls, mature and produce a further generation of eggs, larvae and adults. This process continues until the third and fourth generation when there will be some larvae which migrate down to the roots of the vine. Ordish (1987: 52) has estimated that one winter egg can potentially produce 4,800 million gall-living females by midsummer when the fourth generation emerges.

Stage 2: The larvae that have migrated down to the roots develop as female root living adults, which then lay eggs. From these eggs root living larvae emerge, and mature into female adults, which lay more eggs and produce up to seven further generations of root living larvae in a year. By the late summer, three different types of larvae have emerged: root livers, wanderers, and winged sexuperous larvae. All are still female. The root livers continue to destroy the vine by producing root galls and stopping further root growth. The wanderers climb up through the soil in search of new vines, which they subsequently infest. The winged adults fly a short distance also in search of new vines.

Stage 3: If a wanderer successfully finds a new vine, it overwinters among the roots. When in the spring the temperature reaches 10°C it matures and lays eggs which in turn set in motion the root living cycle described above.

Stage 4: The winged migrants find vine leaves and lay eggs, with each adult laying only one female or one male egg. These eggs hatch and develop into adults which reproduce sexually. The female then lays a single overwintering egg, which completes the cycle.

This is the cycle which is usually found in phylloxera's native habitat in the eastern United States and in other humid areas, where all of the different types of the insect are found. However, in California, although the winged migrants of Stage 4 (Figure 8) lay eggs which give rise to sexual forms, these forms do not mature. Consequently 'the above-ground forms of the life cycle do not occur in California' (Winkler, 1962: 468). In Europe on *Vitis vinifera*, likewise, the leaf stages are also uncommon, and phylloxera is generally found mainly in its root living form. Most American vine species show resistance to the effects of

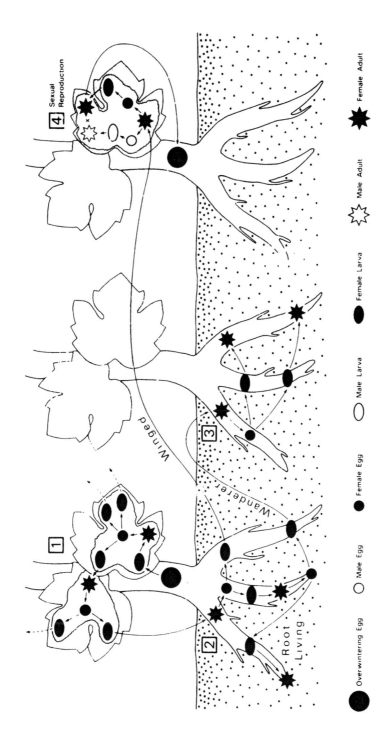

Figure 8 The life cycle of *Phylloxera vitifoliae*

phylloxera, particularly when it is found mainly in the leaf gall form, but *Vitis vinifera* varieties are killed by the root galls and deformities. The solution to the problems facing Europe's vine growers once phylloxera had taken hold was thus to graft scions of *vinifera* varieties on to American rootstocks, but it took a long time for this to be accepted.

The vine in the physical environment

As Rankine *et al.* (1971: 33) have stated so clearly, 'The influences of both soil and climate on the more subtle differences in the composition of grapes, which are so important to wine quality, are little understood'. However, there is little doubt that 'Climate, soil, and slope are the key physiographic factors in viticulture' (de Blij, 1983: 81). This section examines the broad influence of each of these factors in turn.

Climate is the most important factor determining the broad areas of the world where it is possible to grow grapes successfully for wine production (Figure 6) (de Blij, 1985a; Fisher, 1978). Ideally, the average yearly temperature should be approximately 15°C, with a summer maximum of 22°C and a winter minimum of 3°C (Burroughs and Bezzant, 1980), although some vines are cultivated for wine much nearer the equator in countries such as Peru, Bolivia, Kenya and Tanzania (Stevenson, 1988). High summer temperatures are necessary to ensure that the fruit is able to set and then ripen, and cool winters enable the vine to rest. The annual distribution of temperature is also important, and for quality wines it is useful to have temperatures in the order of 30°C in the months of August and September in the northern hemisphere, and February and March in the southern hemisphere. Temperature is also critical in the spring months, since frost at this time of the year can severely reduce the crop by damaging the potential fruiting buds. There are now several ways of reducing the effects of frost, by using heaters in the vineyards, by mixing the air with large fans, or by spraying the vines with water. This last method protects the vines after the initial water has frozen onto the plant by preventing the temperature from falling much below 0°C. However, all of the methods are expensive, and mostly beyond the means of small wine producers. Cold and wet weather later in the spring also gives rise to other problems for vine growers, most notably the conditions known in French as *coulure* and *millerandage*. The former is the dropping of flowers before they are satisfactorily pollinated as a result of a lack of sufficient nutrients, and the latter is when some or all of the grapes in a bunch remain small and hard as a result of uneven flowering (Stevenson, 1988).

Temperature and sunshine are also important in determining the final character of a wine, because of their effects on the ripening of grapes (Kliewer, 1964; Kliewer and Schultz, 1964). Generally, in hot, sunny climates the grapes ripen rapidly, have high sugar levels and low acidity, whereas in cooler climates the acid content of the grapes is higher, the colour deeper, and the sugar content less. Wines from hot climates are thus often more alcoholic and flabby than those from cooler climates, which produce fresher, more acid wines. These differences can to an extent be countered at the vinification stage, but they are also influenced by the variety of vine grown and the method of training used. Thus, for example, in north-west Portugal, the high training of vines on arbours and up trees ensures that the grapes retain high acid levels because little heat is reflected from the soil and the prolific leaves shade the grapes from the sun. In climates nearer the poles and thus with shorter growing seasons, early ripening varieties of grape, such as the Müller-Thurgau, Gutedal or Faber in Germany, or the Chardonnay and Pinot Noir in France, are preferred. Late ripening varieties, such as Nebbiolo, the great grape of north-west Italy, and Cabernet Sauvignon are usually grown in lower latitudes.

Rainfall figures need to be considered in the context of temperatures, which significantly affect the evapotranspiration rate. Most viticultural areas in the world have rainfalls between 400 and 800 mm per year. Where annual rainfall levels are below 400 mm it is usually necessary to provide some irrigation if heat summation figures reach much above 1,400 degree (C) days (Jackson and Schuster, 1981). These heat summation figures provide a general indication of areas suitable for viticulture, and are calculated by subtracting the figure of 10°C from the average monthly temperature, multiplying that figure by the days of the month, and then adding the monthly figures together to produce an annual figure. Thus the northern French vineyards of Champagne have approximately 1,050 degree (C) days a year, whereas the Paarl district of South Africa has between 1,400 and 1,450 degree (C) days, and the northern regions of Chile between 1,600 and 1,700 degree (C) days a year (Winkler, 1962; Jackson and Schuster, 1981; Stevenson, 1988). Seasonal distribution of rainfall is also very important, with most rain being needed in the winter and early spring, provided that the soils are able to store enough for the full growing season requirements. Too much rainfall in the summer and autumn is harmful, since it leads to watery grapes, although in a very dry year some rain just before the vintage can be useful. A reliable supply of water is critical, and irrigation is now quite widely practised, especially in countries ouside Europe, to ensure that water is available at crucial times, particularly when the grapes are ripening. As with frost, hail can be a serious problem in vineyards, destroying leaves and flowers and thus leading to crop failure. High levels of humidity are usually associated with fungus diseases, and are generally to be avoided in vineyards. In

43

certain circumstances, though, as with the Riesling grape along the banks of the Rhine and with the Sémillon in Bordeaux, when levels of humidity are just right they encourage the development of 'noble rot', known as *pourriture noble* in French and *Edelfäule* in German. This is caused by none other than *Botrytis cinerea*, which also causes grey rot (Table 3), but when it attacks the ripe grapes in these specific localities it causes the particular concentration of sweetness that is a hallmark of their finest wines.

Soils provide the interface between the vine and the underlying geology (Bourne, 1980; Wallace, 1972; Pomerol, 1989). As with climate, particular varieties of vine tend to do well on certain soils. Thus the Gewürztraminer seems to prefer fairly deep fertile and loamy soils, whereas the Chardonnay excels on poor quality calcareous soils as in Burgundy (Figure 3) (Robinson, 1986). Generally it appears that soil structure is more important than fertility or chemical composition in influencing vine and grape development. This has been shown particularly clearly for the wines of Bordeaux by Seguin (1965; 1970), who has suggested that it is the drainage capacity of the soils that is of most importance in influencing the quality of a wine. Thus, the best vineyards in Bordeaux are often found on readily drained soils near, but well above, rivers and streams. Here the vine roots penetrate deeply into the ground in search of water and nutrients. The deep rooted nature of vines means that it is essential that soils are not obstructed by the development of hard impervious layers or concentrations of salts, and that the water table is not too high (Bourne, 1980). Loams and gravels, rather than clays and silts, tend to be most suitable for vines, and geological formations with many vertical bedding planes or fissures, such as schists and limestones, are particularly suited to the vine's root systems. Very fertile soils tend to produce vines of high vigour and often result in crops with a high yield but low quality. To some extent this situation can be overcome by restricting the vigour of the vine, through the selection of specific rootstocks and the use of particular trellis systems. In Europe, though, many of the best wines are produced from vineyards with soils of low natural fertility where it would be difficult to cultivate any other crops, as in the Upper Douro valley of northern Portugal (Figure 4). Vines do, nevertheless, require a sufficient supply of basic nutrients: nitrogen and magnesium for the production of chlorophyll; phosphorus, particularly important for the development of flower trusses; potassium, for the production of starch and sugars; and calcium for neutralising excessive acidity. It may well be that high potassium/nitrogen ratios, tending to favour the production of starch and sugar rather than chlorophyll, are important for the successful production of grapes for high quality wines (Hancock, pers. comm., 1989). Some trace elements, most notably boron, also appear to be of significance in influencing the quality of a wine. Further important characteristics of the soils are their colour and reflectivity. Thus unvegetated gravels and rocky soils with high reflectivities tend to lead to a more rapid ripening of grapes than do dark soils. Whatever role the trace elements and mineral contents of wines play in influencing their flavour, it is clearly the case that these characteristics are largely determined by the soils and rocks on which the vines are grown (Pomerol, 1989).

There is, though, considerable debate about the importance of soils in determining the quality and character of specific wines. In Winkler's (1962: 64) words, 'In most of the famous grape-growing and wine-producing regions of Europe and elsewhere, the belief is firm that a particular soil has much to do with the local success. Perhaps it does in some instances, but in others it is largely irrelevant'. The battle lines in this debate have often been too starkly drawn, caricatured by north American wine producers arguing that soils are of little significance (Jekel, 1983a, 1983b), and European, particularly French, producers championing the importance of soils. Part of this disagreement has derived from the frequent mistranslations of the French word *terroir* simply as 'soil'. As Prats (1983: 16) has emphasised, 'When a French wine-grower speaks of a 'terroir', he means something quite different from the chemical composition of the soil . . . The terroir is the coming together of the climate, the soil and the landscape'. This finds its expression in the individual microclimates of specific vineyards, which owe much to their local relief characteristics (Dickenson, 1990). It is the interaction between slope, aspect, soils, altitude, humidity, shelter and drainage, and the way in which these factors influence the critical elements of sunshine, temperature and wind, that distinguishes between the nature of wines made from different vineyards. Generally in the northern hemisphere the best sites are found on gently sloping south or south-westerly hillsides, which attract most of the warm afternoon sunshine, and avoid the problems of frost and excessive humidity. Such considerations, though, are less important in lower latitudes, where frost is not a problem, and where the main consideration is to find locations which are sufficiently cool.

Until recently, the variety of vine grown, the way in which it was trained, and the local microclimate were the main factors influencing the character of a wine. Vinification methods varied little, and wine presses dating from medieval times were very similar to those to be found throughout the world in the nineteenth century. Recent developments in vinification, however, have transformed the wine industry. The most significant of these have probably been the introduction of cool fermentation, the use of active dry forms of cultured yeast, and the development of new methods to stabilise wines before bottling. While poor quality grapes will never make great wines, modern methods of vinification can certainly make drinkable wines from mediocre grapes. Moreover, they can also overcome some of the problems of annual variability, being able to compensate to some extent for the vagaries of the weather.

VINIFICATION: CHEMISTRY AND THE HUMAN MILIEU

Before the mid-nineteenth century the process of alcoholic fermentation was unexplained. In Amerine's (1965: 75) words, it 'remained an empirical process, subject to the vagaries of nature, with man exercising, at best, only partial control'. It was known that pressed grapes would ferment into wine, and that if wines were left open to the air for any length of time they would become unpalatable, but there was no explanation for these processes. It was not until the series of experiments begun in the 1850s by Louis Pasteur that people began to have any real knowledge of why fermentation took place and of how to control it. In 1863 Pasteur began his research on wine, at the express command of the Emperor Napoleon III, who was concerned to establish why so much wine deteriorated before it was consumed. His results were published first in 1866, and then in a second edition in 1875 (Pasteur, 1866; 1875). Pasteur's research revealed that wine deteriorated mainly as a result of the action of micro-organisms, and that these could be killed by heating the wine to 55°C in the absence of oxygen. However, his work went far further than this, and he established the role of yeasts, the distinction between aerobic and anaerobic processes, and the products of alcoholic fermentation. Moreover, he also showed that it was the presence of small amounts of oxygen that enabled wines to mature. From these beginnings, numerous changes have taken place in the way in which wine is fermented and matured, so that much greater control is now possible over all stages of vinification. The basic principles involved in alcoholic fermentation are relatively simple. It is the action of yeasts that converts the sugar in grapes into alcohol. However, many other processes take place in the vinification of wines, and in order to comprehend the historic development of the wine industry it is important that the main features of these processes are understood.

Yeasts and bacteria

In the making of wine there are three stages at which yeasts and bacteria are of importance: the conversion of sugar to alcohol by yeasts; the conversion of ethanol (ethyl alcohol) to acetic acid by acetic acid bacteria; and the conversion of malic acid to lactic acid by strains of lactic acid bacteria.

Yeasts can be differentiated from other types of fungi by usually maintaining a unicellular growth. Until the identification of individual

46

strains of yeast by microbiologists, alcoholic fermentation was a rather haphazard affair set in motion by the naturally occurring yeasts on the skins of grapes. These yeasts are generally found near the pedicels, the stomata and points of damage on the grape skin, and are of many different varieties (Kunkee and Goswell, 1977). Amerine and Joslyn (1970: 331-4) thus note 147 species of yeast reported on grapes and in wines, all of which vary in their ability to ferment sugars, their temperature tolerance (Jacob *et al.*, 1964), their alcohol tolerance, and their effects on flavour. Alcohol tolerance is particularly important, since under normal wine making conditions yeasts are only able to produce a maximum of approximately 16 per cent alcohol by volume. In early experiments, Castan (1927) demonstrated the great variation in the fermentation capacity of yeasts, which when grown in sterile musts of 31.71 per cent sugar produced levels of alcohol ranging from 11.1 per cent by volume to 17.2 per cent by volume. In practice it is usually found that the rate of fermentation decreases appreciably with increases in alcohol content and sticks at levels of between 13 and 15 per cent alcohol, at which point the yeast is inhibited from further activity (Amerine and Joslyn, 1970). Different yeasts, acting at different stages, thus make up the overall process of alcoholic fermentation. Two yeasts are found extensively throughout the main wine regions of the world, *Saccharomyces cerevisiae* and *Kloeckera apiculata*, but they vary in their incidence, and many other yeasts also exist, including species in the genera *Hansenula*, *Candida* and *Brettanomyces* (Amerine *et al.*, 1980). In a classic study of the role of yeasts in fermentation, Domercq (1957) showed how fermentation in red wine musts from Bordeaux was initiated by apiculate yeasts such as *Kloeckera apiculata*. As the ethanol content of the musts built up, these were soon taken over by the more ethanol-tolerant wine yeasts, *Saccharomyces cerevisiae*, and by the end of fermentation several other varieties of yeast, including *Saccharomyces bayanus*, *Saccharomyces chevalieri* and *Hansenula anomola* were found to be present. In recent years, stemming partly from work in California, there has been a tendency to prevent such 'natural' mixed fermentation from taking place through the use of sulphur dioxide, and instead to inoculate musts with particular starter cultures of yeast, designed to ensure the production of wines of a particular quality and character. As Amerine and Joslyn (1970: 334) comment, 'This practice, when properly used, results in a fermentation that begins promptly, proceeds regularly, and goes to completion in a relatively short time'.

Once fermentation has begun, the musts and eventually the young wine are then subject to the action of bacteria of two kinds. Acetic acid bacteria, belonging to the genus *Acetobacter*, are able to oxidize ethanol to acetic acid, and lactic acid bacteria, belonging to the genera *Lactobacillus*, *Leuconostoc* and *Pediococcus* (Amerine *et al.*, 1980) convert malic acid to lactic acid and can also lead to possible spoilage in the wine. Acetic acid bacteria require oxygen for growth, and can therefore be inhibited by minimising the access of oxygen to the wine. During fermentation of red wines, a cap, consisting of the skins, pips and any stalks, naturally forms on the top of the must, and this can readily be subject to the action of the acetic acid bacteria, which will convert the ethanol (C_2H_5OH) to acetic acid, more commonly known as vinegar, ($C_2H_4O_2$) according to the following equation:

$$C_2H_5OH + O_2 \rightarrow CH_3COOH + H_2O$$
ethanol + oxygen → acetic acid + water

This process is usually prevented by keeping the cap moist at all times during the fermentation, either by breaking it up or by pumping the must continuously over it. The addition of sulphur dioxide during and after fermentation is also important in preventing acetic spoilage. When fermentation is over, acetification is also likely unless oxygen is excluded, and it is therefore essential to exclude air by keeping casks or other containers of wine well filled at all times.

Unlike the *Acetobacter*, lactic acid bacteria grow best in the presence of small amounts of oxygen (Kunkee, 1984). While they can lead to an unpleasant smell in a wine and also inhibit the growth of *Saccharomyces cerevisiae*, their main role in vinification is through their action in converting malic acid ($C_4H_6O_5$) to lactic acid ($C_3H_6O_3$). Since this produces a monocarboxylic acid from a dicarboxylic acid it causes a decrease in the sourness of the wine (Fornachon, 1957), and the carbon dioxide gas (CO_2) produced in the reaction can also lead to a slight sparkle in the wine if the fermentation takes place in the bottle. Both processes typically take place in traditionally made *Vinhos verdes* from the Minho region of northern Portugal, which 'owe both their palatibility and their effervescence to this bacterial fermentation' (Fornachon, 1957: 121). In cool wine producing areas, where there is a high natural acidity in the wine which can usefully be softened somewhat, the malo-lactic fermentation can reduce the titratable acidity of the wine by as much as one-third (Amerine *et al.*, 1980), but in warmer areas which naturally produce wines with lower acidities the fermentation may not be so desirable. If malo-lactic fermentation is desired, the wine should be left on its lees at a temperature between 18° and 21°C, but if it is to be inhibited the wine should be removed from the lees as soon as possible, filtered, treated with sulphur dioxide, and increased in acidity to a pH of approximately 3.3.

In several places in the above account, sulphur dioxide (SO_2) has been mentioned in connection with the prevention of undesirable processes during fermentation. The preservative powers of sulphur were known to Roman wine makers, but it has only been during the twentieth century that it has been used extensively to protect barrels, musts and wine. Sulphur dioxide is a normal product of yeast fermentation, but it is also added at various stages in wine making to prevent the activity of bacteria. Yeasts also vary in their resistance to sulphur dioxide, enabling the selective use of certain yeast strains in the fermentation process. Furthermore, the addition of sulphur dioxide can also be used to stop fermentation, thus allowing the retention of some residual sweetness in the wine, although with the current tendency towards wines with lower sulphur dioxide levels chilling or the use of a centrifuge are now more common methods of achieving the same result.

Alcoholic fermentation

The essential role of yeasts in alcoholic fermentation has been noted above. However, the chemistry of fermentation and the products resulting from the transformation of grapes into wine also have important ramifications for an appreciation of the development of the wine trade.

The overall process of the conversion of sugar into alcohol, can be summarised in the Gay-Lussac equation as follows:

$$C_6H_{12}O_6 \rightarrow 2C_2H_5OH + 2CO_2$$
$$\text{glucose/fructose} \rightarrow \text{ethanol} + \text{carbon dioxide}$$

However, alcoholic fermentation is actually much more complicated than this would suggest, and involves twelve separate reactions, with no fewer than twelve enzymes and at least three sets of co-factors as well as several inorganic ions being necessary (Kunkee and Goswell, 1977; Amerine et al., 1980). Theoretically, the products of fermentation should be distributed so that alcohol accounts for 51.1 per cent and carbon dioxide 48.9 per cent by weight, with heat energy also being produced. In practice, though, the alcohol yields are generally somewhat lower than this, with the remainder of the sugar being converted into other products.

The production of heat energy has important ramifications for the progress of fermentation, since it causes a rise in the temperature of the must. In general the optimum temperatures for yeast activity are between 22°C and 27°C (Amerine et al., 1980), but different yeasts tend to be active within different temperature ranges. Moreover, if temperatures are very low fermentation will not commence, or having begun may stop, and high temperatures kill the yeasts leading to the alcoholic fermentation sticking. This latter situation can be particularly dangerous, since the must is especially susceptible to attack by bacteria under such conditions. In practice the critical range for successful yeast activity seems to be between about 15°C and 35°C. Temperature also has effects on the alcohol yield and flavour of the resultant wine. At higher temperatures it seems that small amounts of alcohol are lost by evaporation, thus lowering the potential alcohol yield from a given sugar content. White wines benefit considerably from fermentation at lower temperatures, and Amerine *et al.*, (1980: 211) have summarised these benefits as follows: 'less activity of bacteria and wild yeasts, less loss of volatile aromatic principles, greater alcohol yield, more residual carbon dioxide, and less residual bitartrate'. It is now generally accepted among oenologists that white wines resulting from slow, cool fermentations have a much better bouquet and a finer quality than wines made from the same musts but fermented more rapidly at hotter temperatures. For these reasons there has been a widespread shift during the 1970s and 1980s to methods of cool fermentation in many of the warmer wine producing areas of the world, often using stainless steel fermentation tanks with refrigeration to control temperatures. Must chilling is also important in warm climates where the grapes are often at temperatures above 30°C prior to crushing. While such techniques were initially mainly introduced in California, Australia and South Africa, wine makers in many parts of southern Europe, and particularly in Italy, are now developing cool fermentation techniques and as a result are producing fresher wines of a very different character from those of their predecessors.

Since alcoholic fermentation in wines is basically the conversion of grape sugar to alcohol, wine makers encounter serious difficulties if the grapes are insufficiently ripe. Not only will the grapes be acid, but they will also not have enough sugar to convert to a sufficient degree of alcohol. This is particularly a problem in northern areas of Germany and parts of northern France, where a solution, known as chaptalisation after its French inventor Chaptal, has been in use since the nineteenth century. This involves the addition of sugar to the must at the start of fermentation, thus increasing the final alcohol content of the wine. When used in small amounts this can achieve a better balance between the alcohol and other constituents in the wine, but it is a practice which can clearly be abused and when excessive chaptalisation takes place the resultant wines tend to be unbalanced and of poor quality.

Methods of vinification

It is the task and art of the wine maker to turn good grapes into fine wine. Spurrier and Dovaz (1983: 38) thus comment that 'The oenologist's mastery of vinification techniques is just as important as the grape variety, the soil and the climate . . . A good oenologist can stamp a wine with distinction; a bad oenologist can produce a bad wine even from excellent grapes'. Each *vigneron*, or wine maker, builds on local traditions, legal requirements, and his or her own skills and experience to create a particular style of wine. Despite this diversity, there are common elements to all wine making.

Figure 9 The traditional method of treading grapes: the vintage at Quinta da Foz, Pinhão, northern Portugal

Source: author 4th October 1983

Figure 10 The construction of a modern winery, with stainless steel fermentation tanks, Cave des Hautes Côtes, Beaune
Source: author 27th April 1987

Red wine production generally begins with black grapes, although in some areas, such as Chianti in Italy, white grapes are also used. The grapes must be picked with just the right balance between sugar and acidity, and they are then usually destalked and sometimes partially pressed, before being taken to fermentation vats or tanks. Traditionally, wines were pressed by foot in great stone or wooden containers (Figure 9), but modern methods of vinification use enclosed tanks (Figure 10). Here the natural yeasts, or cultured strains where used, start the process of alcoholic fermentation. Sulphur dioxide is now also usually added prior to the commencement of fermentation in order to protect the must against the actions of acetic acid bacteria, and to inhibit wild yeasts, allowing the selected yeast strain to dominate the fermentation. The length of time and temperature of fermentation are critical for determining the final character of the wine. Short fermentation of one or two days produces light wines, and long or hot fermentations of several weeks or more are used to extract the maximum colour and tannin from the skins for wines that are to be kept and matured. For wines designed to be drunk young, the must is sometimes taken off the skins, seeds and stalks after a couple of days of fermentation, and it is then allowed to complete its fermentation in a separate tank. During normal fermentation the cap has to be continually reincorporated into the must. This is done increasingly by mechanical means, such as autovinification, where the carbon dioxide produced as a by-product of fermentation operates a valve system which regularly sprays must on the cap, or, more commonly, by *remontage*, which is simply the pumping and spraying of must over the top of the cap by means of a sprinkler system. When fermentation is complete, and the sugar has been converted to alcohol, the new wine, known in French as *vin de goutte*, is run off. The remaining skins and stalks left at the bottom of the fermenting vat are then pressed in a vertical press to produce a darker more tannic and more acid wine, known as *vin de presse*. These two wines can be blended together in various proportions to suit the particular taste of the wine maker. The new wine then usually undergoes malo-lactic fermentation before being left to mature, with the better quality wine being aged in wooden casks. The pressed skins and stems are then usually used as fertiliser, having previously been distilled to make a cheap spirit, known in French as *marc*, in Italian as *grappa*, and in Portuguese as *bagaçeira*. In earlier centuries water was also sometimes added to the skins and stems and refermented to produce a drink, low in alcohol content, and known in French as *piquette*, which provided a low quality drink for the poor.

Methods of making white wine are broadly similar, with the major difference being that the grapes are usually pressed first, so that the juice is fermented without being in contact with the skins. Most white wine is made from white grapes, although it should be noted that in some wine making districts, most notably Champagne, black grapes are also used in white wine production. After the grapes have been crushed and destalked they are usually drained in a horizontal membrane press. A light pressing may be added to the free run, after which further pressings are kept separate. The use of sulphur dioxide at this stage should be kept as low as possible, with oxidation being kept to a minimum through the use of cold temperatures by, for example, chilling the must. Temperature control of fermentation and the use of selective yeast strains are essential for the production of clean, fresh white wines. If residual sweetness is required in the wine, fermentation can be stopped by the addition of sulphur dioxide before the sugar has been completely converted into alcohol. Moreover, some producers are now experimenting with the fermentation of part of their white wine in contact with the skins for a short time in order to give it added flavour. In some circumstances, when the grapes have very high sugar levels, the yeasts are killed by the level of alcohol before they have converted all of the sugar. This also gives rise to wines that are sweet, but in this instance they additionally have high levels of alcohol. Following the alcoholic fermentation, the wine maker has to decide whether or not to allow malo-lactic fermentation to take place, and if this is not desired the lactic acid bacteria are killed, again through the use of sulphur dioxide.

While widely differing wines can be made through variation in these basic techniques, there are also a number of more specific vinification techniques used to produce certain other types of wine. Traditional champagne, made using the *méthode champenoise*, thus involves the addition of a sugar/wine solution, known as *liqueur de tirage*, and yeasts to each bottle after the first fermentation is over. This sets in motion a secondary fermentation within the bottle, the carbon dioxide from which is dissolved in the wine. Fortified wines, such as port, are made by the addition of alcohol during fermentation. This stops the natural fermentation of the must, with the degree of sweetness being influenced by the point at which fermentation is stopped. The development and introduction of such techniques is examined in detail in Chapter 8 in the context of the changes that occurred in the wine trade during the seventeenth and eighteenth centuries. More recent changes in the technology of vinification resulting from advances in biochemistry and microbiology, and leading to the introduction of new wines designed to create new markets, are also analysed in the penultimate chapter on the wine industry in the twentieth century.

ELEVAGE: THE AGEING AND MATURING OF WINE

The variety of grape and methods of vinification chosen do much to determine the character of a wine, but the treatment of a young wine, its maturation, and the way in which it is 'brought up', its *élevage*, also play an important role in developing its character. As the previous section has highlighted, a young wine is subject to considerable potential deterioration, not only through the action of bacteria, but also from oxidation as the wine ages. If wine is left open to the air, either in an open bottle or in a partly filled cask it rapidly degrades to acetic acid, acetaldehyde and sometimes ethyl acetate. However, very slow oxidation, such as that achieved in a full cask, leads to the development of the complex aromas and flavours so typical of a good wine. These are created through the slow oxidation of the tannins and flavour compounds in the wine. The harshness of the tannin lessens, and in a red wine the bright ruby colour of its youth fades to a mellow red-brown, with perhaps a mahogany tinge visible at the edge when a glass of the wine is tilted gently on its side. Eventually, as a wine ages and oxidises further, these qualities fade and it goes past its best. However, the length of time necessary for these processes varies depending on a wide range of factors, including grape variety, the state of the grapes at the time of the vintage, methods of vinification, and the subsequent treatment of the wine.

The critical variables influencing the speed at which a young wine will age are access to oxygen, temperature, and light. Throughout the medieval and early modern period all wine was drunk young, usually within a year of the vintage. This was because it was stored in barrels, which gave access to the air, and the wines therefore rapidly became unpalatable as they turned to vinegar. Although the Romans had discovered the art of sealing their wine amphorae, and thus of preserving wine for a longer period, this skill appears to have died out with the introduction of barrels towards the end of the first century AD. It was not until the use of corked glass bottles beginning in the late seventeenth century that vintage wines, as they are known today, began to be developed.

While it is difficult to generalise, wines designed for drinking young will usually have been made from grapes with relatively low tannin content, for example Gamay or Merlot rather than Nebbiolo or Cabernet Sauvignon, and they will also normally have been fermented for a relatively short time in contact with the skins. The resultant wines will be low in tannin, with little potential to improve with age, and they will be bottled soon after vinification. *Macération carbonique*, the vinification of wine under the pressure of carbon dioxide, is also a useful method for the production of soft wines for early drinking, as in Beaujolais. Any suspended matter or impurities in a young wine will first be removed through a process known as fining in which a coagulant, such as gelatine, egg-white, bentonite or isinglass, is added to the wine. This reacts with the proteins and tannin to produce a sediment which can then be separated from the clear wine. As well as fining, or sometimes instead of it, the new wines are also filtered off their lees. Furthermore, all young wines are supersaturated with respect to potassium acid tartrate (Amerine *et al.*, 1980), and in order to avoid this precipitation in the bottle at low temperatures, wines are usually refrigerated for between one and two weeks before fining at temperatures of between -5.5°C and -3.9°C. This precipitates the tartrate crystals which can then be removed during filtration. Some wines, particularly those designed for the mass consumer market, are also pasteurized in order to kill any surviving micro-organisms, but using modern methods of vinification and hygienic wine making conditions this is increasingly becoming unnecessary.

Fine wines designed for ageing are treated very differently. These are usually matured in wood for several years before being fined and bottled as described above. Ageing in wood achieves two main objectives: it enables the slow oxidation and development of the wine to take place by allowing small amounts of air to come into contact with the wine during the process of racking, and it also adds tannin and some flavour. Wines for ageing are usually left to settle for a short time in large tanks, enabling the heaviest sediments to be removed, before being put into wooden casks or another large tank where, if desired, the malo-lactic fermentation can take place. The wines are then put into new casks to mature, usually for two years for fine wines from Bordeaux, but up to perhaps ten years for a Barolo from Italy in an exceptional vintage (Belfrage, 1985). While in cask the wines are regularly racked, by transferring them from one cask to another, usually about once every three months. This separates them from the deposits and lees, which mainly consist of dead yeasts, colouring matter and tartrates, and it also aerates the wine, thus assisting in the process of oxidation. The second main benefit of ageing in wood results from the properties of the barrels or casks themselves. Young wines will have some tannin in them from the skins of the grapes, but further tannin is derived from the wood of the barrels, and through its action as a fining agent it precipitates colouring matter in the wine and assists in its clearing. Depending on whether new or old wood is used, the barrels also add varying amounts of flavour to the wine, and the wood tannin provides further structure to the wine. Traditionally oak has been used for the finest quality wines, although other woods, such as chestnut are also used, with each imparting something of its flavour to the new wine.

After ageing in wood most wines are then bottled and left to mature. For wines that have been pasteurized no further improvement takes place in the wine, and in general it loses its freshness and colour quite rapidly. Other wines, particularly good quality reds, improve appreciably with bottle age. Most white wines are best drunk quite young, usually well under five years, but the great red wines of Bordeaux from the best vintages do not reach their best until over a decade in the bottle. The chemical processes involved in bottle ageing are not well understood. They are partly due to the oxygen already present following bottling, but in the long term, the bouquet of a bottle aged wine owes more to its being aged under reductive conditions. Tannins and anthocyanins condense, gradually forming larger molecules which settle out as a deposit in the bottle, the wine becomes more garnet in colour, and reductive esters are formed from the combination of acids and alcohols in the wine (Peynaud, 1987).

WINE TRADITIONS: ANCIENT AND MODERN

In parts of the world it is still possible to find vines grown, and wines made, much as they must have been for centuries. However, these corners of the viticultural landscape are fast disappearing. They can still be found in isolated districts of northern Portugal or occasional farms in hidden parts of Greece and Italy, but they are fast being overtaken by the demands of modern capital. The way in which viticultural and vinification practices have changed through the centuries forms the focus of attention for this book. In seeking to explain these changes, it explores the structural interaction between ideologies and economies, between social organisations and political entities. At different periods, different conditions and processes have dominated this involvement of people in environments of their own making; wine and the vine have meant different things to people of different eras. However, deep within the cultural identity of southern European societies lies a symbolic imagery in which wine and the vine have a special role. As economic considerations have come to dominate the minds and imagination of capitalist society, this role has been suppressed, but it remains of considerable significance in any understanding of the importance of viticulture throughout the world in the late twentieth century.

3

THE ORIGINS OF
VITICULTURE:
SYMBOLS AND MYSTERIES

The ὠμοφαγία and the bestial incarnations reveal Dionysus as
something much more significant and much more dangerous
than a wine-god. He is the principle of animal life, ταῦρος and
ταυροφάγος, the hunted and the hunter – the unrestrained
potency which man envies in the beasts and seeks to assimilate.
His cult was originally an attempt on the part of human beings
to achieve communion with this potency. The psychological
effect was to liberate the instinctive life in man from the bondage
imposed on it by reason and social custom: the worshipper
became conscious of a strange new vitality, which he attributed to
the god's presence within him.

(Dodds, 1960: xx)

By the end of the Tertiary era plants belonging to the genus *Vitis* were to
be found scattered throughout Japan, eastern Asia, north America, and
Europe (Billiard, 1913). During the subsequent cold and warm spells of
the Quaternary this wide extent was reduced, so that all of its species
came to lie approximately within the latitudinal band between 30° and
50° north (Figure 6) (Amerine and Wagner, 1984). At the same time it
seems that the predecessor of most modern wine grape varieties gradually
evolved under the influence of human selection from wild vines, with
early evidence for its existence being found in various parts of southern
France, Italy, Greece and the eastern Mediterranean (Billiard, 1913;
Younger, 1966; Núñez and Walker, 1989). However, as Levadoux (1956)
has argued, it is not easy to differentiate between the wild *silvestris* and
the cultivated *sativa* forms of vine by morphology alone. Consequently, it
is extremely difficult to identify the place, or places, within their
extensive natural distribution where people first began to cultivate vines
for the production of wine. Negrul (1938, 1960) has suggested that the
hearth of the domestication of the grape vine was probably Asia Minor

and Transcaucasia, where gradual selection of vines with small juicy fruits took place from about 8000 BC. However, there is little direct evidence for particularly early cultivars of *Vitis vinifera* in this region, and Núñez and Walker (1989: 220) argue that

> Doubt must therefore be cast on assigning the origin of viniculture to such regions as Transcaucasia on the basis of supposing that present-day diversity there of spontaneous *V. vinifera* necessarily implies either a greater time-depth there during which the genus developed, with regard to other regions of western Eurasia, or a greater likelihood that it harboured the ancestor of all early prehistoric cultivars.

Moreover, increasing evidence from south-east Spain (Stevenson, 1985) suggests that the western Mediterranean may also have developed a separate indigenous viticultural tradition by 2500–2000 BC.

At a global scale, the development of viticulture in Eurasia, and the observation that none of the other *Vitis* species became used at all widely for wine production, suggests that the cultivation of vines and the origin of wine was not purely a result of the natural distribution of this particular genus. Instead, it is likely that it was closely related to the social, economic and ideological stuctures that emerged in the Caucasus and northern Mesopotamia in prehistoric times. It is indeed remarkable that, despite the wide global distribution of species belonging to the genus *Vitis*, its use for wine production was so limited until the early modern period. Thus, no alcoholic beverage made from the grapes of vines appears to have been produced in the Americas before the arrival of Europeans in the sixteenth century, although numerous species of the genus *Vitis* were to be found there. Instead, the indigenous peoples of Meso-America made alcoholic drinks such as *pulque* from the *maguey*, *tesgüino* from the sprouted kernels of maize, and *balché* from honey mead flavoured with the leaves of *Lonchocarpus* (Bruman, 1940). Interestingly, as Bruman (1967) points out, many of these drinks were produced using largely wild materials, and it is highly likely that wine itself was initially made in the Caucasus and Mesopotamia from the grapes of wild vines.

Viticulture and wine production on a global scale were thus initially quite localised. In Isaac's (1970: 69) words, 'Wild grapes, *Vitis silvestris*, grew widely in Europe and Asia but the original domestication of wine grapes, *V. vinifera*, seems to have taken place between the Caucasus, eastern Turkey, and the Zagros range'. The reasons for the development of this hearth, and for the particular pattern of subsequent spread of viticulture and wine consumption away from this region form the focus of attention in this chapter. In seeking to explain these distributions, it is argued that the symbolic requirements of religion were critically important for the spread of viticulture. Indeed, the origin of viticulture and wine production can be interpreted as being part of the development of a particular religious and ideological conscience that emerged in, and later spread from, the mountainous region near the borders of the modern states of Iraq, Syria, Iran, the USSR and Turkey. The subsequent transfer of aspects of this religious experience westwards into the Mediterranean established the foundation for the eventual spread of viticulture and commercial wine production into other parts of the world where its symbolic significance was previously unknown. While it is possible that vines were cultivated for domestic wine consumption elsewhere in the Mediterranean region prior to contact with this cultural influence, and that Transcaucasia may not have been the only hearth of viticulture (Núñez and Walker, 1989), it was nevertheless in this region that wine production first came to achieve major symbolic and economic significance.

Following partly from the work of Hahn (1896; see also Kramer, 1967), who saw the domestication of herd animals such as cattle as being closely linked with the development of a complex of religious ideas, Isaac (1970) has argued that the development of agriculture was underlain by a religious revolution which replaced the widespread mother goddess cult with a new set of beliefs developed around the idea of the death and rebirth of a primeval deity. In Isaac's (1970: 107) words, 'The new world view rested on the belief that the world emerged from the death of a primeval being or god, whose slain and severed body came to constitute the existing world'. This belief was paralleled, and then replaced in importance by myths concerning the emergence of plant life from the slain body. Subsequently, the development of fertility myths, which must be re-enacted to ensure continued agricultural productivity, led to the emergence of a range of symbolic rituals associated with the agricultural seasons.

The vine and wine can both be seen to have had four important roles in this evolving religious experience, reflecting the metaphorical relations between the symbol and the symbolised:
(a) the vine itself, in its apparent 'death' in the form of a leafless cane in winter, and in its dramatic rebirth and growth in the spring, became a highly pertinent symbol for the death and rebirth of the god and for the whole agricultural cycle;
(b) in turn, the products of the vine came to take on symbolic and ritual significance, holding within them the secret of rebirth, since they could survive beyond the autumnal and winter 'death' of the parent vine in the form of dried fruit or wine;
(c) moreover, wine with its ability to intoxicate and engender a sense of 'other-worldliness' provided a means by which people could actually come into contact with the gods. At one and the same time it could also

represent the dichotomy between 'good', when taken in small quantities, and 'evil', when taken in excess;

(d) fourthly, the cycle itself linked back to human fertility, through wine's ability to break down reason and social customs, and its consequent role as a catalyst for human intercourse.

While each of these symbolic roles is explored in detail within this chapter, particularly in the context of the development of Dionysiac and Bacchic ritual, it is also important to recall Bourdieu's (1977) observation that symbols may be used by those in power in order to retain that political or economic power. The ownership of vines, and the knowledge of how to turn grapes into the wine which enabled the followers of a religious belief to come into contact with the gods, may well initially have been a closely guarded secret held by a religious and political elite as a way of reinforcing their power and control over the people that they ruled.

SOURCES AND EVIDENCE

The sources available for a reconstruction of the practice and spread of viticulture and wine making are numerous and varied, and much of the controversy surrounding the spread of viticulture has been the result of differences in their interpretation. In any resolution of these debates, it is essential to distinguish between evidence concerning the growing of grapes, their vinification, the transport and trading of wine, and its final consumption. Thus, evidence for the cultivation of grapes does not necessarily mean that these grapes were turned into wine, and the presence of amphorae used to transport wine need likewise only indicate that the place where they were found participated at some stage in the wine trade. Moreover, as Williams (1977) has pointed out, the only really reliable archaeological evidence for the cultivation of vines is the survival of vine wood or leaves; the presence of grape seeds, which is often accepted as proof of *in situ* viticulture, may in practice actually derive from imported raisins or even from wine imported in barrels. Likewise, the only reliable archaeological evidence for wine making is the discovery of evidence of vinification, usually in the form of a wine press. Furthermore, there are fundamental problems in trying to interpret the evidence of vine pollen, which by itself can not be used as an indication of wine production (Stevenson, 1985; Núñez and Walker, 1989)

The evidence for viticulture and wine can broadly be grouped into two types, the archaeological and the literary. Within the former, three broad categories of source can be distinguished: that showing direct evidence of viticulture or vinification; that illustrating the production or use of wine; and that recording aspects of the wine trade and viticulture on tablets or papyri. Direct archaeological evidence of viticulture, vinification and the wine trade consists of surviving vine material in the form of wood, leaves and seeds (Núñez and Walker, 1989); the remains of wine presses and cellars for storing wine (Rossiter and Haldenby, 1989); the presence of containers for wine, either in the form of amphorae, drinking cups or wine barrels; and also the labels on such containers, indicating the type and origins of the wine they once held. Of all the sources, this direct archaeological evidence is the most reliable, but it is variable in quality and survival, and its quantity also varies considerably depending on the period under investigation. Thus, direct evidence of vinification is comparatively rare, although numerous classical amphorae have been discovered throughout the Mediterranean region enabling a detailed reconstruction of the Roman wine trade to be undertaken (Tchernia, 1986). The second type of archaeological evidence, that of illustrative material, is generally more widespread, certainly with respect to its temporal coverage, but it is also more difficult to interpret. Illustrative material concerning all aspects of the cultivation of vines, the production and consumption of wine, and its use in a range of rituals is to be found in wall paintings, frescoes, mosaics, vase paintings, bas-reliefs, and statuary. However, the presence of such material at a particular location need not in itself imply that viticulture or wine drinking were to be encountered there. Particularly when such practices had taken on religious significance, their depiction might well reflect their ideological role rather than any practical evidence for vine growing or wine making. Nevertheless, illustrative evidence for the worship of wine gods, for example in the form of statues or wall paintings, does seem likely to indicate that such rituals were practised in the locations where the evidence was unearthed. The third category of archaeological evidence is that of tablets and papyri recording details of vineyards and the trade and manufacture of wine. These include, for example, tablets from Ur dating from about 2400 BC which recorded wine among the temple stores, contract tablets from Nippur in the fifth century BC which gave details of various wine transactions (Younger, 1966), and a papyrus dated 170 BC consisting of a lease of a vineyard near Philadelphia in Egypt (Turner, 1958). The main difficulty with these sources, as with all literary material, is that it is not always possible to determine whether they are describing wine from grapes or from other fruits.

Literary works referring to wine and viticulture generally date mainly from the period after about 500 BC and can be divided into four broad groups: epics, histories and geographies, specific agricultural treatises, and miscellaneous poems and religious writings. The great epic poems, such as the Sumerian *Epic of Gilgamesh* and Homer's *Odyssey*, dealing with tales of heroes, and the relationships between the gods and humanity, tell us much about the religious significance of wine, but they also in passing provide considerable additional information about the use of wine in the daily activities of the people to whom they refer. Historical and geographical works, such as *The Histories* of Herodotus dating from the fifth century BC, and Strabo's *Geography* dating mostly from the early first century AD, also provide a wealth of material concerning the production, consumption and redistribution of wine. However, as with all literary sources, their evidence is not always reliable, and indeed it sometimes disagrees quite considerably with that of the archaeological record. Among the most useful literary works concerning viticulture and vinification are specific Roman agricultural treatises, such as the *De Agricultura* of Cato (234-149 BC), and Columella's *Res Rusticae* written in AD 65. These provide a wealth of information concerning all aspects of the cultivation of vines and the production of wine. There are also a large number of other literary works, mainly in the form of poetry, which either discuss viticulture and wine drinking in detail or give important mentions of them in passing (Dalmasso, 1933). To these must also be added the vast corpus of religious writings, including perhaps most importantly the Bible, which provides a major source for our understanding of the Jewish and Christian symbolism of wine and the vine.

Taken together these literary and archaeological sources provide a wealth of evidence concerning the emergence of viticulture and wine making, and their subsequent spread throughout the Mediterranean region and Europe. Not all of the evidence is complementary, and much still remains unknown about the origins of viticulture. Nevertheless, they provide a basis upon which the reconstruction of the practices of viticulture and vinification, and their symbolic ritual role in the ideology of the peoples of south-west Asia can be undertaken.

THE ORIGINS OF VITICULTURE AND VINIFICATION

Mesopotamia

The cultivation of vines for the making of wine originated some time before 4000 BC and possibly as early as 6000 BC in the mountainous region between the Black Sea and the Caspian Sea (Figure 11), on the borders of the modern states of Turkey, Syria, Iraq, Iran and the Soviet Union (Billiard, 1913; Lutz, 1922; Levadoux, 1956; Negrul, 1960; Younger, 1966; Ramishvili, 1983; Hyams, 1987; Johnson, 1989). It can only be conjecture, but it seems probable that the first wine was discovered by chance, when someone drank the fermented juice of wild grapes that had been collected and stored, perhaps in a pottery vessel. This occurrence must have predated the deliberate cultivation of the vine, and may well have taken place as early as 8000–10000 BC. This probable hearth of viticulture lay to the north of the great plains of the Tigris and Euphrates, which formed the core of the Sumerian, Akkadian, Assyrian and Babylonian empires, and also to the east of the heartland of the Hittite empire in what is now Turkey. It is in the records of these civilisations that the first real evidence for viticulture and vinification is to be found.

By the beginning of the third millennium BC, with the establishment of the Sumerians in southern Mesopotamia in the plains between the Tigris and the Euphrates, there is clear evidence for the cultivation of vineyards. However, wine does not appear to have been common, and both barley and dates were probably the main sources of alcoholic beverages for the majority of the population. Tablets discovered in excavations in Lagash and Ur (Genouillac, 1909; de Thureau-Dangin *et al.*, 1910–14; Woolley, 1934) indicate that in the first half of the third millennium BC vines were grown in a number of small irrigated vineyards, often forming part of temple complexes, and that additional wine was also imported to Sumer from mountains to the east (Younger, 1966). The great cities of the Sumerians, particularly Ur and Lagash, lay at a latitude of about 31° north, and were therefore at the southernmost limit of successful vine cultivation if the climate then prevailing was similar to that of today. These climatic conditions might well therefore be one of the main factors explaining why viticulture never really flourished in these lands which were in any case much better suited to the cultivation of the date palm. The mountains to the east from where the Sumerians apparently imported some of their wine, were possibly the Zagros range of western Iran, and the higher altitudes to be found there will have provided the cooler winters essential for the succesful cultivation of the vine. The great city states of Sumer lay well to the south of the original hearth of viticulture, and it is probable therefore that the land nearer the sources of the Tigris and the Euphrates was more extensively cultivated with vines. The evidence for climatic change in the Tigris and Euphrates valleys is unclear, but it does seem that the region was somewhat wetter than it is today *c.* 3000 BC, and that there was then a period when it became drier up to about 1500 BC before there was another wet peak *c.* 1000 BC (Bowden *et al.*, 1985). Viticulture in southern Mesopotamia could therefore have been more successful at the beginning of the third and first millennia BC, than at other times in the past, but it seems that even then it would have been easier to cultivate vines in the hills and mountains to the east and north.

With the collapse of the Sumerian empire at the beginning of the second millennium BC., and the rise of the Assyrians whose power base lay further to the north-west, initially in the upper valley of the Tigris and then at Babylon, the centre of political and economic power moved towards lands more suitable for the cultivation of the vine. As the Code of Hammurabi dating from *c.* 1700 BC (Gordon, 1971) makes clear, economically the most important drink in Babylon was beer (Hyams, 1987). However, wine was essential for certain religious ceremonies, and was

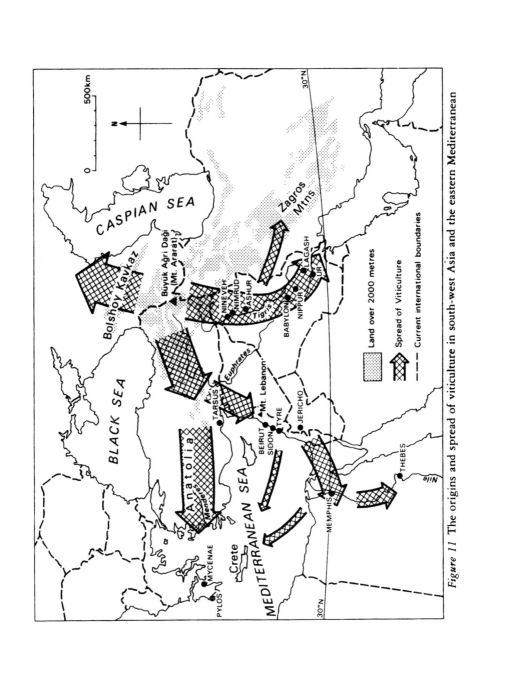

Figure 11 The origins and spread of viticulture in south-west Asia and the eastern Mediterranean

also consumed by the ruling elite. Some controversy surrounds the existence of wine shops in Assyria at this time. Lutz (1922: 128) suggests that such premises were also brothels, but the evidence for this is by no means clear. Based on the Code of Hammurabi, it seems that such wine shops, despite their name, sold mostly beer and were always run by women. Interestingly, the Code also records that if priestesses were to set up such wine shops they were to be burned, again establishing a link between the religious elite and wine. It is possible that, at a time when many of the vineyards appear to have been in the possession of a temple priesthood, such a law may have been designed both to prevent the sale of temple wine in common wine shops, and also to preserve the religious precedence of wine over beer as a sacrificial symbol.

By the first millennium BC., with the rise of the new Assyrian empire, the balance of power shifted further north still to the towns of Ashur, Calah (Nimrud) and then Nineveh. By this date extensive evidence of vine cultivation and wine production is to be found, and Nineveh, near the foothills of the mountains where vines were probably first domesticated, became renowned for its wines. Carved reliefs in stone from earlier centuries illustrate Assyrian processions usually set against a background of date palms, but among the reliefs from Nineveh at the time of Ashurbanipal (668-26 BC) are illustrations of the king seated with his wife under a bower of vines, drinking what is almost certainly the wine of grapes (Figure 12). By the eighth century BC, tablets from Calah (Nimrud) provide detailed lists of the men and women entitled to a daily ration of wine by virtue of their employment in the king's service, with 1 $q\hat{u}$ apparently being the daily ration for ten people (Wilson, 1972). During the first half of the first millennium BC numerous vineyards seem to have been planted in this northern region of the Assyrian empire, and Hyams (1987: 38) has noted that 'the fact that the best wines come from mountain or at least from hilly situations was already established'. As the focus of Assyrian power moved northwards, away from the earlier dominance of the southern Sumerian cities, it came into closer juxtaposition to the heartland of viticulture, and the agricultural expertise built up by the peoples living in the valleys of the Tigris and Euphrates was put to good use in the cultivation of the vine.

This northern power base, though, did not last. With the destruction of the Assyrian cities in the seventh century BC a new Babylonian empire was formed under Nabopolassar (625-05 BC) and then Nebuchadnezzar II (604-562 BC), before Babylon too was to fall to the Persians under Cyrus II in 539 BC. Herodotus (1954: 92), writing soon afterwards in the fifth century BC noted that 'As a grain-bearing country Assyria is the richest in the world. No attempt is made there to grow figs, grapes, or olives or any other fruit trees'. From the previous evidence of numerous vineyards in northern Assyria, it seems that Herodotus is here describing the region

Figure 12 Relief from royal palace at Nineveh illustrating Ashurbanipal and his
queen drinking under a bower of vines
Source: British Museum (124920; BM K24729)

around Babylon. Moreover, he goes on to relate that in Babylonia 'date-palms grow everywhere, mostly of the fruit-bearing kind, and the fruit supplies them with food, wine and honey' (Herodotus, 1954: 92). This reinforces the view that southern Mesopotamia was never an important home for the vine, and that much of the wine drunk there was date wine. Younger (1966) has noted that at this time Babylon imported much wine from Syria, Iran, Armenia and Elam, and Herodotus has provided a clear account of the way in which at least some of this wine was imported. He thus records that boats were built in Armenia, and that 'men fill them with straw, put the cargo on board - mostly wine in palm-wood casks - and let the current take them downstream' to Babylon, where the cargoes were sold and the boats broken up, since it was impossible to paddle them upstream against the current (Herodotus, 1954: 92-3). This account is particularly interesting, since it not only shows that wine was being imported to Babylon by river from Armenia, but it also reveals that it was being transported in wooden barrels.

On the western borders of Nebuchadnezzar's great empire lay the Mediterranean, and on the hills of western Syria and Palestine, where temperatures were cooler and altitudes higher than in the great plains across the desert to the east, further vineyards were to be found. It is likely that it was down this warm temperate corridor, between the desert and the sea, that the art of viticulture first reached Egypt. However, like the Mesopotamian cities of Ur and Babylon, northern Egypt lies at the southernmost limit of modern viticulture in the northern hemisphere, with Memphis being just to the south of latitude 30° north. It was therefore only through the cooling effects of the Mediterranean that viticulture was enabled to flourish in the Nile delta.

Egypt

The inhabitants of Dynastic Egypt have left us with a magnificent pictorial record of viticulture, not only in their tomb paintings, but also in their countless statues and papyri. As in Mesopotomia, beer was the main alcoholic beverage in ancient Egypt, but from the end of the fourth millennium and beginning of the third millennium BC wine appears to have been used by the kings and priests, with large wine cellars being found in association with the temples of First Dynasty kings (Stanislawski, 1975). Much of the wine consumed in Egypt was produced from dates, but there is good evidence that true wine made from grapes was also to be found there from an early date, alongside wine made from other fruits such as pomegranates (Lutz, 1922; Younger, 1966; Stead, 1984; Manniche, 1987; Hyams, 1987).

The Egyptian practice of using stamped seals on their pottery wine jars has provided a considerable amount of information concerning the places of origin of their wines. Most seem to have come from the north of the country in the Delta region (Figure 11), but others were made near Memphis and wines are recorded from both Upper and Lower Egypt (Lutz, 1922). Among the most famous were those from the region near Lake Mareotis and from Tanis. As early as the third millennium a number of different kinds of wine were recorded, classified by colour and quality as well as by location of production (Lutz, 1922; Younger, 1966). Memphis lies at the southernmost limit of viticulture in the northern hemisphere today, but there is good evidence from ancient levels of the Nile that c. 3000 BC the weather was somewhat wetter in this region (Lamb, 1982; McGhee, 1985), and therefore possibly more suitable for viticulture. Subsequent dessication may have made irrigation more necessary for viticulture, but the critical climatic variable for the vine at these latitudes is the winter temperature which needs to be cool. Unfortunately most of the climatic evidence is not concerned directly with temperatures, and it is therefore extremely difficult to make any firm statements concerning the effects of climatic change on Egyptian viticulture.

During the third millennium and the first half of the second millennium it seems that vineyards were mainly owned by the king, the priesthood and some great officials. Vines were grown both in gardens adjacent to houses, and also in separate vineyards. Typical of such vineyards was probably that belonging to Methen, a state official during the Third Dynasty *c.* 2600 BC who had a fine house, measuring 200 cubits by 200 cubits, together with a walled vineyard of 2,000 *stat* (Murray, 1963). The tomb paintings from Thebes during the middle of the second millennium, indicate with great clarity the types of vineyard and methods of wine making that were then common (Wilkinson, 1878). Typically, vineyards were walled, with the vines being cultivated among other fruit trees, plants, and small lakes. The nature of the climate, as in Mesopotamia, necessitated the use of irrigation, and Lutz (1922: 46) suggests that vineyards were also 'generally planted on an artificially raised plot whenever the district lacked hills or mountains'. The hieroglyph sign for a vineyard consists of a vine, from which bunches of grapes hang down, supported by two notched sticks, and this indicates the method by which vines were most commonly trained. Nearly all of the paintings illustrate vines supported on trellises, which not only provided grapes for eating and turning into wine, but also made pleasant shady walkways for their owners.

The tomb of Khaemwese at Thebes dating from *c.* 1450 BC has provided us with a fine description of the vintage (Figure 13). In the top right corner the grapes are being picked by hand by men kneeling on the ground. To the left the grapes are then put into wicker baskets, and below and to the right they are trodden by foot in a wine press, typical of those to be depicted from the earliest days of Dynastic Egypt (Younger, 1966). Above the treading tank there is a pole, supported by two sticks, from which ropes hang down to provide the men with something to hold on to while treading. The wine is then run off, inspected, and to the left put into jars before eventually being stoppered and sealed. From evidence in the tomb of Nebamun, dating from the end of the fifteenth century BC (Younger, 1966: 44) it seems that such amphorae were also used to ferment the wine, and that they were only sealed after fermentation was complete. Figure 13 also illustrates the transport of the sealed amphorae of wine to the tomb, where they are placed as an offering alongside various birds, animals and other fruit. Other aspects of the vintage in Dynastic Egypt are shown in Figure 14 which reproduces illustrations by Wilkinson (1878) from the tombs at Thebes. Figure 14a, from the wall-paintings at Beni Hasan (Lutz, 1922), indicates the use of a wine press to extract more juice from the residue left after the initial treading. Here the skins and other remains of the grapes are put into a cloth bag attached to ropes stretched between two poles. The ropes are then twisted, and further juice extracted. Already in the second millennium BC we therefore

Figure 13 Vintage scene depicted on the walls of the tomb of Khaemwese (Thebes No. 261) *c.*1450 BC (Copy of a painting by Mrs Nina de Gavis Davies)
Source: British Museum (BM K89048)

have good evidence of the use of both *vin de goutte* and *vin de presse*. Figure 14b illustrates in detail the pouring of wine into jars, and these are then labelled and stored in a secure building (Figure 14c). The powerful effects of the wine are all too clearly shown in this last picture, with a man sitting down inside the building, overcome either by the fumes or by having consumed too much of the contents of the jars.

During the reign of Ramses III (1197–65 BC) numerous vineyards were apparently constructed throughout the country, and the Papyrus Harris dating from his reign lists 513 vineyards which were then temple property. Although Egypt therefore clearly produced its own wine, it is also evident that much wine was imported. If the writings of later Greek and Roman authors are to be accepted, Egyptian wines were probably rarely of good quality. As Younger (1966: 47) has pointed out, 'fermentation in the hot Egyptian climate may often have been too rapid, and would have produced a difficult wine which could easily turn sick'. Some Egyptian wines were certainly labelled as being good, or even very good, but there was undoubtedly also a demand for wines from elsewhere in the eastern Mediterranean. Herodotus (1954: 131), writing in the fifth century BC, was of the opinion that the Egyptians had 'no vines in their country' and thus resorted to drinking 'a wine made from barley'. While the evidence cited above, and the continuance of viticulture in Egypt well beyond Herodotus's days (Turner, 1958), indicates this statement to be mistaken, it does nevertheless emphasise that indigenous wine was probably not widely available in Egypt, and that the drink of the mass of the population was probably beer. In discussing the customs of the Egyptians, Herodotus (1954:176) notes that 'Throughout the year, not only from all parts of Greece but from Phoenicia as well, wine is imported into Egypt in earthenware jars' with the jars then being used for the storage of water. He dates this particular practice to the Persian invasions of the sixth century BC, but the importation of wine from the eastern Mediterranean was well established centuries earlier, and probably well before Egyptian conquests in the region in the second millennium.

The evidence of tomb paintings and papyri indicates that wine was used for both social and religious purposes from the earliest days of the Old Kingdom in Egypt. Wine was certainly widely used for libations and offerings to the dead, but again it is not always certain whether date wine or wine made from grapes was the more usual offering. Socially, all forms of alcohol, as in Assyria and Babylonia, were commonly drunk to a state of intoxication (Hyams, 1987; Lutz, 1922). Wine was particularly popular at banquets and feasts; in Younger's (1966: 51) words, 'At the feasts of the upper classes wine was the crowning glory'. However, it is clear that date wine, pomegranate wine, beer and other drinks were also consumed, and there is no real evidence that wine had achieved the position of religious

a.

From Tomb of Baot at Beni Hasan
c. 2000 B.C.

b.

From Tomb at the Pyramids

c.

From Tomb of Antef at Thebes
c. 1450 B.C.

d.

From Tomb of Kynebu at Thebes
c. 1450 B.C.

e.

From Tomb at Thebes

f.

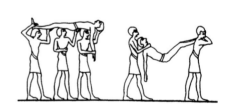

From Tomb at Beni Hasan

Figure 14 Vintage scenes from Dynastic Egypt
Source: based on Wilkinson (1878)

72

and social dominance in Egypt that it was later to assume in Greece and Italy. The banquet scene from the tomb of Nebamun at Thebes, dating from around 1450 BC (Figure 15), indicates what must have been a typical scene. Wine is offered to the guests, at this stage sitting sedately on their chairs; dancing girls and musicians perform before them; and a plentiful supply of food and drink is set out ready to be consumed. Figure 14d illustrates the common practice of mixing wine before it was served, with the use of siphons being essential in order to transfer the wine into smaller serving vessels from the large storage jars without mixing it with the lees. Inscriptions on some of these banquet paintings indicate that drinking to excess was commonplace. Lutz (1922: 100) thus quotes numerous examples of the sentiments of participants at such feasts, including the following spoken by a woman, Nubmehy, from the tomb of Paheri: 'Give me eighteen cups of wine . . . don't you see I want to get drunk! My insides are as dry as straw!' The after-effects of such banquets are also to be seen in the Theban tomb paintings, with Wilkinson (1878) illustrating two such scenes, a woman vomiting, and two men being carried away unconscious by their servants (Figures 14e and 14f). These scenes of feasting, though, also had a deeper meaning. In Manniche's (1987: 42) words, 'the banquet scenes and the details consistently rendered point to a specific interpretation of the scenes as attempting to convey the initial stages in the rebirth proceedings in a language understood by those who knew the key to it'. The symbolic erotic tension in the paintings (Figure 15) is readily apparent: the ladies wear diaphanous garments; they hold lotus flowers and mandrake fruits, the symbols of love; scents and hair, both of erotic significance, were commonly emphasised; and the servant girls pouring drinks represented a play on words, with the hieroglyphs for 'pouring' and 'ejaculating', *seti*, both sounding the same (Manniche, 1987). Such examples of wine drinking associated with love making, reflect a theme that was to become even clearer in later Greek and Roman religious expression.

Viticulture in Syria and Anatolia

Mesopotamia and Egypt have preserved the best early evidence concerning viticulture, vinification and wine consumption. However, it is clear that the population of these countries imported large amounts of wine from elsewhere. In particular, by the middle of the third millennium BC, much of the wine consumed in Egypt came from Palestine and Syria (Breasted, 1906–7), and it seems that, despite the lack of direct evidence, viticulture here must have predated its introduction further south into Egypt (Figure 11). Goor (1966) thus suggests that viticulture may already have been established in Syria as early as 5000 BC, although the earliest firm evidence of grape seeds is not found until 3000 BC at Jericho. Records

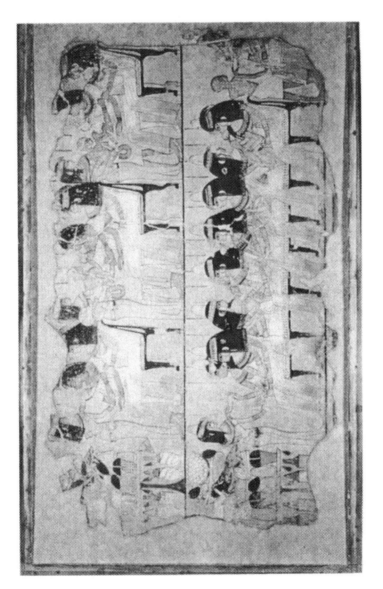

Figure 15 Banquet scene from the tomb of Nebamun (Thebes) *c.*1450 BC
Source: British Museum (BM K225)

of the Egyptian military expeditions to the north-east during the second millennium BC indicate that vineyards were common in Syria and Palestine at that time, and both the Talmud and the Bible treat vines and wine as being a typical part of the scenery and way of life of the Jewish people. Lutz (1922: 22) goes as far as to suggest that 'Syria was the wine country par excellence of the Ancient Near East'. Climatically and environmentally it was well suited to viticulture, with its hills and mountains providing both the cool winters and long hot summers required by the vine. By the beginning of the first millennium BC, with the rise to power of the seafaring Phoenicians, based on the city states of Aradus, Byblos, Beirut, Sidon and, most important of all, Tyre (Figure 11), the wines of Lebanon were being traded extensively in the Mediterranean and elsewhere. Phoenician wine was exported most importantly to Egypt, but also, according to Diodorus Siculus (1939) writing much later in the first century BC, apparently as far afield as north Africa, Arabia, East Africa, India and Spain (Lutz, 1922). By 814 BC, when Carthage was established by the Phoenicians fleeing from Tyre after its conquest by the Assyrians, vines were probably already grown in north Africa, and it is likely that some of the other cities which had traded with Tyre and Sidon had also begun to develop their own viticulture. If the evidence of high levels of *Vitis* pollen in parts of southern Spain is to be accepted, it is even possible that viticulture had been taken to the far west of the Mediterranean well before the end of the second millennium BC, but it is more likely that Iberia represented a separate area of vine domestication and viticultural development (Stevenson, 1985; Núñez and Walker, 1989).

To the north-west of Syria lay Anatolia, the heartland of the Hittite kingdom, which emerged as a major regional power during the second millennium BC (Gurney, 1964; Macqueen, 1986). Its close proximity to the hearth of viticulture (Figure 11) suggests that vines may have been cultivated here in eastern Turkey at a date similar to the development of viticulture in Syria, but there is little direct evidence to support this suggestion. Lloyd and Mellaart (1958) have discovered grape pips, together with barley, wheat, lentils and pulses, in a shrine at Beycesultan in western Turkey on the Menderes (Meander) river dating from the Early Bronze Age (c. 2600 BC–2300 BC), and stains in some of the vessels found in the shrine indicate that they once held liquid, but there is no firm evidence that this was wine. Elsewhere at Beycesultan a wine shop with two storage jars and a pile of drinking cups dating from the Late Bronze Age (c. 1450 BC–1000 BC) has also been found (Lloyd and Mellaart, 1962–1972; Macqueen, 1986). Viticulture was apparently quite widely established in Anatolia by 1000 BC, evidenced by a bas relief discovered at Ivriz, near Tarsus, illustrating the existence of a deity festooned with a vine (Figure 16) (Billiard, 1913; Goodenough, 1956), and a stele from Maras which shows a god holding a bunch of grapes in one hand while

Figure 16   Carved image of a Hittite god at Ivriz
Source: Hirmer Verlag, München

his other arm is around a goddess holding a mirror (Barnett, 1980).

Viticulture must also have spread eastwards at an early date, enabling the Sumerians to import wine from 'mountains to the east' in the third millennium BC (Younger, 1966: 60). Firm evidence of viticulture and vinification in Iran at such an early date is presently lacking, but it seems highly probable that it had spread into the foothills of the Zagros mountains by at least the second millennium BC (Figure 11). From here viticulture spread further eastwards, and de Planhol (1977) suggests that it was already well established in parts of Afghanistan by at least the fourth century BC. Further east, vines were cultivated in China by the second century BC (Laufer, 1919), and possibly as early as the eighth century BC (Weinhold, 1978), and it seems that the knowledge of viticulture thus travelled east from Persia and Afghanistan along the central Asian caravan routes.

Contemporary with the Middle Kingdom in Egypt, and providing another source of demand for wine, was the emergence of Minoan culture in Crete. Here it seems that beer brewed from barley was initially the most popular drink, but that by 1700 BC in late Minoan times it had become supplemented by wine (Evans, 1935: 628). Much of this wine may initially have been imported from Syria to the east or from Egypt to the south, but by the fifteenth century BC, with the Mycenaean conquest of Crete, viticulture had become firmly established on the island, evidenced by the discovery of a wine press at Palaikastro dated to the Mycenaean period (Dawkins, 1902-3). This remarkable discovery revealed that the press had a sloping bed, which was designed to let the must flow easily into a pithos built into the fine hard plaster of the press floor. Other evidence, particularly a gold signet ring discovered at Mycenae illustrating a goddess seated beneath a vine from which hang clusters of grapes (Evans, 1935: 394), as well as references to wine in Linear B tablets, suggest that under Mycenaean influence wine was quite widely consumed in Crete and on the Greek mainland at this time. Nevertheless, as Younger (1966: 83) points out, 'One of the really striking things is the rarity of the vine motif in Minoan art'. Compared with the evidence for viticulture and wine production from the same period in Egypt, it is indeed surprising that the Minoans should apparently have paid so little heed to the vine. However, with the emergence of Mycenae as a regional power and the introduction of the vine and wine to Greece, a new era dawned in the history of viticulture, that of the Heroic Age, in which the symbolism of wine was to take on new significance.

THE PREHISTORIC SYMBOLISM OF WINE AND THE VINE

In neither of the two great hearths of civilisation in the Nile and Tigris-Euphrates valleys was wine made from grapes the most common alcoholic beverage. However, before the end of the fourth millennium BC, it was extensively used for purposes of libation and sacrifice, and it was widely sought after by the ruling classes as a drink to accompany their meals. In both Mesopotamia and Egypt wine and the vine were already associated with specific divinities, but in neither society did these divinities take on the role assumed in later centuries by Dionysus/Bacchus.

The vine and wine: goddesses, gods and divine fluid

The concept of the 'Earth-mother' was one of the earliest divinities of the eastern Mediterranean and south-west Asia, being represented in Greek mythology by Gaea, and in Mesopotamia by Ga-Tum-Dug at Lagash, and by Bau and Innini at Der and Kish (Guirand, 1968a). Later, a number of specialised earth divinities emerged, such as the Assyrian-Babylonian goddess of the vine Geshtin, or Geshtin-an-na (Barnett, 1980; Guirand, 1968a), and the vine god Pa-Geshtin-Dug (Hyams, 1987; Lutz, 1922). At an early date, though, Geshtin ceased to be the 'mother vinestalk' and instead became Nina, goddess of the waters. Hyams (1987: 38–9) has suggested that this transfer indicates the move of a northern goddess southwards to a land where water was of greater significance than wine, but it seems more likely that this is but one example of the underlying importance of the concept of a divine fluid which could be represented in different forms. Goodenough (1956) has emphasised that this divine fluid was intimately connected with the religious concepts of death and rebirth, and thus with a fundamental concern with the maintenance of fertility which underlay most religions in the region. Water provided both the rain necessary for the rebirth of crops, and also the drink essential for the maintenance of human life. Milk from divine breasts provided life for infant mankind, and divine semen provided a symbol for fertility and rebirth. In Goodenough's (1956, vol. 6: 126) words,

> the symbols of divine fluid kept an amazing similarity of value throughout the civilizations of the ancient world. The basic desires were to get the fluid which represented the life, and so the life-giving power, of the god or goddess; to get this fluid primarily for the crops but soon also for the personal security and greater life here and hereafter of the devotees as individuals.

Critically, he also comments that 'Most of these symbols of fluid united in the later symbolism of the vine and of sacramental wine' (Goodenough, 1956, vol. 5: 113). This divine fluid is perhaps most clearly represented not in Mesopotamia, Egypt or Europe, but instead in Indian mythology by the polymorphic deity Soma, which according to Masson-Oursel and Morin (1968: 331)

> is first and foremost a plant, an essential part of the ancient sacrifical offerings. It is also the juice of the plant, obtained by squeezing it between two mill-stones. Then it is golden nectar, the drink of the gods – a precious ambrosia which symbolises immortality and ensures victory over death to all who drink it.

From being relatively unimportant, the deities of wine and the vine gradually came to the forefront of religious consciousness, as symbols concerned with rebirth and the life hereafter. Neither in Mesopotamia nor in Egypt did a specific wine or vine deity achieve particular prominence. In part this no doubt reflected the difficulties of cultivating vines in these regions, but it also resulted from the critical importance of water to the survival of their societies. In Egypt the dominant rhythm was determined by the waters of the Nile: if they failed Egypt would have famine; if they came in too large quantities there would be floods; but if they could be tamed to produce a regular and guaranteed supply they would ensure fertility and prosperity.

The development from a god of nature to a god of human death and rebirth, is perfectly expressed in the ideological developments associated with the Egyptian god Ousir, known to the Greeks as Osiris. According to Viaud (1968: 16), 'At first Osiris was a nature god and embodied the spirit of vegetation which dies with the harvest to be reborn when the grain sprouts. Afterwards he was worshipped throughout Egypt as god of the dead, and in this capacity reached first rank in the Egyptian pantheon'. Osiris was the Nile, the grain, the trees, and even the light vanishing in the evening to reappear as the dawn. As part of his overall identity with nature, Osiris also represented the vine, but it is important to stress that he was never purely associated with or symbolised by the vine. It was in the legend of his murder by his brother Set, and his restoration to life by Isis, his wife, that Osiris rose to prominence in the Egyptian pantheon as the god of the underworld, and it was in this form that he later came to be identified with the Greek gods Hades and Dionysus.

Elsewhere in south-west Asia, however, particularly in areas close to the hearth of viticulture, there is good evidence for the existence of specific vine deities. The image on the bas-relief discovered at Ivriz in Turkey (Figure 16) and dating from c. 1000 BC has been variously interepreted as a wine god (Billiard, 1913: 32), a peasant god (Seltman, 1957: 23), a vegetation god (Hyams, 1987), and a fertility god (Goodenough, 1956, vol. 5: 137), but, whoever he symbolised, the importance of the vine in his representation is clear. Here the god, festooned in a vine and holding a bunch of grapes in his right hand, presents grain stalks to some unknown king standing in homage before him. What the scene seems to represent is the deity's gift of the new life of grain and the vine, thus again reinforcing the connection between fertility and agriculture. Another vine deity is revealed on an electrum plaque dating from the ninth or eighth century BC which was probably made in eastern Anatolia or north-east Syria (Barnett, 1980). This illustrates a nude, four-winged goddess holding a bunch of grapes in each hand.

The imagery connecting fertility with wine and the vine is also illustrated by the Sumerian *Epic of Gilgamesh*, dating from the first centuries of the second millennium BC, but probably existing in much the same form many centuries earlier. In this, the hero Gilgamesh goes in search of the secret of eternal life from Utnapishtim who was set by the gods to live in the land of Dilmun, and on his journey he encounters Siduri, 'the woman of the vine, the maker of wine' (*Epic of Gilgamesh*, 1960: 97). Dilmun has widely been identified as Bahrain (Bibby, 1970; Lloyd, 1978), but in order to get there the Epic records that Gilgamesh had to pass through the mountain of Mashu, which is usually identified as Mount Lebanon (*Epic of Gilgamesh*, 1960: 122). Largely on the basis of this latter identification, Younger (1966: 32) has suggested that Siduri's home was in western Syria on the shores of the Mediterranean, thus reinforcing the arguments for this area being considered to be near the original hearth of viticulture. These identifications, however, appear to be contradictory. If Dilmun lay in the direction of the rising sun to the east of Gilgamesh's homeland in Uruk (Bibby, 1970; *Epic of Gilgamesh*, 1960), and Gilgamesh met Siduri on the sea shore after he had travelled through the mountain following 'the sun's road to his rising' (*Epic of Gilgamesh*, 1960: 96) then she must have been on the eastern side of the mountains rather than to their west. It therefore seems more likely that the mountains where he killed a number of lions and through which he journeyed, were to the east of his homeland and may have been part of the Zagros range (Tigay, 1982). If this is so, an alternative interpretation would be that Siduri lived on the southern coast of Iran on the shores of the Gulf, beyond which lay Dilmun. While the Epic does not specifically mention that Siduri actually taught the skill of wine making to mankind, Younger's (1966) suggestion that her home symbolised the hearth of viticulture is not unreasonable. However, the difficulty with such a suggestion is that it would imply that the origins of viticulture were in southern Iran. On balance it therefore seems most probable that the Epic merely indicates that viticulture was widespread in this general region at that time. Gilgamesh encounters Siduri by the garden of the gods where 'there was fruit of carnelian with the vine hanging from it' (*Epic of Gilgamesh*, 1960: 97), and she tries to pursuade him from continuing his journey with the following words:

> where are you hurrying to? You will never find that life for which you are looking. When the gods created man they allotted to him death, but life they retained in their own keeping. As for you, Gilgamesh, fill your belly with good things; day and night, night and day, dance and be merry, feast and rejoice. Let your clothes be fresh, bathe yourself in water, cherish the little child that holds your hand, and make your wife happy in your embrace; for this too is the lot of man.
>
> (*Epic of Gilgamesh*, 1960: 99)

Gilgamesh refuses, but in Siduri's pleas it is again possible to see the symbolic association between the themes of wine and the vine, and human fertility. In Tigay's (1982: 98) words, 'The barmaid's "hedonistic" counsel befits one of her profession'.

The Epic of Gilgamesh provides us with two other pieces of evidence about the use of wine in early Sumerian society. The first concerns the taming of Enkidu, the wild man of nature, later to become Gilgamesh's inseparable companion. Enkidu is seduced by 'a harlot from the temple of love' (*Epic of Gilgamesh*, 1960: 99), who later also introduces him to the pleasures of wine, saying to him:

> 'Enkidu, eat bread, it is the staff of life; drink the wine, it is the custom of the land.' So he ate till he was full and drank strong wine, seven goblets. He became merry, his heart exulted and his face shone. He rubbed down the matted hair of his body and anointed himself with oil. Enkidu had become a man.
>
> *(Epic of Gilgamesh*, 1960: 65-6).

Enkidu thus becomes a man by eating bread and drinking wine, symbolising the development of agriculture which raised humanity above nature. Moreover, Enkidu is introduced to these symbolic elements following his seduction by a temple prostitute, illustrating not only the apparent widespread religious practice of prostitution in ancient Sumer, but also once again the symbolic connection between wine and human sexual activity.

The final use of wine mentioned in *The Epic of Gilgamesh* is following the hero's death, when bread offerings are made and libations of wine poured out (*Epic of Gilgamesh*, 1960: 117-18). Here is a good example of the frequent use of wine in libations and sacrifices throughout the ancient world (Haas, 1977). Its precise significance in this Sumerian context is uncertain, but wine in its sacrificial role served at least five important symbolic functions: it symbolised the 'divine fluid'; red wine, similar in colour to human blood, indicated or represented human sacrifice; together with bread it represented the first-fruits of agriculture; it symbolised the hope of rebirth through its metaphorical connections with fertility rituals; and the drinking of wine induced that sense of other-worldliness associated with being in the presence of the gods.

Wine and the vine in Jewish symbolism

One religion, that of the Jews, stood out noticeably from amongst others in the region, both in being monotheistic and also in terms of its proscription of idols. Two of the ten commandments given by God to Moses are thus:

"You shall have no other gods before me.
"You shall not make for yourself an idol in the form of
anything in heaven above or on the earth beneath . . ."

(Exodus, XX: 3-4).

Despite these proscriptions, Jewish ritual reflects a number of symbolic
themes connected with wine and the vine that are similar to those of other
religions in south-west Asia during the first and second millennia BC.
The Old Testament thus provides us with much evidence concerning
both the general distribution of viticulture within the lands neighbouring Israel and Judah, and also the symbolic significance of wine and the
vine, much of which was later to be incorporated into Christian religion
(Busse, 1922; Zapletal, 1920; Goodenough, 1956; Schreiber, 1980).

The account of the origins of viticulture in the Book of Genesis largely
coincides with that already established on palaeobotanical grounds.
Thus Noah, after his ark had come to rest on the mountains of Ararat
(Genesis, VIII: 4), being 'a man of the soil, proceeded to plant a vineyard'
(Genesis, IX: 20). If Mount Ararat is taken to be the mountain also known
as Büyük Agri Dagi in eastern Turkey, and this first mention of a
vineyard in the Bible is assumed to represent the beginnings of
viticulture, then this would reinforce the argument that the origins of
viticulture indeed lay in this region.

The Book of Genesis also provides evidence that viticulture was
practised in Egypt during the period when the Israelites resided there (c.
1700 BC-1300 BC), as shown in the dream of Pharaoh's cupbearer (Genesis,
XL: 9-15). Moreover, the numerous references throughout the Books of
Leviticus (XXVI: 3-5), Judges (IX: 7-15; XIII: 2-24), Isaiah (V: 1-7),
Psalms (LXXV: 8; LXXX: 14-16; CIV: 14-15) and Proverbs (IV: 14-17;
IX: 1-6; XXIII: 29-35; XXXI: 16) to wine and viticulture in Canaan and
the land that became Israel, suggests that viticulture and the making of
wine were common on the eastern shores of the Mediterranean during the
second and first millennia BC. Moreover, when men are sent by Moses at
God's command to explore Canaan in the Book of Numbers (XIII: 23, 26)
they specifically cut off a vine branch bearing a cluster of grapes which
they bring back to the Israelites together with pomegranates and figs. In
Goodenough's (1956: 129) words, 'Wine was from early times a proverbial
part of the richness of the promised land, and it naturally appears to have
been an important aspect of the life of Jews at all periods afer they came
into Canaan'.

Much has been written about the symbolic significance of wine and the
vine in Jewish religion (Busse, 1922; Goodenough, 1956; Goor, 1966), and
there is good evidence that the popular ritualistic blessing and drinking
of wine only entered Judaism relatively late in the Greco-Roman period.
Likewise, the emphasis given to the grape cluster as a symbol of

leadership in Jewish society also seems only to date from late antiquity (Porton, 1976). However, wine and the vine were of symbolic importance to the Jewish people in a number of other ways from a very early date. One of the favourite symbols used for the nation of Israel in the Old Testament is thus the vine or the vineyard. Thus in Psalm LXXX: 8-9 the Psalmist calls out to God,

> You brought a vine out of Egypt;
>   you drove out the nations and planted it.
> You cleared the ground for it,
>   and it took root and filled the land.

This imagery is particularly used by some of the Old Testament prophets in contrasting the ideal and obedient Israel, which is seen as a well cultivated vineyard, with a people who have disobeyed God, and who are consequently represented as a degenerate vine or vineyard. For example, the Book of Isaiah (V: 7) thus observes that

> The vineyard of the LORD Almighty is the house of Israel,
> and the men of Judah are the garden of his delight.

Further, Jeremiah (II: 21), in prophesying against Israel, quotes the following statement from God:

> I had planted you like a choice vine of sound and reliable stock.
> How then did you turn against me into a corrupt, wild vine?

A second type of symbolic reference connected with viticulture was the use of wine by the Jewish people for sacrificial purposes. Wine was one of the first-fruits to be offered as a drink offering, but as Goodenough (1956, vol. 6: 129) points out, 'wine is never singled out as of especial and unique importance. Yahweh increases the growth of the vine as he does of other crops but is never a wine god, or especially to be approached through wine.' Indeed, it is noticeable that wine plays no part in most of the major sacrifices about which God gave Moses details at Mount Sinai, such as the burnt offering, the fellowship offering and the sin offering, all of which involved the ritual sacrifice of animals (Leviticus, I–V). This is in marked contrast to other religions in the region where wine does appear to have been associated with the sacrifice of animals. In this context, Busse (1922) usefully distinguishes between two kinds of offering made by the Jews: an older one consisting of an animal and its blood, and a later one in which wine is also used. He suggests that this reflects a change in ritual following the exile of the Jews c. 700 BC, when they came under Persian and Babylonian influence and, unable to make blood sacrifices, substituted wine instead.

A third symbolic reference to wine in the Old Testament is found in the ordinance that those serving God as priests, or undertaking a special vow of service to God, should abstain from wine. Thus, in Leviticus (X: 8-9), God said to Aaron, the chief priest,

> "You and your sons are not to drink wine or other fermented drink whenever you go into the Tent of Meeting, or you will die . . ."

Likewise, the Book of Numbers (VI: 1-4) specifies the conditions which a Nazirite must fulfil as follows:

> The LORD said to Moses, "Speak to the Israelites and say to them: 'If a man or woman wants to make a special vow, a vow of separation to the LORD as a Nazirite, he must abstain from wine and other fermented drink and must not drink vinegar made from wine or from other fermented drink. He must not drink grape juice or eat grapes or raisins. As long as he is a Nazirite, he must not eat anything that come from the grapevine, not even the seeds or skins . . ."

These restrictions can, in part, be interpreted as a logical extension of the general exhortations against excessive drinking to be found throughout the Old Testament, but which are particularly well expressed in the following two quotations from the Book of Proverbs: 'Wine is a mocker and beer a brawler; whoever is led astray by them is not wise' (Proverbs, XX: 1), and

> Who has woe? Who has sorrow? Who has strife? Who has complaints? Who has needless bruises? Who has bloodshot eyes? Those who linger over wine, who go to sample bowls of mixed wine. Do not gaze at wine when it is red, when it sparkles in the cup, when it goes down smoothly! In the end it bites like a snake and poisons like a viper.
>
> (Proverbs, XXIII: 29-32)

This symbolism is also taken a stage further when the Psalmist associates the wicked with the dregs of a wine cup: 'In the hand of the LORD is a cup full of foaming wine mixed with spices; he pours it out, and all the wicked of the earth drink it down to its very dregs' (Psalm LXXV: 8). This is also the symbolism that recurs in the New Testament in the Book of Revelation (XIV: 10), where it is stated that anyone who 'worships the beast and his image and receives his mark on the forehead or on the hand, he, too, will drink of the wine of God's fury, which has been poured full strength into the cup of his wrath'.

Not all of the symbolism concerning wine in the Old Testament, though, is negative. Thus Psalm CIV: 15 refers to 'wine that gladdens the heart of man, oil to make his face shine, and bread that sustains his heart', repeating an imagery very similar to that already encountered in the *Epic of Gilgamesh*, when Enkidu first drank strong wine. Such symbolism is also to be found in the Book of Ecclesiastes (II: 3; IX: 7; X: 19), where wine is seen as making the heart merry, and in the Book of Proverbs (XXXI: 6-7) another nuance is reflected in the exhortation to 'Give beer to those who are perishing, wine to those who are in anguish; let them drink and forget their poverty and remember their misery no more'.

A further theme in the symbolism of wine in the Old Testament concerns its association with fertility and human relationships, again revealing parallels with the religious symbolism of Israel's neighbours. The clearest example of this is in the story of Lot and his two daughters told in Genesis (XIX: 30-2):

> Lot and his two daughters left Zoar and settled in the mountains, for he was afraid to stay in Zoar. He and his two daughters lived in a cave. One day the older daughter said to the younger, "Our father is old, and there is no man around here to lie with us, as is the custom all over the earth. Let's get our father to drink wine and then lie with him and preserve our family line through our father."

This they did, and in time gave birth to sons, who were seen as fathering the Moabites and Ammonites. A more mystical and poetic imagery associating the vine with human sexuality is to be found in the numerous references to the subject in the Song of Songs, of which the following are but two examples: 'Let us go early to the vineyards to see if the vines have budded, if their blossoms have opened, and if the pomegranates are in bloom – there I will give you my love.' (Song of Songs, VII: 12), and 'I said, "I will climb the palm tree; I will take hold of its fruit." May your breasts be like the clusters of the vine, the fragrance of your breath like apples, and your mouth like the best wine.' (Song of Songs, VII: 8-9).

Considerable emphasis has been given here to the symbolism of wine and the vine in the Old Testament for two main reasons: first, it can be seen as reflecting several of the broader symbolic representations of wine and the vine in the ideologies of most of the religions in south-west Asia in the first and second millennia BC, and secondly, and more importantly, much of this symbolism was then taken over and developed in Christianity, which in time became the dominant ideology of societies in which wine was to be the most important alcoholic beverage. A second crucial ideological influence on these societies was that of Greece, and it is therefore to a discussion of the symbolism of the Greek god Dionysus, that this chapter now turns.

Dionysus

By the fifteenth century BC, with the extension of Mycenaean power eastwards and to the south, viticulture had become established on mainland Greece. The evidence of the Mycenaean wine press at Palaikastro on Crete (Dawkins, 1902-3), the wine cellar discovered at the palace of Pylos (Blegen, 1959; Lang, 1959), mentions of wine and wine merchants on Linear B tablets (Seltman, 1957; Younger, 1966: 88-9), and the goddess seated under a vine discussed by Evans (1935), all suggest that the Mycenaeans made and drank wine. Moreover, among the gods named on the Linear B tablets there is mention of a servant or priest of the god Dionysus (Seltman, 1957), who was later to become known throughout the Greek world as the god of wine.

It was these Mycenaeans who featured in Homer's epic poems, the *Iliad* and the *Odyssey*, which were probably written during the eighth century BC. The content of both poems owes much to Homer's own age, but they referred to a time four centuries earlier during and following the fall of Troy, which is traditionally dated to 1184 BC based on the date given by Eratosthenes (Lattimore, 1951). Both poems reflect a widespread use of wine for libations and feasting, and this is generally supported by the discovery of elaborate Mycenaean cups and drinking vessels from the twelth and thirteenth centuries BC. Significantly, Dionysus is but rarely mentioned by Homer, suggesting that while viticulture and wine were common in Greece during the eighth century BC the cult of Dionysus was not yet widespread. In the *Iliad* Homer refers to the mythical occasion when Lykourgos drove out Dionysus and was blinded by the gods as a consequence (VI: 132-40) and to his parentage as the son of Zeus and Semele (XIV: 325), and in the *Odyssey* Dionysus is mentioned in connection with the death of Ariadne (XI: 325) and with his gift of a golden urn made by Hephaestus in which were placed the bones of Achilles (XXIV: 74). The mentions of vines and wine, though, are numerous (Allen, 1961; Seltman, 1957; Younger, 1966). In particular, Homer describes in detail scenes from a vineyard at the time of the vintage on the shield of Achilles (*Iliad*, XVIII: 561-72), the use of wine in libations (*Iliad*, I: 463; VI: 258-60; XXIII: 220; *Odyssey*, XIII: 38-40), the way in which Odysseus made the Cyclops, Polyphemus, drunk on strong wine given to him by Maro of Ismarus (*Odyssey*, IX: 346-64), the transport of wine in amphorae (*Odyssey*, II: 349-51), and the drinking of wine at banquets (*Odyssey*, IV: 622; *Iliad*, XI: 637-9).

The suggestion that Dionysiac religion and ritual entered Greece some considerable time after the first vines were cultivated there and made into wine (Nilsson, 1940) is further supported by Euripides' play the *Bacchae* and also by the evidence of paintings of the god on Greek vases. Dodds (1960: xi) has observed that 'Unlike most Greek tragedies the *Bacchae* is a play about an historical event – the introduction into Hellas of a new religion'. Quite when this religion entered Greece is uncertain, but by the time Euripides was writing *c.* 408-7 it lay far in some mythical past. The first appearance of Dionysus in Greek art is not until about 580 BC, when he is represented as 'a humble, barefoot figure who holds a branch of a grape-vine and walks by himself in a procession of deities' (Carpenter, 1986: 124). As Carpenter (1986: 124) goes on to note,

> it is difficult to see any connection between this insignificant figure and the Dionysos of the frenzied maenads in Euripides' *Bacchae* – the god who has invaded Greece with his cult – or with the god who tears a fawn in half with his bare hands on red-figure vases from the middle of the fifth century.

On this evidence, it seems that Dionysian imagery is a sixth century Attic invention, in which the earlier 'light-hearted symbol of the pleasures of wine' (Carpenter, 1986: 126) is transformed between 540 and 520 BC into the much more serious and dangerous god referred to by Dodds in the quotation with which this chapter began (Evans, 1988).

According to Euripides, the original home of Dionysus was in the mountains of Lydia and Phrygia, and this is strongly supported by the recognition that the word βακχος is the Lydian equivalent of Dionysus (Dodds, 1960). However, it is widely assumed that the cult of Dionysus first emerged in Thrace, from whence it was brought to Boeotia, the Greek islands and eventually to Attica and the Peloponnese. In origin Dionysus (Figure 17) was simply the god of wine, but as a result of traits borrowed from other foreign gods, particularly the Cretan god Zagreus, the Phrygian god Sabazius and the Lydian god Bassareus, his sphere of influence and significance grew (Guirand, 1968b), and he eventually became incorporated into the Olympian pantheon. As Ruck (1982: 233) notes, 'The name of Dionysus himself contains the Indo-Europeans' word for their chief god, Zeus/Dios, and testifies to an attempt to incorporate the alien deity into Indo-European mythology as a son of Zeus'.

There are many myths and legends associated with Dionysus (Dodds, 1960; Guirand, 1968b, Kerényi, 1962, 1976; Nilsson, 1940, 1975; Ruck, 1982), but the key elements in his mythology are that he was the son of Zeus and Semele, daughter of Cadmus, the King of Thebes; that he was brought up by the nymphs on the slopes of a fabled mountain called Nysa, where he was educated by the Muses and other secondary gods; that on reaching maturity he discovered how to make wine from grapes; and that he then took his discovery to Attica, Phrygia, Thrace and elsewhere. Significantly, though, there was a Greek tradition that Dionysus turned away from Mesopotamia because its inhabitants preferred beer (Hyams, 1987: 37; Younger, 1966: 55), and this reinforces the arguments made earlier in this chapter that viticulture and religious symbolism associated with wine and the vine were generally less important than the production and consumption of beer in Babylonia and Sumeria. In recognition of Dionysus' upbringing on the mountain of Nysa, he is often represented in the company of the secondary divinities associated with the fertility of

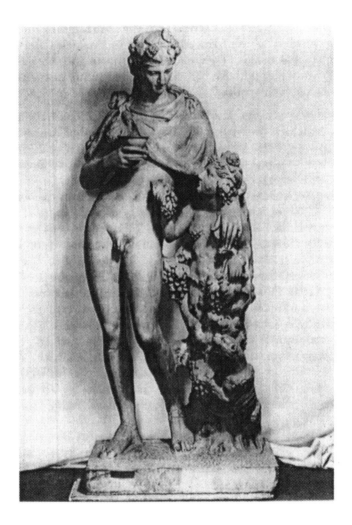

Figure 17 Statue of Bacchus with a personification of the vine (AD 150–200)
Source: British Museum (SC 1636)

nature who reputedly gave him his education. The most usual of these were the Satyrs (spirits of the forests and mountains), Silenus (a drunk and obese old man who had been his tutor), Pan (the half goat shepherd god of woods and pastures), the Centaurs (savage cattle herders who were half horse and half man), the Nymphs (youthful and beautiful deities of forests and mountains), and Priapus (the god of garden and field). When anyone rejected his cult they were struck by madness, and in the case of the inhabitants of Argos, the women tore up and devoured their own children. Euripides' play the *Bacchae* relates the story of the death of Pentheus, King of Thebes, who having rejected Dionysus was torn to pieces by the female members of his family when he unwisely followed them to a mountain where they held their ritual frenzied dance.

From being a god of the vine and wine, Dionysus rapidly also became the essential power of nature (Ruck, 1982). In Dodds' (1960: xii) words, his domain was 'not only the liquid fire in the grapes, but the sap thrusting in a young tree, the blood pounding in the veins of a young animal, all the mysterious and uncontrollable tides that ebb and flow in the life of nature'. His followers sought communion with the god which transformed them into a βακχος or βακχη, and there were two main ways in which this could be achieved: through the drinking of the deity in the form of wine, and through the ορειβασια, frenzied dancing by women on mountain tops in imitation of the Maenads. The culmination of this dancing, in which only women were allowed to participate, was the tearing to pieces of an animal and the ritual eating of its raw flesh, which then empowered them with its own vitality.

From being the essential power of nature, Dionysus also became closely associated with the afterlife. Under the influence of Orphic mysticism he became the god who was killed, who descended to the underworld of death, and was then born again, reflecting the widespread symbolism of the death and rebirth of the vine that has already been referred to in the context of earlier religions in south-west Asia. Moreover, it is in this role that Dionysus came to be identified with the Egyptian god Osiris, and with a range of fertility rituals.

Following their introduction into Greece, the rituals associated with the worship of Dionysus eventually came under state control, and in Attica by the time of Euripides they had lost much of their original character (Dodds, 1960). Among the earliest festivals associated with Dionysus was the Agrionia, which appears to have been first celebrated in Boeotia. At this festival the Bacchantes immolated a young boy, and in Chios and Lesbos it was also associated with human sacrifice, later to be substituted by flagellation (Guirand, 1968b). In Attica by the fifth century BC Dionysus was worshipped at a number of different rural festivals throughout the year associated with the different seasons of the viticultural calendar. In December–January when the vines were pruned he was celebrated with Demeter and Kore at the festival of Haloa; in January–February there was the Lenaea, the festival of the wine press; at the end of February came the Anthesteria, the festival of spring fertility, celebrated with revelry and wine drinking; at the end of March were the most brilliant festivals of the Greater or City Dionysia, during which numerous plays were performed; and then in September or October there were the Eleusinian Mysteries, an old agrarian cult into which elements of the cult of Dionysus became introduced, and at which mystical visionary experiences were probably induced, not only by wine but also by the ingestion of psychoactive alkaloids in the form of a water-soluble extract of ergot (Dodds, 1960; Guirand, 1968b; Nilsson, 1940; Ruck, 1982; Osborne, 1987).

Dionysus, under the name of Bacchus, was familiar in Italy from well before the second century BC, with Dionysus being identified from an early date with the Roman god of the fertility of the fields, Liber Pater (Bendinelli, 1931). However, Livy, writing at the end of the first century BC, relates that the senate soon came to regard the Bacchic rituals as being a threat to public security, and banned them in 186 BC. In his *History of Rome* (XXXIX: viii-xix), Livy provides a record of the official condemnation of these nocturnal rituals:

> To the religious content were added the pleasures of wine and feasting, to attract a greater number. When they were heated with wine, and all sense of modesty had been extinguished by the darkness of night and the commingling of males with females, tender youths with elders, then debaucheries of every kind commenced: each had pleasures at hand to satisfy the lust to which he was most inclined. Nor was the vice confined to the promiscuous intercourse of free men and women, but false witnesses and evidence, forged seals and wills, all issued from this same workshop; also poisonings and murders of kin, so that sometimes the bodies could not even be found for burial.
>
> (Livy, in Lewis and Reinhold, 1951: 469)

To stamp out these rites, the consuls decreed that all future Bacchic ceremonies should be prohibited, that all the priests and participants in the cult should be apprehended, and that all places of Bacchic worship should be destroyed. Furthermore, those guilty of debauchery, murder, forgery and other fraudulent practices were sentenced to death. It was not primarily, though, a fear of the new religion that prompted this action, and as Lewis and Reinhold (1951: 469) note

> Roman conservatism in religious matters and aversion to orgiastic religious practices played a part in the senate's decision to suppress the Bacchic societies. But fundamentally the senate was motivated by the fear that the growth of secret organizations (which were prohibited by Roman law), especially among the lower classes, might harbor or foster revolutionary movements.

Although several thousand people were put to death, the worship of Bacchus was still to be permitted on an individual basis, so long as not more than five people participated together in any ceremonies.

Subsequently the cult appears to have gone underground, to reappear in popular form once again during the last years of the Republic.

A variety of paintings illustrating aspects of this later Bacchic ritual has survived, and using this evidence Nilsson (1975: 21) has argued that 'The liknon filled with fruit among which a phallus rises, often covered with a cloth, is the characteristic symbol of the Bacchic mysteries of the Roman Age'. These symbols can readily be understood in terms of Dionysus/Bacchus' role as a god concerned with fertility and new life. The liknon was a winnowing basket, representing the harvest, but it was also used as a container for fruit, particularly grapes which were always associated with Dionysus as the god of wine. As Nilsson (1975) points out, Dionysus himself was not initially directly associated with the phallus, but Priapus and the Satyrs, both in his retinue, were ithyphallic. This combination of the grape cluster with the phallus, symbolising the life power which the dead lack and need, is clearly represented in depictions of the god Priapus (Figure 18).

The paintings preserved on the walls of the Villa dei Misteri at Pompeii, destroyed in the eruption of Vesuvius in AD 79, provide a particularly clear image of the basic rite of initiation into these Bacchic mysteries in the first century AD. These show that the rites were still mainly concerned with the initiation of women, that the ceremony involved flagellation and dancing, that it was associated with wine drinking and banqueting, and that it culminated in the revelation of the phallus, the hope of new life. It seems that, for the Romans, the Bacchic mysteries were one way in which fear about the afterlife could be allayed, with belief and initiation into his rites offering a promise of revelry after death. However, as Nilsson (1975: 146) suggests, 'the sexual symbols of the cult were certainly not, as among simple people, thought of merely as bringers of fertility; for the well-to-do townspeople they had a piquant attraction'. He goes on to conclude that 'The mysteries of Dionysos appealed to well-to-do people who loved a pleasant and luxurious life . . . These people were not in earnest about religion' (Nilsson, 1975: 146–7).

WINE AND THE VINE: SYMBOLS OF FERTILITY AND LUXURY

From its origins in the mountains of Kurdistan and eastern Turkey, viticulture and wine making spread to all of the major centres of prehistoric civilization in south-west Asia and the eastern Mediterranean. In all of these societies the constant need to ensure the continued fertility of the land engendered a complex of ideological beliefs and symbolic rituals designed to enable the societies to reproduce themselves. In such circumstances, the vine, with its apparent death in winter and its

Figure 18 Roman statue of Priapus (first–second century AD)
Source: British Museum (GR1772.3-2.99; BM k95125)

vigorous rebirth in the spring, became a particularly pertinent symbol for
the annual cycle of the death and rebirth of nature as a whole. Gods of the
vine are therefore found associated with agrarian fertility rituals in all of
the major religions where viticulture was practised. In southern latitudes,
where it was difficult to grow vines successfully, and where beer
dominated as the alcoholic beverage of the mass of the population, the
gods of the vine never played a major role in the pantheon. However,

where vines flourished, as in Canaan and then in Greece and Italy, the gods of the vine took on a new and more significant role. Wine, the product of the vine, both in its symbolic and in its physiological effects, brought people into communion with their gods. In the words of Dodds (1960: xx), 'The psychological effect was to liberate the instinctive life in man from the bondage imposed on it by reason and social custom: the worshipper became conscious of a strange new vitality, which he attributed to the god's presence within him'. From here, wine and the vine also became part of the ritual celebration of human life and fertility. It is wine above all that features as the drink at banquets and feasts; it is wine that accompanies the revels and facilitates the love making.

However, wine was also of particular social significance. As the Nimrud wine tables make clear, wine as a relatively scarce commodity was initially the preserve of the chosen few, the servants of the royal court and the priests. In both Egypt and Mesopotamia beer was the common alcoholic beverage, and the drinking of wine was restricted to the elite, those who either controlled the production of wine or the trade networks which permitted it to be imported. In contrast, in countries such as Greece and later Italy, where the vine flourished, wine was to become the alcoholic beverage for the mass of the population. Here, in the social distinctions concerning the use of wine, lies a theme which will recur throughout this book. Moreover, once wine had entered the ritual symbolism of societies, those who controlled its production were also able to maintain an element of economic control over the rest of the population. This survey of the symbolic and ideological significance of wine and the vine must therefore be complemented by an analysis of the economics of viticulture and vinification in antiquity. It is thus to the ownership of vineyards, the labour relations involved in the production of wine, and the economics of the wine trade that the next chapter now turns.

4

WINE IN THE
GRAECO-ROMAN ECONOMY

Experto mihi crede, Silvine, bene positam vineam bonique generis et bono cultore numquam non cum magno faenore gratiam reddidisse. [Believe me, Silvinus, that a well-planted vineyard, of a good kind and maintained by a good vine-dresser, has never failed to provide returns with considerable interest.]

(Columella, *de Re Rustica*, IV.iii.5)

From the establishment of viticulture on the Greek mainland, the vine soon came to prominence alongside wheat and olives as one of the three fundamental products of the agrarian economy of the Mediterranean. The symbolic and ritual significance of wine and the vine was of undoubted importance in the early spread of viticulture, but this role has often been overstated in the past (Allen, 1961; Younger, 1966; Hyams, 1987; Johnson, 1989) with insufficient attention being paid to the importance of economic factors in the development of viticulture and the wine trade. It is therefore important to focus attention in this chapter on the role of wine and the vine in four broad aspects of the economy of the classical world: the *system of agrarian production*, including the structure of land ownership, the labour relations of production, the cultivation techniques, and the methods of vinification; the *location of different types of economic activity*, represented by the distribution of vineyards, wine presses, and potteries manufacturing amphorae; the pattern of *demand*, ranging from the use of wine in religious rituals and at banquets to its role as an exchange commodity; and the resultant network of *trade linkages*, whereby wine was distributed throughout the Graeco-Roman world.

The relative lack of archaeological and literary evidence concerning the agrarian economy of pre-Hellenistic Greece, compared with the wealth of information on the Roman economy in later centuries, makes it difficult to identify processes of economic change during the Classical and Archaic periods. It is indeed tempting to see many elements of the

Graeco-Roman economy as generally static and unchanging. Garnsey (1988: 43) has thus suggested that 'the essential structure of Mediterranean society and the character of its economic base remained relatively stable throughout the period of classical antiquity'. However, while some elements of the system of agrarian production undoubtedly altered little over many centuries, other aspects saw considerable change. This chapter therefore begins by outlining the basic organisation of viticulture and the wine trade in Greece, and then, where the evidence permits, it traces in more detail the nature of viticulture in the Roman world, the organisation of the Roman wine trade, and the eventual spread of viticulture throughout the Mediterranean.

VITICULTURE AND THE WINE TRADE IN THE GREEK ECONOMY

Agrarian production

By the eighth century BC, when literary evidence concerning agriculture first becomes available, it is apparent that viticulture and wine making were quite widely established in Greece. Hesiod's *Works and Days*, which was written probably in the late eighth century BC, provides the first detailed account of Greek agriculture and it combines advice on agriculture with social and political comment, superstitions, and moral maxims (West, 1978). In his account of a farmer's year, Hesiod records among other tasks the pruning of the vines and, in the following quotation, the timing of the vintage in his native Boeotia:

. . . when Orion and the Dog Star move
Into the mid-sky, and Arcturus sees
The rosy-fingered Dawn, then Perses, pluck
The clustered grapes, and bring your harvest home.
Expose them to the sun ten days and nights
Then shadow them for five, and on the sixth
Pour into jars glad Dionysus' gift.
(*Works and Days*, 609–17; Hesiod, 1973: 78)

Here, not only is Perses given guidance that he should pick the grapes in September, but there is also a brief indication of the method of vinification. The technique recommended by Hesiod, involving the drying of the grapes and thus the concentration of sugar in order to produce sweet wines, appears very similar to that used in Italy today to produce *passito* wines. The finished wine was then to be stored in jars, or *amphorae*, remains of which have been found in the archaeological record throughout the classical world.

In the rural communities of Hesiod's Boeotia, Garnsey (1988) has suggested that a relatively uniform peasant subsistence agriculture prevailed. However, with the emergence of the *polis*, or city-state, organisation in Greece, the nature of land ownership became intrinsically connected with the provision of food for the urban community. According to Cooper (1977/78) citizens of the *polis* held family farms, which were theoretically of a sufficient size to sustain all the members of their household. With the growth of these cities, it then became necessary to increase their food supply if hunger and famine were to be avoided, and Garnsey (1988) has suggested five ways in which this might have been done: through the extension of domestic production either by the cultivation of new lands or the use of increasingly intensive methods of cultivation; through the extension of a city's territories as a result of conquest; through the colonisation of new lands; through the importation of staple foods; or through the more equitable distribution of the food available.

The standard family farm probably produced little more than the subsistence needs of its owners, and all the evidence suggests that, while there was some agrarian specialisation influenced in part by the nature of the physical environment, mixed farming based on wheat, olives and vines predominated. Dispersal of landholdings and diversification of crops produced appear to have been two of the main strategies adopted by farmers in their efforts at risk minimisation (Garnsey, 1988). However, by the fifth and fourth centuries BC there is some evidence of increasing agricultural specialisation, partly influenced in Attica by Solon's sixth century legislation in favour of olives and vines at the expense of cereals (Cooper, 1977/78: 162). While many small farms were cultivated by family members, it is evident that slavery also played a fundamental role in the Greek rural economy. In essence the ownership of slaves enabled the citizen of a *polis* to undertake his military and civic duties without having to spend all of the time cultivating his fields and orchards. Jameson (1977/78: 125) has thus argued that in the Classical period slaves 'extended the reach of the family's work force and that this permitted forms of intensification that enabled the farmer to be fully a citizen'. While cereals provided the basis of subsistence, olives, vines, vegetables and legumes were also generally cultivated, and of these it was probably the vines that required the most labour (Jameson, 1977/78). It therefore seems that where vineyards were cultivated for any more than the production of sufficient wine for a household's annual requirements, they were usually cultivated by slaves.

Hesiod's *Works and Days* illustrates that the practice of pruning vines was already well established in the eighth century, but it is in the work of Theophrastus (*c.* 370–285 BC) that the first systematic account of viticulture is to be found. In his *Enquiry into Plants* Theophrastus (1916) discusses the physiology of the vine, methods of its pruning and propagation, its environmental requirements, and the pests and diseases which attack it. Within his account, he shows a clear awareness of the need to prune vines in order to improve their growth and fruitfulness (II.vii.2), and he recommends that they should be cultivated on low ground, with particular varieties being chosen to suit particular soils (II.v.7). Typically vineyards appear to have been planted on plains in regular rows (IV. iv. 8).

The extent of viticulture

While much can be discovered about the practice of viticulture in Greece, it is much less easy to be certain about its overall extent. It seems evident that by the time of Hesiod and Homer in the eighth century BC, viticulture was practised in most parts of Greece from Thrace to Boeotia and Arcadia. Moreover, it also appears that most farmers grew at least a few vines in order to make their own wine. The development of the major city states, such as Athens, Thebes, Argos and Sparta, provided a focus for demand, and it is reasonable to conjecture that it was around such cities that specialist vineyards and wine producers first emerged. By the fifth century, with increased urban demand, certain areas much further away also came to specialise in wine production, and typical of these may have been the island of Thasos at the north of the Aegean. About a dozen kilns producing amphorae have been identified on the island, and Osborne (1987) has suggested that many farmers were then specialising in wine production for the town of Thasos. It is also possible, though, that much of this wine was then exported further afield.

There has been much debate about the types of wine favoured by the Greeks, and their places of origin. One of the most famous of these wines is the Pramnian wine mentioned by Homer in the *Iliad*. According to the *De Materia Medica* of Dioscorides (1934: 604), written in the first century AD, this sweet wine appears to have been made from dried grapes in a manner similar to that described by Hesiod, but Allen (1961) suggests that it may have been produced from the juice of the grapes that emerged from under their own weight before being pressed. It is likely that the word Pramnian indicated the style of wine, rather than its place of production, although Dioscorides (1934: 606) records that that the island of Lesbos was renowned for the production of another wine made from sun-dried grapes and known as Omphacites. Elsewhere in the Aegean, other islands such as Chios and Cos, also produced famous wines, but of a very different kind. Cos wine was reputed to be sour, and the wines of Chios were heavily resinated (Younger, 1966). Typically, Greeks added a wide range of flavouring and diluting substances to their wines, including sea water, spices, honey and resin. Some of these were probably designed to cover the taste of wine that was turning to vinegar, while others acted as preservatives. Wines produced according to Hesiod's prescription would have been both sweet and high in alcohol content, and the addition of water will have served to reduce their strength, making them lighter and easier to consume in quantity.

From the eighth century BC the establishment by Greek city states of new colonies throughout the eastern Mediterranean, partly in response to demographic pressure but also as a result of political unrest, social conflicts, economic imperialism, and an expansion of maritime trade, was a further important factor in the spread of viticulture. Cumae near Naples was thus founded c. 750 BC, and Syracuse in Sicily in 734 BC, with other cities such as Sybaris and Crotona in southern Italy, and Naxos and Messina in Sicily, also being established later in the eighth century (de Beer, 1969; Cornell and Matthews, 1982), and it seems that their inhabitants rapidly developed the practice of viticulture in these colonies. In the first instance it is probable that the colonists brought vines with them from their mother cities, but it is likely that in other cases indigenous wild vines were also cultivated. By the fifth century BC, Greeks such as Herodotus (de Blij, 1983), called southern Italy Oenotria, the land of vines, which is indicative of the importance that viticulture had attained in the region at that date.

To the north in Etruria, modern Tuscany, peoples probably originating in Asia Minor, and later known as the Etruscans, had settled the land between the rivers Tiber and Arno during the ninth and eighth centuries BC (Cornell and Matthews, 1982; Bonamici, 1985). There is much debate as to whether the Etruscans introduced viticulture to Etruria from their homeland, whether they began cultivating the local indigenous vines under their own initiative, or whether they learnt of the necessary skills through their later contacts with the Greeks in southern Italy (Bouloumié, 1981). If they indeed originated from Lydia, as suggested by Herodotus (1954: 52), it is likely that they would have been familiar with both wine and viticulture in their homeland, and consequently they may have begun to cultivate vines in Etruria as early as the eighth century BC. However there is no firm evidence of Etruscan viticulture before the seventh century BC, and it appears that contacts with the Greeks in southern Italy played an important part in the development of Etruscan viticulture (Settis, 1985: 50). In this context it is significant to note that it was the vineyards of southern Italy, rather than Etruria that were of most renown during later centuries, suggesting that it was the activities of the Greeks in Calabria and Campania that were of most significance for the development of viticulture in Italy. Further migrations by Greek peoples led to the establishment of additional colonies in the western Mediterranean. Thus, in about 600 BC Massilia (Marseille) was founded by the Phocaeans in southern France, for the first time bringing the commercial cultivation of the vine to the land that was to dominate the production of quality wine during the medieval and modern periods.

The demand for wine

Much has been written on the uses of wine in the Greek world (Billiard, 1913; Allen, 1961; Younger, 1966; Hyams, 1987; Johnson, 1989), but it is important to stress that there were demands other than simply the ritual use of wine for libations and in the worship of Dionysus. From at least Mycenaean times it is evident that wine had become part of the basic triad of staple foods of people living in Greece. As urban populations increased, and particularly the populations of those without access to land, so too did the demand for wine. This in turn increased the potential for farmers with sufficient resources of land and capital to profit from the production and importation of wine to the cities.

Typical of the drinking parties common to Athens in the fifth and fourth centuries BC are the accounts of such symposia by Plato (1951) and Xenophon (1922). Both commentators describe the way in which discussion and debate, accompanied by liberal quantities of wine, took place following a dinner. The drinking itself was determined by rules, chosen by the host or the president of the symposium, who decided upon the amount of water that was to be mixed with the wine. In Plato's *Symposium* the guests chose to drink their wine very weak, following a heavy bout of drinking the previous night, but even so by the morning most of them were incapably drunk. In general such symposia were accompanied by flute-girls and dancing, but in Plato's account Eryximachus suggested that such entertainment should be dispensed with and that the members of the company should instead amuse themselves by taking it in turns to render a speech in praise of love. Scenes of such drinking parties are to be found represented on numerous fifth century BC red figure vases from Attica (Figure 19). Another common illustration on such vases is the game of *kottabos*, in which the dregs from a glass of wine were dextrously thrown at some mark.

Widespread drinking of wine, certainly among the rich, was thus a common feature of Greek life, but it is also evident that wine was the preferred drink at all levels in Greek society. Unlike the situation in Mesopotamia or Egypt, where beer was the drink of the poor, the arrival of the vine in Greece brought with it the consumption of wine throughout the social order. Not all of this wine was of high quality, and for many of the urban poor their wine may well have been equivalent to the French *piquette*, made from the addition of water to the remains of the grape skins and pips following their final pressing. In rural areas, though, wine was made and consumed by most farmers, as part of an

Figure 19 Red Figure Kylix (480 BC) depicting wine drinking
Source: British Museum (GR 1843.11-3.15; BM K60415)

agrarian system which was later, under the Romans, to achieve dominance throughout the Mediterranean.

The Greek wine trade

There were two basic elements to the Greek wine trade: a rural–urban flow of wine designed to supply the urban market, and a long distance trade, often of higher quality wines, which provided quantities and qualities unavailable locally. Two fifth century BC inscriptions from Thasos (Osborne, 1987) provide an insight into some of the legislation concerning the wine trade at that time, and suggest that much of this may have been designed to protect the consumer from unscrupulous merchants. Thus the purchase of wine in the large-mouthed jars known as *pithoi* was legal only if these jars were sealed, small quantities of wine were not to be sold from large containers, and wine was not to be watered down before sale. In addition, trade in a forward market was prohibited, with anyone buying grape juice or wine from a crop still on the vines before the first day of the month Plynterion being liable to a fine. Together with a prohibition preventing Thasian ships from carrying foreign wine, Osborne (1987) sees this as part of an attempt by the authorities to ensure the true origin of wines claiming to be Thasian.

With the establishment of colonies westwards in the Mediterranean during the eighth century BC a new market for Greek wine was opened up, and the discovery of Ionian amphorae in France indicates that by the sixth century Greek wine was probably reaching the Upper Saône valley and the Jura (Blakeway, 1932–33). Wine was also exported to the towns along the shores of the Black Sea as well as to Egypt (Garlan, 1983). Prisnea (1964) further notes how Greek wine and oil exports had extensively penetrated the Danube region by the fifth century. Once the colonies in southern Italy and Sicily had become firmly established, they too developed their own vineyards and began trading in wine. While wine from the Aegean maintained its prestige, it seems that by the third century the development of vineyards in Italy had led to a reduction in the westerly exports of wine from the Aegean and Greek mainland. Not only was Italy producing its own wines, but it was by then also exporting them northwards and westwards into Gaul.

VITICULTURE AND VINIFICATION IN THE ROMAN WORLD

The importance of viticulture in the Roman agrarian economy is emphasised in the agricultural works written in Latin from the second century BC onwards (Dalmasso, 1937). The earliest known treatise on agriculture, though, was written in Punic by the Carthaginian Mago.

Following the destruction of Carthage in 146 BC this twenty-eight book treatise was translated into both Latin and Greek and provided one of the standard works on the subject, frequently being referred to by such subsequent writers as Varro and Columella (Mahaffy, 1890). Unfortunately Mago's treatise does not survive, and its precise date is also unknown. It is therefore in Cato's notes on agricultural practice in the second century BC that the first comprehensive survey of Roman viticulture is to be found.

Cato: *de Agri Cultura*

The *de Agri Cultura* of Marcus Porcius Cato (234–149 BC), known as Cato the Censor, the Elder and the Orator in order to distinguish him from his grandson, is the earliest surviving prose work in the Latin language. As a collection of agricultural precepts, it provides a wealth of information on subjects ranging from advice on farm purchase and absentee management to sections on recipes and the year's supplies. In so doing it emphasises the importance that the cultivation of vines had attained in the agrarian economy of Italy by the second century BC, a point well illustrated at the beginning of the treatise when Cato (I.7) provides the following advice on the purchase of a farm:

> If you ask me what sort of farm is best, I will say this: One hundred *jugera* of land consisting of every kind of cultivated field, and in the best situation; [of these] the vineyard is of first importance if the wine is good and the yield is great . . .
>
> (Cato, 1933: 4)

In Cato's work there is clear evidence of the profit motive in farming, and Brehaut (1933) has suggested that the importance Cato accords to vines and olives is indicative of a transition that was then taking place in Roman agriculture, with the traditionally subsistence based economy, heavily dominated by cereal production, being gradually replaced by a more commercially oriented agriculture in which the cultivation of vineyards and olive trees played a dominant role (Duncan-Jones, 1974; Spurr, 1986). Within the 162 chapters of *de Agri Cultura* Cato discusses the nature of a vineyard, the equipment required, the calendar of work including details of vinification, wine allowances for slaves, the maintenance of presses, recipes for different types of wine, regulations concerning share-cropping, and procedures for selling wine.

In discussing the laying out of a vineyard, Cato (VI) shows a clear awareness of the need to cultivate different vines in different environmental conditions. Thus, his advice on the setting out of a vineyard includes the following exhortation: 'On the ground that is called best for wine and is exposed to the sun plant the small Aminean grape, the twin

Eugenean and the little yellowish grape' (Cato, 1933: 17). He also recommends that on farms near the city vines should be trained on trees (VII), and he gives a comprehensive list of the infrastructure required for a vineyard of one hundred *iugera* (*c.* 62 acres, or 24.1 hectares), including sixteen labourers, various livestock, three presses, storage jars, strainers, cauldrons, pruning and cultivation equipment, and a range of other smaller items (XI). Within his calendar of work for the year he provides information on the method of pruning vines (XXXII), on their propagation by layering and grafting (XXXII, XL, XLI), on the need for good vineyard care and maintenance (XXXIII), on the use of nurseries for the young vines (XLVII), on the possibilities of transplanting old vines (XLIX), and on methods of vinification for different wines (XXIII, XIV, XXV, XXVI). Wine for slaves was to be made from the addition of water to the refuse of the wine press (XXV, LVII), and Cato adds the following instructions for the amount of such wine to be allocated for each slave:

> the total of wine for the year for each man, seven *amphorae*. For the slaves working in chains add more in proportion to the work they are doing. It is not too much if they drink ten *amphorae* of wine apiece in a year.
>
> (Cato, 1933: 80)

Estimates of the capacity of Dressel 1A and 1B wine amphorae vary from about 17 litres to 27 litres (Tchernia, 1986), but if an average figure of 22 litres is taken, the ten amphorae allocation to slaves in chains would be equivalent to 293 0.75 litre bottles.

The importance of the wine press is emphasised throughout Cato's writing, with details of its maintenance following the vintage being described in chapter LXVIII. In chapters CXI to CXV, and CXX to CXXV he gives a number of recipes for wine, including precise instructions for the preparation of Coan wine, for improving the bouquet of the wine, and for the preparation of wines as medicines for various ailments. His recipe for Coan wine (CXII) closely follows earlier Greek accounts, with the grapes being left on the vines to ripen well, then being dried in the sun for two or three days, before being added to sea water. For medicinal uses wines with various additives were particularly recommended for moving the bowels, for easing the passing of urine, for indigestion and for lumbago.

His instructions concerning the working of a vineyard by a share tenant were as follows (CXXXVII): the share tenant 'is to take good care of the farm, the vineyard trained on trees, and the grainland. The share tenant gets hay and fodder enough for the work oxen that are on the place. All the rest is owned in common' (Cato, 1933: 118). While this account does not specify the precise shares to be received by the tenant, it does indicate a tenurial arrangement for vineyards which came to be prevalent throughout much of the Mediterranean region in later centuries. As in Thasos in the fifth century BC, Cato records strict rules concerning the sale of wine, both as grapes on the vine, and also as wine in storage jars (CXLVII, CXLVIII and CLIV). Particular emphasis is placed on conditions concerning the tasting, measurement and storage of the wine, designed to ensure that the wine that has been sold will be stored preferably by the purchaser rather than the producer.

Cato's account of viticulture and vinification in the second century BC thus provides a comprehensive description of the cultivation of vineyards, which he saw as being the most profitable form of agriculture at that time. Writing over a century later, Varro was less confident about their profitability.

Varro: *Res Rusticae*

Marcus Terentius Varro (116-27 BC) commenced his manual on husbandry, *Res Rusticae*, when he was eighty, and began the section on vine training (VIII) with the comment that 'An objection sometimes made to vineyards is that their costs eat up the profit' (Varro, 1912: 30). He goes on to suggest that the costs of viticulture depend in part on the type of training method used. The cheapest method was to cultivate vines on the ground without supports as in Spain, either allowing the vines to trail along the earth itself or to support the fruiting canes with small forked sticks, but Varro also distinguishes four kinds of yokes used to support vines on uprights in different parts of Italy. Thus he records that 'Poles are used in Falernum, reeds in Arpinum, cords in the country about Brundisium, and withes in the district of Mediolanum (Milan)' (Varro, 1912: 30). Two methods were then used to train the vines: the one along lines at right angles to the trees, the other in slanting lines representing the sloping roofs of the *compluvium* of a Roman house. In referring to the best situations for different types of vine, Varro reiterates Cato's instructions, but he also comments that the supports for the vines should be protected to the north (XXVI). This would seem to indicate that vines were normally grown to the south of their supports, thus permitting the grapes to gain maximum access to sunlight for ripening and also protecting them from any northern winds.

Varro's discussion of viticulture and vinification is much more cursory than that of Cato, but he does illustrate three particular practices of interest: he recognised that different types of grape ripened at different times; he commented that during the vintage good farmers selected their grapes for different purposes; and he also observed that some people press the stalks and skins a second time, with the result tasting like iron (LIV). Finally, in chapter LXV Varro discusses aspects of the élevage of wine, noting the potential of different grapes to produce wines of differing longevity. Thus he commented that some grapes produce wine that rapidly turns acid and requires drinking within a year, whereas others will mature with age. Of these, he particularly identifies Falernian wines, 'which are the more valuable when you bring them out, the more years you keep them in the cellar' (Varro, 1912: 117).

Written at a similar date to Varro's *Res Rusticae*, but of a very different nature, is the *Georgics* of Virgil–Publius Vergilius Maro, which was first published in 29 BC. Although not designed as an agricultural manual, Virgil's account of an 'agricultural golden age' (Johnston, 1980) does, nevertheless, provide additional glimpses of Roman viticultural practices in the first century BC. While Putnam (1979: 7) is correct in stressing that 'only when we eradicate from our minds any lingering notions that the poem is utilitarian, fostered though they be by the poet's overt subject and by the genre, can we begin to come to grips with the *Georgics'* extraordinary qualities', Virgil's descriptions of agricultural practice were based on the agrarian landscapes with which he was familiar. As a didactic poet, his audience would have expected him to be accurate in his descriptions, and his comments on Roman viticulture can therefore be used in the expectation that they reflected current practices. It is in Book II of the *Georgics* that Virgil (1974) addresses Bacchus and viticulture in most detail. In particular, he compares the neatly aligned vineyard with the deployment of a Roman legion (II. 273–88), observing that on rich level ground it is possible to plant vines closer together than on sloping ground, and that vineyards should not slope towards the setting sun (II. 298). He goes on to identify particularly suitable times of the year for the different tasks associated with planting, ploughing and maintaining a vineyard, noting the importance of dressing the vines on which he suggests that never enough time can be spent (II. 397–400). In so doing he emphasises the need for careful pruning of the vines, and the importance of regularly breaking up the soil in the vineyard and clearing away the unwanted vegetation.

Columella: *de Re Rustica*

The most comprehensive account of Roman viticulture is to be found in the twelve books of the *de Re Rustica* of Lucius Junus Moderatus Columella (1941, 1954, 1955), written in the middle of the first century AD, c. AD 65 (Rossiter, 1981; Purcell, 1985; Spurr, 1986). Columella was a native of Cadiz, in southern Spain, but he spent most of his life near Rome owning a number of farms in Latium, and he therefore wrote about Roman agriculture from his own practical experience (Ash, 1941). *De Re Rustica* has provided the context for an extensive debate on the profitability of Roman viticulture (Duncan-Jones, 1974; Carandini, 1983; Tchernia, 1986; Garnsey and Saller, 1987), but it discusses much else besides concerning Roman vine cultivation and vinification. Thus Book III covers the soils suitable for vines, different vine varieties, vine nurseries, vine cuttings, soil preparation, vine planting and vineyard design, while Book IV goes into more detail on the depths of trenches for vines, methods of training and pruning vines, vine supports (White, 1975), methods of layering, the restoration of old vineyards, duties of the vine dresser, and it also includes a section on general precepts for the vineyard owner. Further information on the laying out of vineyards is given in Book V, and the final Book XII gives information on different wines and methods of vinification. A further work by Columella (1955), *de Arboribus*, provides additional details on the selection of land and plants for a vineyard, and on methods of vine cultivation and pruning.

Of particular interest are Columella's statements concerning the laying out of vineyards, the pruning and training of vines, and methods of vinification. He shows a clear awareness of the great diversity of vine types, and of the environmental factors suitable for each kind. In so doing he observes that

> the wise farmer will have discovered by test that the kind of vine proper for level country is one which endures mists and frosts without injury; for a hillside, one which withstands drought and wind. He will assign to fat and fertile land a vine that is slender and not too productive by nature; to lean land, a prolific vine; to heavy soil, a vigorous vine that puts forth much wood and foliage; to loose and rich soil, one that has few canes. He will know that it is not proper to commit to a moist place a vine with thin-skinned fruit and unusually large grapes, but one whose fruit is tough-skinned, small, and full of seeds.
>
> (III.i.5; Columella, 1941: 229)

Moreover, he also recognised that the character of a given vine variety could change according to where it was cultivated. Thus he commented that some vines 'have so far departed from their peculiar character, through a change of place, as to be unrecognizable' (III.ii.30; Columella, 1941: 251). Following Graecinus, Columella recommended for vineyards soils that were warm rather than cold, dry rather than wet, and loose rather than compact (III.xii.4), and in contrast to Virgil he advocated that in warm regions vineyards should face east, while in cooler regions they should face southwards. Vine cuttings were to be planted in trenches (IV.iv), and for best results they should be leaf-pruned and trained from a young age (IV.vi). By the third autumn after the planting of a vineyard, Columella recommended that vines should be propped up on strong supports. Two canes were to be allowed to grow from each vine, (IV.xvi), and they were to be pruned in subsequent years so that they developed into the shape of a star (IV.xvii).

In determining the best time to pick the grapes at the vintage, which usually occurred at the end of September and the beginning of October, Columella (1955) noted that some people 'have attempted to test the ripeness of grapes by tasting them so that they may judge thereby whether their flavour is sweet or acid' (XI.ii.68), but he himself recommended the determination of the ripeness of the grapes by the colour of the pips (XI.ii.69). The importance of cleanliness is stressed throughout Columella's writings on vinification, and he advocated the fumigation of wine cellars before the grapes were harvested (XII.xviii.3). The must was to be pressed, and then stored in jars which had been lined with pitch. Interestingly Columella comments that in order to preserve the wine some people boiled the must, thus reducing its quantity by up to one-third or even one-half (XII.xix.1), but ideally he recommends that the best wines should have no preservatives added to them 'for that wine is most excellent which has given pleasure by its own natural quality' (XII.xix.2; Columella, 1955: 229). Other popular methods of preserving musts and wine were through the addition of pitch (XII.xxii–xxiv), and, following the method used by most Greeks, through mixing it with salt or sea-water (XII.xxv).

These practices recommended by Columella illustrate that by the first century AD Roman viticulture was highly developed, and most of the practices other than the concoction of various medicinal wines and the methods of wine preservation, that were then current may still be found in use today. The archaeological evidence concerning vinification generally agrees with the descriptions of Columella, Varro and Cato, but Rossiter (1981; see also Rossiter and Haldenby, 1989) has also illustrated how it adds to our knowledge of Roman practices in three significant ways. He thus notes how such evidence indicates that the use of wine presses, although by no means universal, was common, particularly at farms with a substantial level of production, that grapes were often trodden in the press room, and that it was usual for wine to be run off into a reservoir before being transferred into storage jars.

The role and profitability of viticulture in the Roman agrarian economy

The differences between the views of Columella and those of Cato, concerning the profitability of wine production, indicate, though, that there were considerable changes in the role of viticulture in the broader agrarian economy between the second century BC and the first century AD. Cato's account suggests that viticulture and wine production were highly profitable in the second century BC following the Roman victories over Carthage. This profitability then appears to have declined by the time of Varro, and when Columella was writing in the first century AD, he has generally been seen as giving a 'picture of a wine industry on the defensive or in the doldrums' (Garnsey and Saller, 1987: 59). The viability of viticulture in Italy was closely influenced by developments of viticulture in the provinces, and by changes in the nature of trade, subjects to be discussed in the next two sections of this chapter, but at this juncture it is useful to evaluate the evidence concerning the relative importance of viticulture in the Italian agrarian economy during these three centuries.

In examining this evidence three observations need to be borne in mind: first, most of the literary evidence, and particularly that of Columella, referred specifically to the villa system, and as Carandini (1983: 186) has pointed out his arguments may not have been relevant to the '*cultura promiscua* which might be practised even on large farms, or to the run-of-the-mill wines of the peasant'; secondly, the Roman agrarian economy was much more complex and diverse than has often been recognised; and thirdly, recent research has suggested that cereal farming was ubiquitous during this period, and did not decline dramatically in the face of increased olive and vine production (Spurr, 1986). Indeed, while villa estates based on slave labour did increase in importance during the first two centuries BC, much Italian agriculture at this time continued to be broadly peasant-based and subsistence oriented (Hopkins, 1978; Weidemann, 1981; Rathbone, 1983).

Spurr (1986: ix) has summarised the traditional arguments concerning agricultural change following the wars against Carthage in the following terms:

> Increased investment in Italian land was supposedly characterised by the growth of agricultural villas, which specialized in olives and vines, or by sheep ranches. Both concerns were staffed by slaves and both were profit oriented. Grain, it is asserted, neither profitable, nor suited to slave labour, was relegated to poor soils and allocated minimum attention ... This process continued into the early imperial period. Then, from the end of the first century AD, growing provincial competition in wine and oil, and a reduction in the slave supply, gradually forced arboriculture to cede to extensive pastoralism and cereal cultivation on *latifundia* run mainly by tenants.

Against this view, Spurr (1986) suggests that throughout this period cereal production flourished, often serving local markets, and that slaves could indeed be kept profitably on estates where cereals and legumes were the main crops.

Much of this debate has revolved around differing interpretations of Columella's figures for the profitability of wine production during the mid-first century AD. At face value, Columella's calculations for a seven *iugera* (c. 4.5 acres, 1.75 hectares) vineyard suggest that, even allowing for the very low yields of one *culleus* to the *iugerum* suggested by Graecinus, it was possible to make a profit from viticulture (III.iii.8–15). Columella himself considered that vineyards that were not yielding three *cullei* to the *iugerum* should be uprooted (III.iii.11), and he also sold young vines to his neighbours (III.iii.13), thus further adding to the profits of his enterprise. In making these calculations, though, Columella failed to include all of the relevant costs. In Finley's (1985: 117) words,

> though he allows for the purchase of the land, of the slave vine-dresser, the vines and the props, as well as for the loss of two years' income while the new vines are maturing, he forgets the farm buildings, equipment, ancillary land (for cereal grains, for example), the maintenance costs of his slaves, depreciation and amortization. His implied 34% annual return is nonsense.

Duncan-Jones (1974: 44) has attempted to recalculate Columella's figures, taking account of the items he omitted, and he derives a net rate of yield for income from the sale of wine and nursery plants during the first eight years of the vineyard's existence of 9 per cent, which is well below Columella's figure of 25.3 per cent. Thereafter, Duncan-Jones suggests that the running yield would have been 15.3 per cent instead of Columella's 34 per cent, and if the sale of nursery plants is ignored this would further reduce the running yield to 10.7 per cent. Against these arguments, though, Carandini (1983: 191) claims that Duncan-Jones's corrections 'seem arbitrary, in that they are based on a misunderstanding of the texts and erroneous methodological presuppositions'. He goes on to argue instead that 'the expenses which are missing in Columella's accounts according to Duncan-Jones are all very unimportant . . . except for the villa, the wheat fields and the woodland' (Carandini, 1983: 194). These he then dismisses as costs relevant to the vineyard account since 'they may have been on the estate for over a century and seemed in the end a natural precondition for its existence' (Carandini, 1983: 194). Carandini suggests that once an original nucleus of slaves had been created it would have been accounted for in the domestic sector of the accounts, rather than in the capital sector, and should therefore properly be absent from the vineyard accounts. These arguments, though, are highly speculative, and Finley (1985: 181) has countered Carandini's claims asserting that 'For all this there is neither a shred of evidence nor a shred of probability'.

On balance, while Columella's figures are indeed incomplete, the critical point would seem to be that, in defending viticulture against its enemies (Tchernia, 1986), he showed on his own lands that vineyards were indeed profitable. Even taking Duncan-Jones's lowest figures, farmers were not going to make a loss in cultivating vines for wine. The real problem in interpreting them, though, is that there are no comparable data for earlier times, such as the second century BC, when Cato had, for example, argued that vines were the most profitable of all crops.

These figures must also not be taken out of context, and in order more fully to understand the changing profitability of viticulture during the Republic and Empire it is essential to examine both the wider spread of viticulture through Europe and the Mediterranean and also the changing structure of the Roman wine trade.

THE SPREAD OF VITICULTURE IN THE ROMAN WORLD

The extent of viticulture during the first century AD

While Cato, Varro and Columella provide a wealth of information concerning the practice of viticulture, their primary emphasis is not on the extent and distribution of vine cultivation in the Roman world. Although Columella, in particular, discusses the various qualities of wines made from grapes in different parts of Italy and elsewhere, it is in the works of Pliny and Strabo that the most comprehensive accounts of the distribution of viticulture in the countries around the Mediterranean are to be found.

The *Geographica (Geography)* of Strabo (*c.* 63 BC–*c.* AD 21) gives a detailed account of the 'known world' at the end of the Republic and the beginning of the Empire (Cornell and Matthews, 1982). In so doing, he illustrates that by then viticulture was well established throughout the Mediterranean. Originating from Amaseia on the southern shores of the Black Sea, Strabo shows most familiarity with the wines of the eastern Mediterranean and what is now Turkey. Thus he reported that the Monarite wines of Cappadocia rivalled the Greek wines (XII.ii.1), that the lands of Priapus, the Pariani and the Lampsaceni were abundantly supplied with the vine (XIII.i.12), and that Catacecaumenite wine was inferior to none of the more notable wines (XIII.iv.11). The wines of the Aegean, particularly those of Cos, Chios and Lesbos are particularly praised (XIV.ii.19), although he considered that those of Samos (XIV.i.15) were poor. Further east, Strabo comments that the vines of Hyrcania, south-east of the Caspian Sea, were very productive (XI.vii.2), and that in Aria 'the land is exceedingly productive of wine, which keeps good for three generations in vessels not smeared with pitch' (XI.x.1; Strabo, 1954, V: 279). He quotes Aristobulus as stating that wine was produced in India in the country of Musicanus, but he also notes that other writers said that India produced no wine. Syria, particularly in the area around Laodicea (XVI.ii.19), abounded in wine, and the vine was also apparently grown in Babylonia (XV.iii.1) and in the marshes at the mouth of the Euphrates (XVI.iv.1), although in Arabia he comments that most wine was made from dates (XVI.iv.25).

Turning to north Africa, Strabo considered that the wines of Libya were of poor quality, being over-diluted with sea water, but in Egypt, referring to the area around Lake Mareotis, he observed that 'the vintages in this region are so good that the Mareotic wine is racked off with a view to ageing it' (XVII.i.14; Strabo, 1949, VIII: 59). The Egyptians also apparently imported wines from the north, and Strabo noted in particular that Lesbian wines were sold in Naucratis (XVII.i.33). To the west, in Iberia, large amounts of wine and grain were exported from Turdetania, to the north-west of Cadiz (III.ii.6), and vines were also grown on a large island in the river Tagus near Moron, modern Almeirim, in Portugal (III.iii.1).

To the north and west of peninsular Italy, Strabo mentioned that the Ligurians only produced a small amount of poor quality wine, which was usually mixed with pitch (IV.vi.2), although Rhaetic wine was reputedly as good as much Italian wine (IV.vi.8.). Aquileia, at the head of the Adriatic, served as an emporium for the Illyrian tribes, and it reputedly had wooden jars the size of houses in which wine was stored (V.i.8; V.i.12). Within peninsular Italy itself, the best wines came from Latium and Campania. For Strabo, the top three wines were the Falernian, the Statanian and the Calenian (V.iv.3), although close to these were the Surrentine wines (V.iv.3), and Caecuban, Fundian, Setian, and Alban wines were also widely famed (V.iii.6). By the beginning of the first century AD viticulture was therefore well established throughout the coasts of the Mediterranean, from Iberia to Syria, and from southern France to northern Egypt. However, Strabo gives no indication that the vine was widely cultivated away from the coastal plains of southern Europe.

Half a century later, Pliny the Elder (AD 23/4-79) in his *Naturalis Historiae (Natural History)* (Pliny, 1945-50) not only amplifies many of Columella's observations on viticulture and vinification, but he also provides an extensive list of wines produced in different parts of the empire. He thus lists ninety-one varieties of vine (XIV.iv.20-43), fifty kinds of quality wine (XIV.viii.59-72), thirty-eight varieties of foreign wines (XIV.ix.73-6), together with a range of other salted, sweet and artificial wines. Pliny gives pride of place among the vines to the five varieties of the Aminean vine, with the two varieties of the red-wooded Nomentian vines taking second place, followed by the apiana, all of which he considered to be indigenous to Italy (XIV.iv.21-4). Most of the other types of vine found in Italy were, according to Pliny, imported from Chios or Thasos. Significantly, he also observed that a number of vines, such as the Nomentian and the apiana, were particularly suited to frost and cold climates (XIV.iv.23-4). This indicates that new vine species were being introduced into more northern climates, and he also specifically noted the recent introduction of a vine, which he describes as blossoming on a single day and thus being proof against all accidents, at Alba Helvia in the province of Narbonensis, southern France (XIV.iv.43). This vine, known as *carbonica*, the charcoal vine, was apparently being planted throughout Narbonensis at the time he was writing.

Pliny divided the better quality wines into four main classes (XIV.viii.59-72). Traditionally the best wine was reputed to be the Caecuban from Latium, but this apparently no longer existed when he was writing. Wines of the second rank, and the best that were then available, were the three varieties of Falernian wine from Campania, while the third rank wines included the sweet wines of Alba, and the Surrentine, Massic, Statanian, and Calenian wines, together with those from Fundi, Veliternum and Priverna. The fourth ranked wines, used at public banquets, were mainly those from Sicily around Messina. Below these wines were a range of other lesser ranking Italian wines. Of the foreign wines he gave the highest status to those of Thasos, Chios and Lesbos, although those from Tarraconensis and the Balearic Islands in Spain were reputed to compare favourably with the best Italian wines (XIV.viii.66-ix.76). In describing the many types of wine available, Pliny was careful to point out the great importance of the land and the soil in which they were grown, for he considered that the same types of vine grown in different places could produce very different types of wine (XIV.viii.70).

Pliny, unlike many of his predecessors, gave several important insights into changes that were taking place in the distribution of viticulture. Not only did he comment that new vines were being planted in the province of Narbonensis, but he also observed that at the time he was writing the vineyards of Spain were highly productive (XIV.viii.68,71). Moreover, he bemoaned the degeneration of Falernian wines, which he blamed on the increasing concern of the producers with quantity rather than quality (XIV.viii.62). At Vienne, just south of Lyon in France, he reported that a new type of vine had been discovered, which was being widely planted in the surrounding territories imparting the flavour of pitch to the wines (XIV.iii.18), and in the Alps he noted that wines were kept in wooden containers, which were heated during cold winters, rather than in amphorae (XIV.xxvii.132-3). All of these observations point to the spread of viticulture northwards during the course of the first century AD.

Most of Pliny's observations on viticulture (XIV.v.44-52; XVII.iii.25-41; XVII.xxv.115-7; XVII.xxxv.152-214) and vinification (XIV.xxv.122-30; XIV.xxvii.132-6; XVIII.lxxiv.309-20) build on principles already established by Cato and Columella. However, he does emphasise a number of additional interesting observations, noting that in some places spontaneous secondary fermentation of the wine makes it lose its flavour (XIV.xxv.125; *cf.* Columella, 1954; XII.xxiv.3), and he further describes ways in which vines may be protected from the ravages of insects (XVII.xxxvi.215). Pliny's account also pays particular attention to the uses for wine for medicinal purposes (XIV.xxii.116–18; XXIII.ii–xxvi.6–104).

The impression gained from the writings of Strabo and Pliny is thus that vines were widely cultivated throughout southern Europe and Turkey by the middle of the first century AD. While traditionally the wines of Greece and the Aegean had the reputation of being of the highest quality, Pliny notes that from the middle of the first century BC Italian wines had come to enjoy an equally high reputation (XIV.xiii.87), and that by the time he was writing it was the wines of Italy that were the most renowned. But the seeds of change are already to be found in Pliny's *Natural History*: Spain was emerging as a major producer of wine, the quality of Italian wines was on the wane, and new vines were being planted in Narbonensis.

The spread of the vine northwards into Gaul.

Most of the vineyards cited by Pliny, Strabo and other earlier Greek and Roman authors were situated in coastal locations or adjacent to major rivers. The costs of transporting wine in amphorae were high, particularly over land, and long distance trade was thus mainly by ship. Using Diocletian's price edict dating from AD 301, Duncan-Jones (1974) has estimated that the cost ratios for sea transport, inland waterways and road transport at that time were 1:4.9:28–56. As Peacock and Williams (1986: 64) have illustrated this implies that 'it was cheaper to send goods from one end of the Mediterranean to the other than to transport them overland a distance of 75 miles'. Consequently, if wine producers sought to sell their wines to any other than local markets it was essential for their vineyards to be located in close proximity to navigable rivers or coastal ports. This, to some extent, explains the dominant role of the Aegean islands and the coastal plains of Campania and Latium in the Graeco-Roman wine trade prior to the middle of the first century AD.

The conquest of Iberia, eventually completed in 133 BC, opened up the possibility of wines from Iberian vineyards competing with those of Italy, and Julius Caesar's campaigns between 58 and 52 BC likewise led to the opportunity for viticulture to become more widely established in Gaul. Using literary evidence, together with the discovery of provincial amphorae and evidence of viticultural and vinification practices, it is possible to trace the gradual spread of viticulture northwards away from the Mediterranean (Dion, 1959; Lachiver, 1988). While vines had been grown around Massilia (Marseille) at least as early as the first Phocaean settlement there in the sixth century BC, it seems that most wine consumed in southern Gaul at that time was imported from Greece and southern Italy. With the establishment of the province of Narbonensis in 118 BC, Roman methods of viticulture became firmly established (Figure

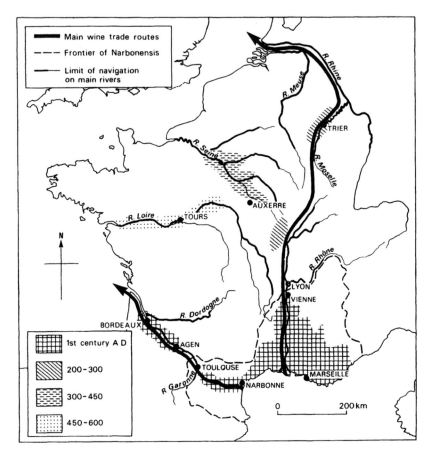

Figure 20 The spread of viticulture in Gaul
Source: derived from Dion (1959) Drinkwater (1983) and Lachiver (1988)

20). Pliny's account, cited above, indicates further that by the middle of
the first century AD new vines were being cultivated near Vienne and Alba
Helvia, and at this date vines were probably to be found in most of the
main river valleys of Narbonensis.

At the end of the first century AD, the Emperor Domitian issued an
edict, which has widely been interpreted as having significantly reduced
the extent of viticulture in the provinces (Dion, 1959; Hyams, 1987; de
Blij, 1983). The main piece of evidence concerning this edict is found in
the *De Vita Caesarum* written by Suetonius, who made the following
observation under his description of the life of Domitian (VIII.Domitia-
nus.7):

On one occasion when there was an extreme abundance of wine

with an extreme scarcity of corn, thinking that the cornfields were being neglected by too great attention to the vineyards, he issued an edict forbidding the planting of new vines in Italy and ordering the cutting down of the vineyards in the provinces, not more than half, at most, of the existing vines to be left; but he did not persist in carrying out this reform.

(Suetonius, 1930: 165)

This edict, which had been dated to c. AD 90–2, has been widely debated, but Levick (1982: 67) would seem to be correct in arguing that 'The main objective of the edict was . . . to increase grain production throughout the Empire'. Concern with the provision of sufficient grain to supply Rome and the legions was always high on the political agenda (Garnsey, 1988), and while 'a secondary purpose may have been to reduce the pressure of competition on the Italian vine-grower' (Levick, 1982: 67), it does not seem that this was the prime intention of the edict. The differentiation between Italy and the provinces was probably nothing out of the ordinary, and would have been expected given the privileged position of Italy within the Empire (Levick, 1982; Garnsey and Saller, 1987). Tchernia (1986: 232) notes how, at the end of the first century AD, there was a crisis in the wine industry of central Tyrrhenian Italy, caused by a period of overproduction following the devastation caused by the eruption of Vesuvius in AD 79. As is clear from Pliny's account, this part of Italy provided much of Rome's requirements for wine, and extensive replanting following the destruction wrought by Vesuvius created a glut of wine. It was the misfortune of the Campanian vineyard owners that in c. AD 92 this overproduction of wine coincided with a poor grain harvest. In the words of Garnsey and Saller (1987: 60) the edict can be seen as representing 'the impulsive reaction of an emperor who knew from the experiences of each of his predecessors, if not yet from his own, the political dangers involved in permitting his subjects, in particular the plebs of Rome, to go hungry'. While a later edict by Domitian apparently also prohibted the planting of vines within city boundaries (Tchernia, 1986), Suetonius's own comments that Domitian did not persist in carrying out his reforms suggest that they did not in practice represent as significant a crisis for viticulture, either in Italy or in the provinces, as was once thought.

The distribution of amphorae within Gaul indicates that Mediterranean wines were widely available there by the end of the first century BC; in Dion's (1959: 103) words, 'La vaste dispersion des fragments d'amphores dans la Gaule extra-méditerranéenne atteste l'ampleur du débit que trouvaient en ce pays, vers la fin de l'époque républicaine romaine et dans les premières anneés de l'Empire, les vins d'origine méridionale.' [The great dispersal of amphora fragments in the interior of Gaul attests

115

to the magnitutude of the flow of wine from southern origins to the country towards the end of the Roman Republic and the first years of the Empire.] Two great trade routes had been established, along which countless amphorae of wine must have travelled. The westerly route, called *le seuil de Narouze* by Dion (1959), left Narbonne in the direction of Toulouse, from whence it went to Agen and eventually to Bordeaux along the river Garonne. The other northerly route, the *couloir rhodanien*, followed the river Rhône to Vienne and Lyon, and then the river Saône before cutting across country to the Mosel and the Rhine (Figure 20). From the edges of the province of Narbonensis, where the vine was well established by the middle of the first century AD, it is then generally argued that viticulture spread northwards and westwards following the directions of these two trade routes (Lachiver, 1988).

However, the traditional vines of the Mediterranean were not well suited to the climate of Gaul, with its colder winters and generally wetter conditions, and, as Pliny noted, new vines were already being introduced in Narbonensis during the first century AD. Some of these may have resulted from the rare chance hybridisation of indigenous vines with introduced varieties, but in most instances a process of selection, by which vines that survived or flourished in the new conditions in Gaul were chosen as the basis for new vineyards, enabled viticulture to spread northwards. Two varieties of vine in particular, cited by both Pliny and Columella, seem to have become particularly important in Gaul. In the west, the *biturica*, which may have originated in Spain, and was the ancestor of the Cabernet varieties of today, came to be the main variety in the region around Bordeaux, whereas around Lugudunum (Lyon) the Allobroges developed an indigenous variety known as the *allobrogica* (Dion, 1959; Lachiver, 1988).

While Pliny mentioned the *biturica* grape, Strabo did not, and it seems likely therefore that it was during the course of the first century AD that viticulture became established in the Bordelais. Tracing the origin of the Burgundian vineyards is, however more difficult (Dion, 1952a, 1952b, 1959; Thevenot, 1952; Tchernia, 1986; Lachiver, 1988). Thevenot (1952) has suggested that the disappearance of amphorae from Burgundy during the second century AD was indicative that the Burgundians were no longer importing wine, but were instead producing their own wines. However, against this Dion (1959) points out that during the second century AD there was a large community of wine merchants at Lyon, and he suggests that their presence was incompatible with the existence at the same time of a flourishing wine producing area in Burgundy. More importantly, though, the disappearance of amphorae during the second century was not restricted to Burgundy, but was instead found throughout the Mediterranean, and this suggests that it was not purely local factors in Burgundy which led to the decline. Firm evidence for the presence of viticulture in Burgundy dates from AD 312 when the Emperor Constantine visited Autun. In a panegyric delivered to Constantine by the people of Autun they bemoan the poverty of their countryside, and in so doing note the existence of a region of well established vineyards known as the *Pagus Arebrignus* (Galletier, 1952; Dion, 1959). This has been identified as the Côte d'Or in the vicinity of Beaune and Nuits St Georges (Dion, 1952a, 1959; Lachiver, 1988), and it can therefore be asserted with some certainty that viticulture was established here by the beginning of the fourth century at the latest. In the panegyric, the vines are specifically stated to be old and neglected, and Dion (1959) and Lachiver (1988) argue that the neglect resulted from the invasions of the region by the Franks and Alamanni during the middle of the third century AD. This period of political disorder which in turn generated an economic crisis within the Roman empire (Cornell and Matthews, 1982) would not have been condusive to the establishment of new vineyards in Gaul, and it therefore seems probable that the vineyards of Burgundy were established some time earlier in the third century AD.

Two pieces of legislation during the third century further affected the development of viticulture in Gaul. In an edict of AD 212 the Emperor Caracalla conferred citizenship on most free inhabitants of the empire (Garnsey and Saller, 1987: 16), and according to Dion (1959) this had the effect of removing the privilege of cultivating vines, which until then was the prerogative only of Roman citizens. Further, Domitian's edict was then repealed in the reign of Probus (AD 276–82), during his successful campaigns against the Germanic tribes in which he secured the frontiers along the Rhine and Danube, and this removed any remaining legal restrictions on the development of viticulture. The military crisis of the third century had enforced a shift of resources to the frontier regions of the empire, and associated with this a number of cities, such as Augusta Treverorum (Trier), emerged as imperial capitals. This will have acted as a significant incentive for the development of viticulture in such regions, and certainly by the end of the third century vines appear to have been cultivated in the valley of the Mosel near Trier (Figure 20) (Bassermann-Jordan, 1907). There is as yet no evidence that viticulture had penetrated the Loire or Seine valleys in the west and north of Gaul by this date, and it therefore seems likely that it was the administrative and social importance of Trier as an imperial capital that acted as the major incentive for the development of viticulture in its vicinity. It is indeed possible that viticulture was also introduced near other imperial capitals, such as Sirmium, Naissus and Serdica in the Danubian regions, and Prisnea (1964: 172–3) certainly argues that vine cultivation was widespread in Dacia during the Roman period. In this context, Mócsy (1974: 298) describes the find of an altar in Pannonia dedicated to Liber Pater on the occasion of the planting of a vineyard of 400 *arpennes*, and in addition he notes that the Emperor Probus specifically ordered vines to be planted at the Alba Mons and at the Aureus Mons east of Singidunum. Mócsy (1974: 298-9) also observes that 'Viticulture in Pannonia is attested by finds such as vinedressers' knives, which are also presented in the hands of the deity Silvanus . . . and by a few wine presses', but that according to Dio Cassius the quality of wine produced was poor. It therefore appears that by the late third century vines were being cultivated quite widely in the middle Danube valley.

The extensive cultivation of vines on the slopes of the Mosel during the fourth century is well attested in the poem *Mosella* written by Ausonius *c.* 370-71 (Ausonius, 1919). In this poem he compares the vineyards of the Mosel with those of Campania, Thrace and his own Garonne (lines 157-60), and describes the hills around the river as being overgrown with vines. Elsewhere, in another poem *De Herediolo* (On his little patrimony) written *c.* AD 379, Ausonius (1919) also provides a brief description of his own estate near Bordeaux, which included a hundred *iugera* of vines, and in his account of the city of Bordeaux, which forms part of his *Ordo urbium nobilium*, he again mentions its fame as a wine producing area. Elsewhere in Gaul during the fourth century AD viticulture appears to have become established in the Yonne valley around Auxerre, and by the fifth century vines were also to be found in the vicinity of Paris (Figure 20). The introduction of viticulture to the Loire seems to have been a later development. Gregory of Tours in his *History of the Franks* (IX.18) noted that in the late 580s the people of Nantes owned vineyards and made wine (Gregory of Tours, 1927: 387), and it seems that vines may well only have been planted at the mouth of the Loire towards the end of the fifth century, although they were cultivated further inland somewhat before this. Dion (1959) suggests that the late arrival of viticulture in the region of Nantes might reflect protectionist policies of the Roman authorities in Bordeaux who sought to prevent competition for their own wines from adjacent parts of Gaul.

The northern limit of viticulture in the Roman era is widely considered to have been just to the north of Paris and the Mosel (Cornell and Matthews, 1982). However, there is some evidence that the Romans also introduced viticulture to Britain. Wine was certainly popular in England prior to the Roman conquest in the first century AD, attested by the remains of numerous amphorae found in southern England (Harding, 1974; Cunliffe, 1978), but there is no firm evidence that vines were cultivated at this date. Much of the evidence adduced in support of the cultivation of vines in Roman Britain has been shown by Williams (1977) to be of dubious validity, although there may possibly have been a short-lived vineyard dating from the late third century AD at North Thoresby in Lincolnshire (Webster, *et al.*, 1967). While many of the finds of Roman grape pips and vine produce in lowland England probably resulted from imports of wine or grapes, Williams (1977) suggests that the grape debris and vine stems found in Gloucester and Boxmoor respectively during the nineteenth century may well indicate that viticulture on a very small scale was indeed practised in southern England. If this was indeed so, it implies that vines may also have been cultivated in many other northern parts of the Roman Empire for which firm evidence is at present lacking.

THE ROMAN WINE TRADE

It is impossible to explain satisfactorily the growth of commercially oriented estates worked by slaves in Italy during the second century BC, the spread of vineyards into the provinces, and the changing profitability of viticulture during the Empire without an understanding of the mechanisms and patterns of Roman trade. While the discovery of numerous amphorae provides considerable artefactual evidence concerning the Roman wine trade, much still remains unknown about the significance of this trade within the Roman economy (Duncan-Jones, 1974; Garnsey, Hopkins and Whittaker, 1983; Peacock and Williams, 1986; Tchernia, 1986; Garnsey and Saller, 1987).

Amphorae and traders: objects and actors

Pottery amphorae were the main containers used in seaborne Mediterranean trade before the third century AD, holding a wide variety of products, ranging from wine and olive oil, to fish, dates and fruit. In the second and first centuries BC wine was mainly transported in amphorae of the type known as Dressel 1 (Figure 21), and Tchernia (1983: 87) has seen the diffusion of these particular amphorae as constituting 'the most spectacular evidence of the export of agricultural produce from Italy in the ancient world'. They appear to have evolved from earlier Graeco-Italian varieties around 130 BC (Peacock and Williams, 1986), and were used exclusively for the export of wines from the Tyrrhenian coast (Tchernia, 1983). Peacock and Williams (1986) see the heavier Dressel 1B amphorae, with collar rims, as having developed from the Dressel 1A variety during the middle of the first century BC, at which time foreign imitations from eastern Spain and southern France also begin to appear. The Spanish amphorae probably carried the wines of Tarraconensis, which according to Pliny (XIV.viii.71) compared favourably with the finest wines of Italy, while those from Gaul are indicative of the beginnings of viticulture in that province. The distribution of Dressel 1 amphorae throughout Europe and the Mediterranean (Figure 22) provides a good indication of the extent of the Roman wine trade during the first century BC, and illustrates not only the long distances over which wine was transported, but also the dominant part played by rivers in this trade (Panella, 1981). Thus amphorae from south-west Italy were found as far afield as southern Britain and north-west Africa, and along all of the major rivers of Europe, such as the Rhône, Rhine, Garonne, Ebro, and Guadiana. It is not easy to estimate the level of trade represented by these surviving amphorae, but Tchernia (1983: 92) suggests that somewhere in the order of forty million amphorae may have been unloaded in Gaul during the course of the century during which the Dressel 1 amphorae were in use, representing a flow of up to perhaps 100,000 hectolitres of wine a year.

In the penultimate decade of the first century BC the Dressel 1 amphorae were replaced by a new variety imitating amphorae produced on the island of Cos in the Aegean and known as Dressel 2-4 (Figure 21) (Tchernia, 1983, 1986; Peacock and Williams, 1986). Whereas the Dressel 1 amphorae weighed *c.* 25 kg with a capacity/weight ratio of 0.88 litres/ kg, the newer Dressel 2-4 amphorae were much lighter, weighing between 12.0 and 16.5 kg with a capacity/weight ratio of 1.09-2.04 litres/ kg (Peacock and Williams, 1986: 52). Peacock and Williams (1986) see this change as resulting from new drinking habits which favoured watered down wines such as those from the Aegean, but their more favourable capacity to weight ratio would also suggest that they may well have been introduced partly to achieve a greater amount of wine per shipment than had previously been possible. Significantly, Dressel 2-4 amphorae are extremely rare in Gaul, and Tchernia (1983, 1986) suggests that this reflects the rapid spread of viticulture in Gaul beginning at the end of the first century BC. A further technical innovation concerning the wine trade at this time was the introduction of ships which carried large jars, known as *dolia*, anchored amidships, but these seem only to have been an experiment and were not persisted with far into the first century AD (Garnsey and Saller, 1987). Dressel 2-4 amphorae remained the main carriers of Italian wines until the end of the first century AD, when there is a sharp reduction in their abundance. This decline has widely been used to support arguments for a crisis in Italian viticulture at this time, and a collapse in the wine trade, but, as Tchernia (1980) and Garnsey and Saller (1987) point out, wine from Campania continued to be exported to Rome and elsewhere overseas in containers as yet unidentified.

The widespread distribution of Dressel 1 amphorae attests to the existence of an extensive commercial network of traders dealing in wine during the first century BC. However, much less is known about the people involved in the wine trade at this date. Purcell (1985) has argued that in the old established vineyard regions of Campania and Etruria corporations of traders played a relatively unimportant role, but that in the newly developing vineyards at the end of the Republic, associations of *mercatores* gradually came to dominate the trade, buying up wine from

After Dressel

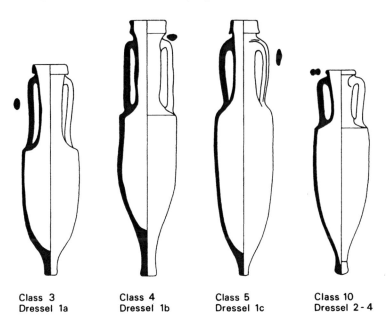

Dressel 1   Dressel 2   Dressel 3   Dressel 4

After Peacock and Williams

| Class 3 | Class 4 | Class 5 | Class 10 |
| Dressel 1a | Dressel 1b | Dressel 1c | Dressel 2-4 |

Figure 21   Roman wine amphorae
Source: derived from Dressel (1899) and Peacock and Williams (1986)

121

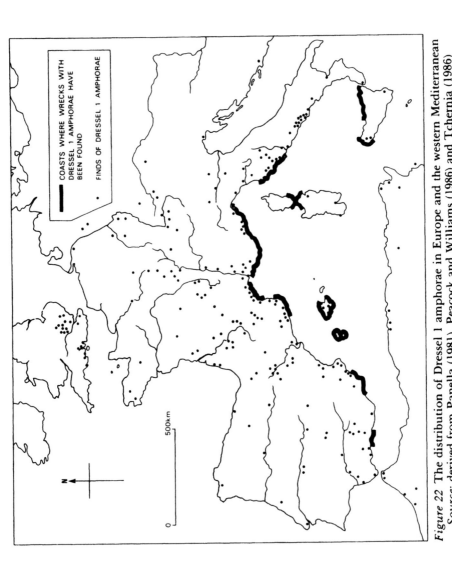

Figure 22 The distribution of Dressel 1 amphorae in Europe and the western Mediterranean
Source: derived from Panella (1981), Peacock and Williams (1986) and Tchernia (1986)

estates, transporting it overland to the sea and then exporting it overseas. Much of the wine trade at this time still seems to have been in the hands of relatively small-scale individual landlords, and it is not until the first century AD that there is firm epigraphic evidence of the existence of corporations of wine merchants. Indeed, Purcell (1985: 12) has suggested that 'the *vinarii* and their organization are, on the basis of this material, above all a phenomenon of the second century'. The discovery close to the port of Ostia of an Augustinian deposit of 181 wine amphorae, 58 of which were from Tarraconensis and Baetica in Spain (Hesnard, 1980), indicates that by the beginning of the first century AD wine was being imported to Rome from a number of foreign destinations, possibly by merchants based at Ostia. The first mention of a *Portus Vinarius* at Rome itself is found in AD 68, and by the end of the first century AD purpose-built wine warehouses, *cellae vinariae*, are found along the banks of the Tiber (Purcell, 1985). By the beginning of the second century AD a large *forum vinarium* had been established at Ostia, and this provided the base for two guilds of wine merchants, the *negotiatores fori vinarii* and the *corpus splendidissimum inportantium et negotiantium vinariorum*, as well as being the location of periodic wine auctions (Bendinelli, 1931; Hermansen, 1981; Purcell, 1985). Elsewhere in the Empire in the second century, other groups of wine merchants had also become established, most notable of which were the *Negotiatores vinarii* based at Lyon, who played an important role in distributing wine northwards into Gaul (Dion, 1959: 129)

The changing structure of demand and supply

The distribution of amphorae provides an indication of where wine was traded, but to explain the alterations in types of amphorae and the organisation of the wine trade it is also essential to consider the changing structure of wine demand and supply.

The popularity of wine during the centuries either side of the birth of Christ was in part influenced by its religious significance, but it is evident from the writings of Roman agronomists that from at least the second century BC vines were also being grown in Italy in the expectation of financial profit. The creation of large slave estates transformed the agrarian economy, displacing numerous subsistence holdings and giving rise to a peasant exodus from the land. However, as Hopkins (1978: 3) has argued, 'Many peasant farms remained intact', and the role of free peasants, owning some land but also working as tenants and labourers on the estates of the rich, must not be ignored. Despite the expansion in the number of slave estates producing wine, olives and oil during the second and first centuries BC, slaves accounted for only 35 to 40 per cent of the total population of Italy at the end of the first century BC (Hopkins, 1978), and therefore in much of the country peasant subsistence agriculture must have continued to survive.

For slave estates to have been profitable it was essential for there to have been a substantial market for their products, particularly wine, olive oil and wheat. This was provided in part by the growth of Rome and other prosperous Italian towns. By the end of the first century BC it is probable that the population of Rome was somewhere in the order of one million inhabitants, having risen from perhaps 100,000 in 300 BC (Hopkins, 1978; Cornell and Matthews, 1982: 88), and this vast increase in population generated a substantial demand for cheap wine, much of which came to be supplied by the expanding number of slave estates in Campania and Latium. The high density of bars in Roman towns such as Ostia, Pompeii and Herculaneum (Figure 23) testifies both to the popularity of wine drinking among all levels of Roman society, and also to the substantial market for wine that existed in other Roman towns (Bendinelli, 1931; Kleberg, 1957). Such demand could not have been satisfied by traditional peasant subsistence agriculture alone, and necessitated the creation of large specialist estates, such as those described by Cato and Varro.

While urban wine demand was one factor fuelling the expansion of commercially oriented viticulture in Italy during the first two centuries BC, another important incentive was the export of Italian wine to Gaul in exchange for slaves. Writing in the first century BC, Diodorus Siculus (V.xxvi.1) thus recorded that

> The Gauls are exceedingly addicted to the use of wine and fill themselves with the wine which is brought into their country by merchants, drinking it unmixed, and since they partake of this drink without moderation by reason of their craving for it, when they are drunken they fall into a stupor or state of madness. Consequently many of the Italian traders, induced by the love of money which characterizes them, believe that the love of wine of these Gauls is their own godsend. For these transport the wine on the navigable rivers by means of boats and through the level plain on wagons, and receive for it an incredible price; for in exchange for a jar of wine they receive a slave, getting a servant in return for the drink.

> (Diodorus Siculus, 1939: 167)

Tchernia (1983) sees these inequitable terms of trade as being the result of the system of potlatch practised by the Gauls, rather than commercial exchange, but this text is of particular significance since it indicates not only the practice of exchanging slaves for wine, but also something of the trading methods used. In addition, the high concentrations of amphorae at towns such as Toulouse further suggests that it was at such places that

Figure 23 A Roman bar at Ercolano (Herculaneum) (AD 79)
Source: author 15th May 1987

the Italian wine was transferred into barrels for its further distribution into Gaul (Figure 22). Wine must also have been exchanged for a range of other goods, particularly metals, but it seems likely that the slave trade accounted for as much as one third of the imports of Italian wine into Gaul during the first century BC (Tchernia, 1983). The development of slave-based estates in Italy, in which viticulture played an important role, can therefore be seen in part as being enabled by the export of wine to Gaul in exchange for the very labour power needed to cultivate the vineyards.

The collapse of the wine trade to Gaul at the end of the first century BC, evidenced by the disappearance of Dressel 1 amphorae, has widely been interpreted as being connected with a decline in the slave trade from Gaul to Italy (Tchernia, 1983). The precise reasons for this, though, are not altogether clear. While it is possible that Italian wine was still exported to Gaul, but in unidentified containers, and that slaves were still taken to Rome, but in exchange for commodities other than wine, these practices seem unlikely. Instead, the explanation for the dramatic collapse of this exchange trade, in a mere two decades, needs to be sought in changing demand and supply relationships in both Gaul and Italy. Within Gaul at the end of the first century BC it seems likely that there was less demand for Italian wine, not only because the vineyards of Narbonensis were beginning to expand in terms of their own production, but also as a result of increased imports of wine from Spain. Moreover, two particular factors would also appear to have reduced the ready supply of slaves available in Gaul. Tchernia (1983) argues that the social reorganisations introduced by Augustus meant that an unrestricted trade in slaves was no longer possible, and it also probably became increasingly less easy to maintain the slave supply in view of the vast numbers of slaves already exported to Italy over the previous two centuries. Once Julius Caesar's conquest of Gaul was completed in 51 BC, the supply of Gallic slaves would appear to have fallen dramatically. Turning to Italy, Finley (1985: 86-7) has suggested that the structural transformation of Roman society that took place during the principate of Augustus, 'in which orders again became functionally significant' and 'in which a broader spectrum of statuses gradually replaced the classical bunching into free man and slaves', meant that landowners no longer fully exploited slave labour. There may well, therefore, have been less demand for slaves from Gaul, with estate owners coming to rely instead on slaves born on their own properties or obtained from elsewhere. It is also possible that by the end of the first century BC the expansion of large slave estates had increased as far as was possible, given the effects that this had on the peasant economy.

The dominant factor explaining the rapidity of the change in orientation of the Italian wine trade was, however, most likely to have been the vastly increased demand for wine from Rome and other Italian towns, which meant that it became more profitable for Italian wine producers to sell their wines within Italy rather than export them overseas. The expansion of slave estates meant that between 80 and 8 BC approximately half of the free adult male population of Italy left their farms and emigrated to the towns or were settled on new properties (Hopkins, 1978). The expropriation of their land meant that they were no longer able to produce their own wines, and they therefore provided a rapidly growing market for the large wine estates of Campania, Latium and other parts of Italy. At some critical point in the last two decades of the first century BC this market became sufficiently large and profitable to divert the exports of wine which had previously gone to Gaul. Thereafter, during the first and second centuries AD, Italy became a major wine importer, drawing its supplies from as far afield as Spain, Gaul and Greece.

That wine drinking was commonplace in Roman society is well attested by the numerous references to it in the works of Roman poets. Griffin (1985: 87), for example, has shown how

> every nuance of the drinking and appreciation of wine can enter into the work of the poets. Invitations; simple wine and expensive wine; modest parties, philosophical parties, and wild parties; the actual events of the *convivium* – all could be handled, and on all levels of stylisation which vary from something very close to realism, all the way to epic elevation on the one hand, or to Dionysiac splendour on the other.

The *convivium* was the Roman equivalent of the Greek symposium already referred to. Literally, it was a living together, but unlike its Greek equivalent women were admitted more freely, and for the participants wine formed an important element in the pursuit of both pleasure and love. Generally the convivia described by the poets were relatively sober affairs, but as Griffin (1985: 84) points out

> It is none the less possible to see what many of them actually were like or threatened to turn into. The two dangers were that guests would start carrying on with girls who belonged to other members of the party, or that drunken brawls might break out.

From the poets of the first century BC, most notably Horace (65-8 BC), it is evident that wine was considered to be of some considerable importance among the elite for whom they were writing. Griffin (1985) thus observes that both connoisseurship in wine and the invocation of Bacchic imagery and mythology raised the status of the pleasures of wine. The prestige value of wines of certain vintages and from particular vineyards is made readily apparent in Latin poetry. Horace frequently mentions Falernian and Coan wines, and in his satire on the party given by Nasidienus Rufus for Maecenas (*Satires*, II.viii.9–19) he refers to Coan, Caecuban, Chian, Falernian and Alban wines, the latter two being the more prestigious (Horace, 1987). The *Satyricon* of Petronius, dating from the middle of the first century AD, likewise emphasises the prestige associated with vintage wines, well illustrated in the following extract from Petronius's description of Trimalchio's dinner party (*Satyricon*, XV.34)

> Carefully sealed wine bottles were immediately brought, their necks labelled:
>
> ### FALERNIAN
> ### CONSUL OPIMIUS
> ### ONE HUNDRED YEARS OLD
>
> While we were examining the labels, Trimalchio clapped his hands and said with a sigh:
> 'Wine has a longer life than us poor folks. So let's wet our whistles. Wine is life. I'm giving you real Opimian. I didn't put out such good stuff yesterday, though the company was much better class'.
>
> (Petronius, 1986: 56)

Although Trimalchio was either lying or had been cheated, since no wine merchant would have labelled his wine in such a manner, this reference to the classic vintage of 121 BC when Opimius was consul, well illustrates the social importance of vintage wines to the aspiring elite of Rome during the Republic and early Empire.

Nevertheless, from the end of the first century BC, demand for wine from among the mass of the poorer urban population increased considerably, aided in part by the widespread distribution of pastry and sweetened wine, *crustulum* and *mulsum*, by benefactors seeking to solicit the political support of this section of the community (Purcell, 1985). Studies by Kléberg (1957) and Hermansen (1981) clearly illustrate the high density of *tabernae* (taverns), bars, wine-counters and drinking establishments to be found in Roman towns such as Pompeii and Ostia from the first century AD (Figure 23). The growing demand for low quality and spiced wine (*conditum*) by the urban poor throughout the Empire resulted in the changing orientation of Italian vineyards so bemoaned by Pliny (XIV.viii.62). Ancient vineyards which had once produced high quality wine, such as those of Falernum, were by the end of the first century AD concentrating on the lower quality mass market. Likewise, there was an appreciable expansion in the density of vineyards in locations adjacent to towns, most notably in the proximity of Rome, with wine from Alba for example rising to a position of considerable importance. There was thus, to some extent, a shift in emphasis from the coastal vineyards located with specific reference to the export trade, to inland vineyards providing for the rising urban wine demand. This should not, though, be interpreted as a crisis for Italian viticulture in the face of foreign wine imports, but rather as a reorientation induced by wine producers and merchants eager to maximise their profits. The increase in demand for low quality, rather than vintage wines, meant that wine production became less capital intensive. Vintage wines, kept for a number of years, require considerable capital investment in the provision of storage facilities, whereas wines produced for immediate sale have a much more rapid turnover and can therefore be produced and sold more cheaply. Such wine, designed to be drunk within the year, did not require such careful handling, nor did it need to be stored in sealed amphorae. Moreover, since much of it was transported short distances overland by cart, other less fragile containers, such as wooden barrels and inflated hides (Figure 24) could increasingly be used for its transportation. This is therefore likely to provide much of the explanation for the disappearance of the Dressel 2-4 amphorae at the end of the first century AD.

It is difficult to obtain accurate estimates for the price of wine at different periods during the Roman Republic and Empire. In part this is due to the practice by which wines of different qualities were charged at different rates, but it also results from regional price variations, and the tendency for prices to fall during large harvests and to rise at times of scarcity. Columella (III.iii.10) used a minimum figure of 300 *sestertii* per culleus, equivalent to 15 *sestertii* per amphora, in his mid-first century AD calculations for the wholesale price of wine, but as Duncan-Jones (1974) points, out this figure is subject to some doubt. Thus graffiti and bar tariffs from Pompeii and Herculaneum, dating from AD 79 and only a short while after Columella was writing, give retail prices for wines of different qualities as 1, 2, 3, 4, and 4.5 *asses* per sextarius, equivalent to 12, 24, 36, 48 and 54 *sestertii* per amphora. For purposes of comparison, bread typically cost about 2 *asses* per loaf at Pompeii, and 2 *asses* was also the commonest figure for the service of prostitutes in the Pompeian graffiti (Duncan-Jones, 1974). Other first century wine prices noted by Duncan-Jones (1974: 46-67) include a figure of 8 *sestertii* per amphora for the wholesale price for sale on the vine, and a possibly apocryphal retail price also of 8 *sestertii* per amphora. Moreover, Frank (1936) has suggested that the price of imported Spanish wine at this time was usually in the range of 8-12 *sestertii* per amphora. A century later, in AD 153 a capital gift designed to provide cash and provisions for the College of Aesculapius and Hygia, which was situated on the via Appia and less than two Roman miles from Rome, enables a figure for the cost of wine of 61-88.5 *sestertii* per amphora to be calculated (Duncan-Jones, 1974, 364). Duncan-Jones (1974) interprets these figures as implying that the price of wine in Rome was therefore higher than that in Campania, derived from the prices pertaining at Pompeii and Herculaneum, but the figures could also indicate that wine prices had risen somewhat between the two dates. The considerable price advantage obtained by Spanish

Figure 24 Marble relief illustrating the transport of wine by ox cart
Source: British Museum (GR 1805.7-3.458; BM K95124)

wines during the first and second centuries AD, when compared with the prices noted above for Italian wines, goes some way to explaining the vast numbers of Spanish amphorae found at Monte Testaccio, by the banks of the Tiber in Rome (Frank, 1940: 220).

Rapid inflation during the third century AD led Diocletian to issue an edict in AD 301 designed to restrict prices and to halt price-speculation. This was only promulgated in the eastern Empire, and although it failed to limit inflation, it does provide a unique insight into the structure of prices at the beginning of the fourth century. Thus wheat prices in the edict are between 25 and 50 times those of the second century AD (Duncan-Jones, 1974: 366–7), and given that the edict was designed to reduce prices it is probable that inflation was higher still. Wine prices in the edict vary from 30 *denarii* per sextarius for the best quality wines, to 8 *denarii* for ordinary wine (Frank, 1940), and with 16 *asses* being equivalent to 1 *denarius* this represents an increase of 107 to 128 times in the price of wine since the first century AD. For comparative purposes, beer was priced at between 2 *denarii* per sextarius for Egyptian beer and 4 *denarii* per sextarius for Celtic or Pannonian beer, and was thus between two and four times cheaper than wine. The daily wages recorded in the edict varied from 20 *denarii* for a shepherd with maintenance, to 50 *denarii* for a farm labourer, carpenter or baker, and 150 *denarii* for a figure painter. The most common daily wage was set at 50 or 60 *denarii*, and this was therefore just over six times the value of one sextarius of wine.

THE GREEK AND ROMAN CONTRIBUTION TO VITICULTURE

The introduction of the vine to Greece changed forever its economic and social significance. Whereas previously wine had generally been the preserve of the ruling classes, the development of viticulture in Greece made it available to all levels of society. While the emergence of viticulture was closely related to the introduction of Dionysiac ritual and symbolism into the Greek mainland, its later spread owed much to the profit seeking aspirations of Greek and Roman wine producers and merchants. Although wine and the vine had clear symbolic significance, the evidence concerning whether or not this was used by those in power to maintain their position of dominance is ambiguous.

Two particular pieces of evidence do provide some support for the argument that wine was used by the ruling classes to maintain their hold on power. The first is that wine was regularly taxed by the urban and state authorities (Frank, 1936), with the revenues then being used to support their positions of authority. In the case of Thasos in the fifth century BC, Osborne (1987) has shown convincingly how the town of Thasos came to dominate the countryside through its control over the wine trade. The wine laws were promulgated from the town, the taxes and fines went to the religious sanctuaries on the town acropolis, and the stamps for the amphorae were produced in and distributed from the town. Moreover, by deciding to cultivate vines, Osborne (1987: 107) argues that the farmers of Thasos made themselves dependent on the town, which could choose either to buy or not to buy wine from the countryside. The Roman authorities likewise received vast revenues from their taxation of wine imports, attested by the stamp marks on the handles of amphorae and known as *ansarium*. These customs, also levied on oil, 'were collected at barriers set up on the major highways leading into the city' (Palmer, 1980: 223), and provided a considerable source of revenue for the Roman state.

A second way in which the Roman elite used wine to maintain its status and authority during the Empire, was through the regular distribution of *mulsum*, sweetened wine, to the urban poor. Purcell (1985) sees the main significance of this as being to introduce wine-drinking to the poor, and thus to increase the demand for wine, but it was also designed to maintain the political support of the poor for the benfactor who actually paid for the wine. This use of wine for the maintenance and legitimation of political support achieved its clearest expression in the 270s AD, when Aurelian built a new wall around Rome designed to combine the functions of defence and taxation. Aurelian's wall generally followed the circuit of the customs barrier dating at least from the first century AD (Palmer, 1980), and it provided a sound deterrent against smuggling, chanelling Rome's imports through the customs stations situated just outside the thirty-seven gates. Wine duties were levied in kind, and were then used to provide 'systematic doles of wine for the urban populace' administered from the temple of the Sun (Palmer, 1980: 220). In addition, Aurelian also apparently established numerous new vineyards using slave labour as far north as the Maritime Alps, in order to provide wine for his munificence. Here then is an excellent example of the use of wine by the Emperor to maintain the support of the populace of Rome. However, it was not necessarily the symbolic value of wine which enabled him to appear as a benefactor, but rather its economic and physiological significance. Wine had long since entered the cultural consciousness of the mass of the Roman population, and it is difficult to assert that this form of legitimation was enabled purely because of its symbolic role.

The lasting contribution of the Greeks was to introduce vines into the triad of crops that came to be so characteristic of peasant subsistence agriculture around the Mediterranean. From the second century BC the development in Italy of vineyards based on slave labour, and designed specifically for the production of wine for sale, introduced a second important element into the historical geography of viticulture, and firmly established the division between 'commercial' and 'subsistence' viticulture that survives to this day. The expansion of profit oriented, export based vineyards in Italy and eventually in the provinces, at the expense of peasant subsistence agriculture, generated a wealth of new commercial possibilities. The export of wine to Gaul in return for slaves during the first two centuries BC enabled the further development of these slave-based estates, but with the consequent rapid growth of towns, whose population was swelled by the immigration of displaced peasants, the orientation of this trade switched dramatically from Gaul to Italy. From then onwards, Italy became a wine importer, drawing its supplies from as far afield as Gaul and Spain.

The economic and political disruptions of the third century, the conversion of the Empire to Christianity under Constantine at the beginning of the fourth century, and the eventual collapse of the Western Empire in AD 476 transformed the entire economic, social, political and ideological context of European viticulture. It is therefore to the sweeping changes introduced by the Germanic invasions of the Roman Empire, and their effects on viticulture and the wine trade that attention in the next chapter now turns.

5

VITICULTURE AND WINE IN THE EARLY MIDDLE AGES

> While they were eating, Jesus took bread, gave thanks, and broke it,
> and gave it to his disciples, saying, "Take and eat; this is my body."
> Then he took the cup, gave thanks, and offered it to them, saying,
> "Drink from it all of you. This is my blood of the covenant, which is
> poured out for many for the forgiveness of sins. I tell you, I will not
> drink of this fruit of the vine, from now until that day when I drink
> it anew with you in my Father's kingdom."
>
> (The Gospel according to Saint Matthew, VI: 26-29)

Following the collapse of the western Roman Empire in the fifth century
AD, it is widely argued that the survival of viticulture depended upon the
symbolic role that wine played within Christianity; a role derived
primarily from its use by Christ in establishing a new covenant, described
above in the words of St Matthew's gospel, where wine symbolised the
sacrificial blood of Christ. De Blij (1983: 46) has thus argued that 'If an
early identification with religion promoted viticulture in the ancient
world, viticulture's very survival depended on its religious associations
during the centuries that followed the fall of the Roman Empire'. Seward
(1979: 14) has similarly suggested that 'Monks largely saved viticulture
when the Barbarian invasions destroyed the Roman Empire, and
throughout the Dark Ages they alone had the security and resources to
improve the quality of their vines slowly and patiently'. Moreover, the
precise processes by which viticulture was maintained and strengthened
by the Church have been described by Johnson (1971: 14) as follows:

> The Church had been the repository of the skills of civilization in
> the Dark Ages. As expansionist monasteries cleared hillsides and
> walled round fields of cuttings, as dying wine-growers bequeathed
> it their land, the Church came to be identified with wine - not only
> as the Blood of Christ, but as luxury and comfort in this world.

These authors thus view the Church not only as enhancing wine with a
specific symbolic value, but also as maintaining and developing a

viticultural tradition handed down from the vestiges of the Roman Empire. In Dion's (1959: 188) words,

> Autant que par la conservation et la transmission des méthodes de culture héritées de l'Antiquité romaine, l'Eglise a servi la viticulture par le surcroît de prestige qu'elle lui a conféré. Elle a placé la vigne au sommet de la hiérarchie des symboles. [As much as through the conservation and transmission of cultivation methods inherited from Roman antiquity, the Church served viticulture by the increase in prestige which it conferred on it. The Church placed the vine at the summit of the hierarchy of symbols.]

Although such views dominate the literature on viticulture in the early Middle Ages, they have not gone unchallenged. Younger (1966: 234), for example, has argued that

> This theory is patently untrue. The Church had little, if anything, to do with the transmission of viticulture from the Ancient into the Christian worlds. Wine-growing was brought over the Dark Ages by private enterprise, and the traditions of viticulture were continued through the memories of lay *vignerons* rather than through the manuscripts of monastic libraries.

Younger provides little firm empirical support for this argument, and the evidence that he does consider is mainly drawn from the twelfth century. Nevertheless, in countering the traditionally accepted view of the survival of viticulture, he usefully draws attention to the role of economic interests in the maintenance of a viticultural tradition in Europe. In order to evaluate these two contrasting arguments it is essential to understand something of the processes by which the western Roman Empire collapsed, and also the role of wine and the vine in Christian symbolism. It is therefore to these two issues that the first sections of this chapter are addressed, before attention returns to the question of the survival of viticulture between AD 500 and 1000.

CHRISTIANITY AND A DIVIDED EMPIRE

Ever since Edward Gibbons's *History of the Decline and Fall of the Roman Empire* (1776–88), the subject of the fragility of empires has been one which has fascinated historians and geographers alike. No single empire has retained its position of pre-eminence for any length of time. In explaining imperial decline scholars have usually resorted to one of two types of explanation: those based on unique characteristics of individual empires, or those seeking explanations based on general laws of imperial growth and decline. Of all instances of imperial decline, however, it is the collapse of the western Roman Empire that is most often cited as having ushered in a subsequent period of social and economic instability and chaos. Indeed the description of the subsequent era of early medieval history as a veritable Dark Age lends substantial support to this image.

The Christian Roman Empire

The military, economic and social crisis that emerged in Europe during the second half of the third century was in part precipitated by the invasions of the Franks and the Alamanni. However, Diocletian's seizure of power in AD 284, and his division of the empire into two halves in 286, with the west being ruled by the co-emperor Maximian, reintroduced a degree of stability. This was reinforced by the establishment of the Tetrarchy, or Rule of Four, in 293 by which Galerius and Constantius were appointed as Caesars holding rank beneath the two Augusti, or co-emperors. As a result 'the "centrifugal" tendencies in the structure of imperial power, so obvious in the third century, were contained by being legitimised' (Cornell and Matthews, 1982: 172). The maintenance of this relatively stable government until 305 enabled a series of military, monetary and taxation reforms to be introduced, which laid the basis for the subsequent administration of the Empire until the downfall of its western part in the fifth century. Critical among Diocletian's reforms was the division, for the first time in Rome's history, of the administration between civil and military functions. From 303 he also promulgated a series of organised persecutions of the Christian communities within the Empire. Following Diocletian's abdication in 305, these were continued under Galerius in the east, although in the parts of the Empire controlled by Constantius and Maxentius they were allowed to lapse. The vigour with which the persecutions were practised suggests that by the fourth century Christianity had achieved a position of some considerable significance in the Empire, and its proximity to those in power is indicated by the observation that 'Diocletian's own wife and daughter were among those compelled to sacrifice as Christians or Christian sympathizers' (Cornell and Matthews, 1982: 178).

The exclusion from power of Maxentius, son of Maximian, and Constantine, son of Constantius, as a result of the promotion of Maximinus Dia and Severus to the ranks of Caesars following the retirement of Diocletian and Maximian in 305, precipitated a series of civil wars which were in the end to result in the conversion of the Roman Empire to Christianity. In a series of battles, Constantine eventually gained complete control of the west. His victories were reported to have been associated with various dreams and visions, most notably one on the eve of the crucial battle of the Milvian Bridge against Maxentius in 312. In this dream he was told to paint the *chi-rho* Christian monogram on the shields of his troops, and to engage battle under this sign. Constantine's subsequent conversion to Christianity laid the basis for the Edict of Milan, which he issued together with the eastern emperor Licinius in 313. This provided for freedom of worship and the restoration to the Christian Church of all their property that had previously been confiscated.

Soon afterwards, following his victories over Licinius at Hadrianople and Chrysopolis, which made him sole ruler of a reunited Empire, Constantine embarked on the creation of a new Christian capital in the east on the site of the ancient Byzantium. The Christian Constantinople, which was formally dedicated in 330, was thus built as a direct symbolic alternative to the pagan Rome. Moreover, the imperial family also undertook an ambitious programme of church building, not only in Constantinople, but also in the Holy Land and Jerusalem, where work was begun on the Church of the Holy Sepulchre. By the second half of the fourth century Christianity was widely practised throughout the Empire and apparently by all levels of society. The activities of Christian writers and ecclesiastics, such as Jerome (c. 345-420), Ambrose (c. 340-97), and Augustine (354-430), played an important part in defining the relationship between Church and State during this period, but it was not until the introduction of a series of laws in the 380s and early 390s, which abolished pagan sacrifice and led to the closure of the ancient temples, that Christianity became the official state religion.

The division of Empire and the fall of Rome

At Constantine's death in 337 the political organisation of the Roman Empire again became divided, this time between his three sons, with Constantine II becoming emperor of Gaul, Spain and Britain, Constantius emperor of the east, and Constans emperor of Illyricum and Italy. Thereafter, during the second half of the fourth century there were alternating periods of political unity and division, with the one constant factor being the continued battles fought in defence of the frontiers along the Rhine, the Danube and in the east against Persia. Following the death of Valens in battle against the Goths in 378, Theodosius became emperor, and in 394 he again succeeded in uniting the two halves of the Empire. However, a year later he died, leading to the final and irrevocable division of the Empire in 395 with Honorius becoming emperor in the west and Arcadius emperor in the east. Subsequently the eastern and western Empires followed very different fortunes, with the west eventually falling to Odoacer in 476 while the eastern Empire survived until its conquest by the Ottoman Turks in 1453. This division of the Empire into two halves was in part a result of the difficulties encountered in governing such a vast area, but it also served as a catalyst for the eventual collapse of the west. In particular it created a fundamental fiscal problem, since more than two-thirds of the imperial revenues had traditionally been derived from the east, while most of the military power was tied up in the west defending the northern borders against the Germanic tribes.

The seeds of Rome's destruction had already been sown as early as the third century, but as Grant (1976) has emphasised, within less than a hundred years both the immense power of Rome and the army that had sustained it had vanished. In many ways it is indeed surprising that the Empire survived for as long as it did, and Starr (1982: 3) has even suggested that in geographical, economic, psychological and political terms the Roman Empire was actually 'an impossibility'. The reasons for the collapse can usefully be divided into those that were external and those that were internal. Prime among the external factors were the wars with the Germanic tribes, which had begun as early as the third century with the incursions of the Alamanni and Franks across the Rhine, and the Goths into Greece and Asia Minor. These were temporarily delayed by the reforms of the late third century, which provided for the creation of military capitals in the frontier regions behind the Rhine–Danube border. However, from the mid-fourth century, and particularly during the first half of the fifth century, a new wave of incursions, precipitated by the westward movement of the Huns, led to the Vandals, Visigoths, Ostragoths and other Germanic tribes breaking across the frontiers of the Empire. These led to more or less continued warfare, treaties providing for the settlement of certain of the tribes within the Empire as *foederati*, and considerable social and economic disruption. Between 406 and 409 a mass invasion of the north of the Empire saw the Vandals, Suebi and Burgundians sweep south-westwards through Gaul. Rome was then sacked, first in 410 by the Huns under Alaric, and then later by the Vandal king Geiseric in 455. Although the power of the Huns was broken by a combined Roman, Visigoth and Burgundian army in 451 at the battle of the Catalaunian Plains near Châlons-sur-Marne, the eventual outcome of these tribal movements was that the previous unity imposed by Roman authority became replaced by a multiplicity of small kingdoms in Europe most of which were in conflict with each other.

Alongside the growing power and military successes of the Germanic tribes (Ferrill, 1986), though, must also be considered the internal problems of the Roman Empire, which despite large increases in the size of the army prevented absolute resistance to the invasions. Grant (1976) has usefully divided these into six categories: a failure of the army, in which generals struggled for individual power regardless of the fortunes of the state, and in which the military gradually became alienated from the people; an increasing class division, as different social groups came into conflict with each other and with the state; a growing credibility gap, not only between the emperors and the people, but also between the people and the bureaucracy; a disintegration of the policy of incorporation of different peoples and races into the Empire, as allies turned against allies and race against race; a growth in the number of different groups which opted out of their social and military responsibilities; and a gradual undermining of effort directed towards resistance against the invaders, reflected in a division of interests between the church and the state. While the state was concerned with military salvation in this world, the church became increasingly oriented towards spiritual salvation in the world to come. At the heart of any understanding of the decline of the Roman Empire must therefore be the recognition that it disintegrated in part from within and was therefore unable to face the external invasions of the fifth century.

WINE AND THE VINE IN CHRISTIAN SYMBOLISM

Christian symbolism of wine and the vine

The conversion of the Roman Empire to Christianity which proceeded apace from the beginning of the fourth century brought with it new beliefs and a new religious symbolism. However, as Cornell and Matthews (1982: 193) stress, 'the ancestral religion gained a tremendous tenacity when forced onto the defensive by an aggressive Christianity supported by the emperors'. While rejecting outright most types of pagan worship, some forms of Christian symbolism, derived in part as they were from earlier Jewish practice, retained a remarkable similarity with certain pagan rituals. Paramount among these was the symbolism associated with wine and the vine. As Goodenough (1956, vol.5: 99) has commented, 'In Jewish art remains, wine is presented in purely pagan forms, those same pagan forms which reappear as the most important group of symbols of early Christianity, surpassed only by the cross itself'.

The symbolic drinking of wine as part of the Christian Eucharist, Mass or Communion, was to play a fundamental role in influencing the global distribution of viticulture, and the social and ideological significance of wine, but wine and the vine are also to be found in three other main aspects of Christian symbolism. Thus, throughout the New Testament of the Bible, there is a clear continuation of the symbolic theme likening the Jewish people to a vineyard, or vine. This is most clearly expressed in Christ's parable of the vineyard (Matthew, XXI: 33–44; Mark, XII: 1–12; Luke, XX: 9–18), in which a landowner, symbolising God, plants a vineyard, and lets it out to some farmers, representing the people of Israel. At the vintage the vineyard owner sends servants to receive the produce of the vineyard, but the servants are beaten and abused by the husbandmen. The owner then sends his son, who is in turn killed by the husbandmen. Finally, the owner himself returns and destroys the husbandmen, and gives the vineyard to others to tend. Here, the son represents Christ, who, in relating the parable, foretells his own death. Through the vineyard owner's subsequent actions, knowledge of salvation, represented in the vineyard and previously only available to God's chosen people, the Jews, is made available to others.

A second symbolic theme is to be found in Christ's use of the vine to represent himself, and the relationship between his followers and God. This is most clearly expressed in Saint John's Gospel (John, XV: 1-5):

> "I am the true vine, and my Father is the gardener. He cuts off every branch in me that bears no fruit, while every branch that does bear fruit he prunes so that it will be even more fruitful. You are already clean because of the word I have spoken you. Remain in me, and I will remain in you. No branch can bear fruit by itself; it must remain in the vine. Neither can you bear fruit unless you remain in me.
>
> "I am the vine; you are the branches. If a man remains in me and I in him, he will bear much fruit; apart from me you can do nothing. . .".

This symbolism in which Christ is described as a vine has several characteristics in common with the imagery in which Dionysus was also so symbolised, and this may be one reason why the writer of the gospel emphasised that Christ was the *true* vine, to provide a contrast with other extant imagery associated with vines. However, the use of the vine symbolism can also be interpreted as being derived from the Old Testament imagery likening Israel to a vine (Psalm LXXX: 8-9), and the emphasis on Christ as the true vine might thus also have been intended here to represent Christ as the fulfilment of the Old Testament. Early Christians were nevertheless eager to emphasise that their religious symbolism, illustrating a particular relationship between Christ, his followers and God, was fundamentally different from that of the worshippers of Dionysus and other religions, which they saw as being idolatrous and much more overtly associated with fertility rituals.

A third theme concerns the use made of wine by the Christian priesthood. It thus seems that Christian priests were under much less stringent rules concerning wine drinking than were Jewish priests, who were forbidden to drink wine whenever they went into the Tent of Meeting (Leviticus, X: 8-9). Although deacons and priests in the early Christian church were required to be sober (1 Timothy, III: 2-3, 8), it is significant that Paul specifically recommended that Timothy should '. . . use a little wine because of your stomach and your frequent illnesses.' (1 Timothy, V: 23). The development of this theme through the New Testament is of particular interest. In Luke's gospel, an angel is recorded as telling Zechariah that his son John the Baptist, the forerunner of

Christ, ' ". . . is never to take wine or other fermented drink. . ." ' (Luke, I: 15). Elsewhere, Luke reiterates this, noting that 'John the Baptist came neither eating bread nor drinking wine' (Luke, VII: 33). This imagery parallels the vows taken by Israelites who wished to serve God as Nazirites (Numbers, VI: 1–4), but there are no such descriptions of Christ abstaining from wine. Indeed Christ was upbraided by the devout Jewish scribes and Pharisees for eating and drinking with publicans, and was questioned by John's disciples as to why his own disciples failed to fast (Mark, II: 15–28; Matthew, IX: 10–17; Luke, V: 27–35). Moreover, according to John's Gospel (II: 1–11), Christ's first miracle was to turn the water into wine at a wedding in Cana. This story of the marriage at Cana (John II: 1–11) is of particular interest since, as Dodd (1965: 223) has emphasised, while this passage 'is highly theological in character', with the good wine being kept until the time of the incarnation of the Word in Christ, it does also reflect a common motif more usually associated with Dionysus, who throughout the Greek world was frequently credited with the transformation of water into wine.

The most overt incorporation of the element of wine into Christian symbolism, however, is with the ritual eating of bread and drinking of wine, which are taken to represent Christ's body and blood. For Christians, wine symbolises Christ's blood, sacrificed for the sins of the world. Moreover, Christ instituted this new covenant, requiring his disciples to consume the bread, and by inference the wine, in remembrance of him (Luke, XX: 14–22). The parallels between this and Dionysiac rituals should not be drawn too closely. In Dionysiac ritual, wine was drunk in large amounts in order to achieve the state of otherworldliness associated with achieving union with the god and becoming ενθεος, whereas, for Christians, wine is drunk in small quantities as a symbolic memorial of Christ's death and resurrection (Jungmann, 1959). Likewise, in Christian ritual the wine and bread are combined, whereas Dionysiac ritual appears to have been most usually associated with wine alone. Perhaps the greatest contrast, though, lies in the Christian belief that Christ's blood was necessarily sacrificed to atone for human sin, providing the only way in which people are able to achieve salvation. There is nothing in Dionysiac ritual directly to compare with this belief that death results from sin, and that it was therefore essential for Christ to die in order to conquer both sin and death.

While the precise nature of the symbolism thus changed from Dionysiac to Christian ritual, a common thread linking wine to new life remained. Moreover, Dionysus or Bacchus, and his vinous associations, formed only a small part of the Greek and Roman pantheon. In contrast, under Christianity, which became the religion of the whole Roman Empire and subsequently one of the major religions of the world, wine was elevated to a symbolic position of the utmost importance.

Wine and the vine in Christian usage

This symbolism linking Christianity with wine and the vine found its material expression in a wide range of paintings, mosaics and sculpture throughout the late Roman world. Thus the Temple of Clitumnus, a fifth century church just to the north of the source of the River Clitumnus in Umbria, depicts a Christian cross entwined with bunches of grapes and vine leaves on its pediment, thus combining earlier pagan symbolism with that of the newly established Christianity. Likewise, the fourth century church of Santa Costanza, in Rome depicts a magnificent set of images of the vintage in mosaics on its barrel vaulted roof (Figure 25). This early church, built as a mausoleum for Constance, daughter of Constantine, thus illustrates the cutting of grape bunches with small curved knives, their transport in ox carts, and their subsequent treading. Similar scenes are also to be found in a mid-sixth century mosaic from the Church of Sts Lot and Procopius at al-Mukhayyat in Jordan, which depicts bunches of grapes being cut, transported, and trodden in what looks like a screw press. Also from the mid-sixth century, well after the fall of the western Roman Empire, a mosaic in the church of San Vitale at Ravenna illustrates Theodora, wife of the Emperor Justinian, offering a gold chalice, or possibly the wine of Christ's sacrifice, to the newly established church. This artistic symbolism associating wine and Christianity was maintained throughout the medieval period, and is well indicated, for example, by the painting of the Madonna and Child by Mantegna dating from the fifteenth century, which shows the infant Christ eating a bunch of grapes representing his sacrifice and death.

It was the need for Christians to drink wine made from vines (Matthew, XXVI: 29; Mark, XIV: 25; Luke, XX: 18) as part of the Eucharist, Mass, or Communion, however, that was to provide the most important material expression of this new symbolism. Throughout the Christian world some wine, whether locally produced or imported, was always required, and under canon law that wine had to be natural wine made from grapes (New Catholic Encyclopedia, 1967, vol. 14: 959). There is, though, considerable uncertainty over the precise amount of wine involved, since this would largely depend on the regularity with which communicants celebrated the Eucharist. By 1100 the receipt of Communion by the laity at Mass had become a rare event (Jungmann, 1959; Klauser, 1969), and during the next century it became widespread practice for the laity only to receive bread (Knowles and Obolensky, 1969). In 1215 the Fourth Lateran Council firmly defined the theory of transubstantiation and thereafter the practice whereby the laity was excluded from participating in the wine was extended (The New Catholic Encyclopedia, 1967). This was one of the practices condemned by fourteenth and fifteenth century reformers, such as John Wyclif and John Huss, in turn forcing the Catholic

Figure 25 Vintage scene from the roof of the Church of Santa Costanza, Rome
Source: author 13th March 1987

authorities to pronounce definitively on the subject. In particular, during the fifteenth century the Hussites pioneered both the giving of the chalice to the laity, and also the giving of Communion in both kinds to children from baptism. At the Council of Constance (1414-18) this was prohibited and it became 'the general law in the West that all except the celebrant should communicate only under the speces of bread' (Addis and Arnold, 1951: 314). The subsequent history of participation in the Eucharist forms part of the wider conflict between Catholic and Protestant authorities. While Luther developed the basis of the German Mass in 1526, with full lay participation in the wine, the Council of Trent (1545-63) reinforced the teaching of the Council of Constance against this, and in the Tridentine Mass lay down the essential basis of subsequent Catholic rite (Jungmann, 1959). It was only as a result of the revisions of the Tridentine Mass in the Second Vatican Council (1962-65) that it once again became permissible for Catholic laity to participate in the sacrament of wine.

Despite this uncertainty, the critical point which requires emphasis is that a new ideological requirement for wine had been established, which meant that where at all possible Christian believers regularly had to partake of wine as part of their religious practice. Although the amounts involved may have been very small, some grape wine had to find its way to the very outer limits of the Christian world. Where this was not achieved through the sending of gifts of wine, or through wine made from locally grown vines, this ideological requirement therefore provided an opportunity for commercial profit to be gained.

VITICULTURAL CONTINUITY AND SURVIVAL

The survival of viticulture in the west

Arguments claiming that the survival of viticulture after the fall of the western Roman Empire depended on the symbolic significance of wine to the Christian church and the activities of numerous diligent monks (Seward, 1979; De Blij, 1983), are premised on the assumption that pagan Germanic tribes largely destroyed the vineyards of western Europe in the fifth and sixth centuries. The evidence for this, however, is far from conclusive. Viticulture was well established in much of Gaul to the south of a line joining the mouth of the Loire, Paris and Tours apparently well before the middle of the fifth century (Figure 20). The northward expansion of viticulture to the Seine, the Moselle and the Loire had thus continued despite the political and economic turmoil of the earlier Germanic invasions of the third century. Indeed, it seems highly probable that the Visigoths, Suebi, Vandals and Burgundians who settled in Gaul and Spain early in the fifth century continued to cultivate vines, possibly using the enslaved labour of some of the local inhabitants whom they displaced. In Gaul elements of the Gallo-Roman aristocracy survived, and as the letters of Sidonius Apollinaris reveal, a surprising degree of cultural continuity was retained, including the cultivation of vines particularly in the Auvergne (Lachiver, 1988). Sidonius Appollinaris (II.xiv.1) writing to Maurusius between 461 and 467, for example, mentions the latter's vines and vineyard at the village of Vialoscum, which was probably located near Clermont Ferrand (Sidonius, 1915: 62). There is thus little evidence of a mass destruction of vineyards at the time of the final collapse of the western Empire.

To argue that the Germanic tribes consciously destroyed the vineyards of Gaul and Iberia would also imply that they preferred other beverages to wine, but of this there is likewise negligible evidence. Indeed, writing as early as AD 98 in the *Germania* Tacitus was able to describe the drinking habits of the Germanic tribes in the following manner:

> Their drink is a liquor made from barley or other grain, which is fermented to produce a certain resemblance to wine. Those who dwell nearest the Rhine or the Danube also buy wine. Their food is plain – wild fruit, fresh game, and curdled milk. They satisfy their hunger without any elaborate cuisine or appetizers. But they do not show the same self-control in slaking their thirst. If you indulge their intemperance by plying them with as much drink as they desire, they will be as easily conquered by this besetting weakness as by force of arms.
>
> (Tacitus, 1970: 121)

While Tacitus was not proved correct about the effects of their fondness for alcohol, and even allowing for some exaggeration, this extract nevertheless indicates a certain demand for wine among the Germanic tribes, which must have been sustained at the time he was writing in part by exports of wine northwards and eastwards from Gaul.

If the Germanic invasions did not lead to the widespread destruction of vineyards, what they did result in was the eventual collapse of the economic and social structure of the Roman Empire in the west. The division of Italy, Gaul and Iberia into a number of small warring kingdoms by the sixth century effectively prevented the long distance trade of goods such as wine, which had been a hallmark of the Empire during the first two centuries AD. Moreover, the decline in the urban population, particularly that of Rome, dramatically curtailed the demand for such trade. The outbreaks of bubonic plague and epidemic diseases, such as smallpox and possibly diphtheria and cholera, in Europe during the sixth century following the so-called plague of Justinian in AD 540, further reduced population levels (Shrewsbury, 1971; Pounds, 1973), and lowered the overall demand for wine. In such circumstances, patterns of wine production and consumption took on a much more local character, generally becoming oriented primarily towards subsistence needs.

During the late fifth and early sixth centuries, the emergence of the Frankish kingdom in France gradually reintroduced an element of political stability, and provided the context within which viticulture could again expand. The Franks' homeland lay in what is now Belgium and the Netherlands, but under Clovis (482-511) they conquered the lands between the Somme and the Loire in 486, and then gained victory over the Alamanni c. 496. It was during this battle that Clovis is said to have been converted to Christianity. Thereafter he conquered the Burgundians c. 500, and then at the battle of Vouillé in 507 he overcame the Visigothic kingdom to the north of the Pyrenees. Under the sons of Clovis, the Frankish kingdom expanded further southwards and eastwards during the 530s, only to suffer a sequence of partitions and reunifications during the remainder of the sixth and seventh centuries. Eventually under the Carolingians Pepin (751-68) and Charlemagne (768-814) a united Empire finally emerged, with its territories expanded to include the land of the Saxons and the Lombard (Langobard) kingdom in northern Italy.

Most of the evidence concerning viticulture during this period between the fifth and the tenth centuries is found in the writings of ecclesiastics, in charters and in the details of lands held by churches and monasteries. The survival of this evidence tends to suggest that the Church therefore played a dominant role, but it must be treated with some caution since it was the monks and clerics who maintained the skill of writing; peasant wine makers left no written records. What the surviving evidence suggests is that three groups of people continued to cultivate vines, and gradually expanded the area of viticulture in France. These were the princes, bishops and monks, all of whom were generally to be found living in or near towns and cities (Dion, 1959; Lachiver, 1988). During this period, the consumption of wine in France and northern Europe once again became closely associated with the physical manifestation of social status. In Dion's (1959: 171) words, 'Elle reste indépendamment du rôle que lui assigne la religion, un ornement nécessaire à toute existence de haut rang et par là même l'une des expressions sensibles de toute dignité sociale'. [It remained, independently of the role given to it by religion, a necessary ornament for the very existence of high rank, and on the same grounds one of the tangible expressions of all social dignity.] Unlike the cultivation of vines and wine drinking in the Mediterranean lands under the Roman Empire, where wine was the beverage of all levels of society, wine and the vine had become attributes of the social status of the new elites emerging in northern Europe. The heartland of the Carolingian Empire of the second half of the eighth century lay to the north of the Seine and the Moselle, around towns and palaces such as Aix-la-Chapelle (Aachen), Attigny, Reims and Köln (Cologne). This was beyond the limits of Roman viticulture, and is at the very edge of the area climatically best suited to viticulture today. The cultivation of vines and the production of wine in such an area was thus difficult and expensive, being beyond the limits of any but the rich. In southern Europe peasant viticulture survived as part of the Mediterranean subsistence economy, but in most of France it gradually became associated with the palaces, cathedrals and monasteries established by the rulers of the Empire.

There are numerous examples of the establishment and maintenance of vineyards within France by bishops, monks and the secular nobility between the sixth and tenth centuries (Dion, 1959; Younger, 1966; Lachiver, 1988). Venantius Fortunatus (Carminum V.vii.1–12), writing in the sixth century, thus describes the shady vines on the estate at Cariacus adjacent to the Loire belonging to Felix the Bishop of Nantes (Fortunatus, 1961: 118), which Lachiver (1988) identifies as having been located at Saint-Rémy-la-Varenne. Several bishops even moved their episcopal seats so that they would be in places more favourable to viticulture. Gregory of Tours, in his *History of the Franks* (III. 19), thus writes how his great-grandfather Gregory, Bishop of Langres, chose to reside at Dijon, where vines were more readily grown (Gregory of Tours, 1927: ii.103), and in Merovingian times the Bishop of Augusta Veromanduorum (Saint-Quentin) likewise apparently moved his seat to Noyon as a result of its suitability for viticulture (Dion, 1959; Lachiver, 1988). For such prelates wine was needed not only to provide for sacrimental use at Masses held in their dioceses, but also, perhaps more importantly, for their own consumption and for the entertainment of guests.

The expansion of monasticism during the eighth and ninth centuries, has provided us with considerable evidence concerning the extent of viticulture at that period, since many monasteries were granted vineyards among their possessions. Significantly, though, this should not be taken to imply that it was necessarily the monks who saved viticulture during the Dark Ages, as argued for example by Seward (1979). Monasteries were granted lands, including vineyards, by secular princes and lords, and this evidence should more correctly be seen as reinforcing the view that it was among the secular nobility that viticulture survived in the period following the collapse of the western Empire. The great contribution of the monasteries, having been granted the vineyards, was to develop quality techniques of cultivation, which were enhanced by their continuity of ownership. From the accounts of the abbey of Saint-Germain-des-Prés in Paris, and in particular the Roll of Abbot Irminon dating from 814, it is possible to obtain a good idea of both the extent and cultivation practices of such monastic vineyards (Durliat, 1968; Lachiver,

1988). The abbey owned between 300 and 400 hectares of vines, just under half of which was directly exploited, the remainder being cultivated by tenant peasantry in return for rents in kind of wine. The vineyards were all located close to the river Seine and its tributaries, at places such as Morsang, Coudray, Epinay, Maisons, Villeneuve and Palaiseau, thus enabling the wine to be transported easily to Paris. Each year the abbey had approximately 6,400 hectolitres of wine at its disposal, for use in Mass, for general drinking by the monks, and for sale. Moreover, the large amounts of wine produced by the peasantry, over and above the rents they owed to the abbey, suggest that even here in northern France a peasant viticulture and demand for wine had survived.

To the east, monastic foundations in Germany also provide evidence of the expansion of viticulture beyond the Rhine (Bassermann-Jordan, 1907; Schreiber, 1980; Weiter-Matysiak, 1985). In particular the *Codex Laureshamnis* from the abbey of Lorsch, which was founded in 764 near Heidelberg, provides extensive details of the vineyards granted to the abbey in the eighth century (Minst, 1966-70). Thus, in the year of its foundation it received vineyards at Oppenheim from Folcred and Berticus, in 765 it received vineyards at Bensheim from Udo, son of Lando, and in 766 it received a further vineyard in Bensheim from Stalan. In 773 it was granted a large estate including vineyards by Charles, the future Charlemagne, and between 765 and 864 it received nearly a hundred vineyards at Dienheim (Seward, 1979). The abbey of Lorsch's early vineyards therefore lay on both sides of the Rhine, in Rheinhessen and the Bergstrasse, but the significant point to note about these grants is that they again emphasise that German vineyards were in the possession of the secular nobility in the eighth century, prior to their donation to the abbey.

The nobility were the third great force in establishing and developing viticulture in the period between AD 500 and 1000. The above examples well illustrate that most monastic vineyards were probably initially derived from secular estates, but it also seems that kings, princes and the secular nobility of northern Europe were keen to have wine from their own vineyards. Primarily this was required for consumption at banquets and in the entertainment of guests. As Lachiver (1988: 54) has argued, 'Comme les gens d'Eglise, les Grands reçoivent d'autres Grands, parfois même le roi, et l'offrande du vin est toujours de rigueur'. [Like the clergy, the nobles received other nobles, sometimes even the king, and the offering of wine was always obligatory.] The vineyards of the nobility formed parts of the estates surrounding their palaces and castles, so that on their frequent peregrinations they would always have wine at hand without having to go to the expense of transporting it with them.

Viticultural continuity in the eastern Roman Empire

While the political collapse of the western Empire based on Rome engendered considerable social and economic change, the political continuity in the eastern Byzantine Empire centred on Constantinople provided the stability within which viticultural traditions continued virtually uninterrupted. Under Justinian who ruled in the east from 527–65, the African provinces were retaken from the Vandals in 533, and Italy was eventually also reconquered by 553. As a result of this conquest, Italy, and particularly Rome, became impoverished, and the senatorial class which had survived the establishment of barbarian rule during the first half of the fifth century disappeared. Byzantine rule in Italy centred on Ravenna, and by the 560s 'little could be done to prevent the rapid occupation of Italy by the Lombards' (Cornell and Matthews, 1982: 223).

Around the shores of the eastern Mediterranean, in Greece, Turkey, Syria and Palestine, viticulture continued to flourish, much as it had done centuries earlier. There is some evidence that the destruction of the Temple at Jerusalem during the Jewish revolt of AD 66–70 had led to the abandonment of a number of vineyards in Judaea, and the Jerusalem Talmud (Damai, Ch.1, Halacha 1) records that 'Formerly grapes were plentiful . . . [but] nowadays there are not many', suggesting an element of decline. Roman writers from the fourth century refer to numerous vineyards in the eastern Mediterranean, and Jerome, for example, comments on the practice of growing vines up elaborate terraces in the Negev (Goor, 1966). In Greece and Turkey wine was produced both on small peasant holdings, and also commercially for sale to the urban markets, particularly in the capital, Constantinople. Much of this wine was sweet, made from the Muscat vine, with the grapes often being dried to give them a higher sugar content, and later in the medieval period when this wine began to be exported to northern Europe in larger amounts it often travelled under the generic name 'Romania', derived from its source in the eastern Roman Empire.

During the fifth and sixth centuries the wines of Gaza, in the Negev in Palestine, became particularly famous (Mayerson, 1985), with Gregory of Tours (*History of the Franks*, VII.29) noting that they were stronger than the wines of Gaul (Gregory of Tours, 1927: 306). The development of vineyards in this region seems to have been closely related to the rise of monasticism there during the fourth century, and to the activities of the proto-monk Hilarion. If Jerome's life of Hilarion (*Vita Hilarionis*) is to be believed, it would appear that one of the main activities of the monks in the monastery that he established was the cultivation of vines and the production of wine (Schaff and Wace, 1954: 303–15). The expansion of these vineyards, evidenced also by the discovery of remains of numerous wine presses in the vicinity of Gaza and Elusa, was further enhanced by the pilgrimages to the Holy Land undertaken by people from France, Italy and Spain during the fifth and sixth centuries. Gaza was the southernmost port of Palestine, and Mayerson (1985) has suggested that it became a favourite staging centre for pilgrims who having tasted the wine then took some back with them to the western Mediterranean. Once the taste for wines from Gaza had become established, the same routes were then plied by merchants. This, though, was only to be a short lived efflorescence, for with the Arab conquests of the seventh century, and the imposition of Islamic rule, the production of wine in the eastern Mediterranean was severely curtailed.

THE ISLAMIC CONQUESTS AND THEIR EFFECTS ON VITICULTURE

Islam, wine and the vine

During the course of the seventh century, the remarkable outbursting of vitality that began in the Hijaz in the Arabian peninsula, and found its expression in the extraordinarily successful Arab conquests undertaken in the name of Islam, was to have a profound effect on viticulture and wine production. In a mere fifty years between 630, when Mecca was occupied by the forces of Muhammad, and 680, at the death of the Caliph Muawiya ibn abi Sofian, the whole Persian Empire from the Taurus mountains to Samarkand had been conquered, as had much of the Byzantine Empire along the southern and eastern shores of the Mediterranean. The imposition of a new ideological structure built on the writings and sayings of Muhammad and forming the basis of the Islamic religion, was theoretically to transform the social and economic foundations of viticulture through the prohibition of wine consumption.

The basis of the Islamic prohibition on wine derives from the following three references within the Koran (Dawood, 1974: 355, 397; Ali, 1983: 86, 270, 271):

> They ask you about wine and gambling. Say: 'There is great sin in both, although they have some profit for men; but their harm is far greater than their profit'.
>
> (Sura II. 219)
>
> Believers, wine and games of chance, idols and divining arrows, are abominations devised by Satan. Avoid them so that you may prosper.
>
> (Sura V. 90)
>
> Satan seeks to stir up enmity and hatred among you by means of wine and gambling, and to keep you from the remembrance of Allah and from your prayers. Will you not abstain from them?
>
> (Sura V. 91)

Upon these sayings, implying that gambling and wine are sinful since they keep the devout from prayers, the rulers of the Islamic world forbad the consumption of alcohol. Strictly translated, the Arabic word *Khamr*, translated as wine in the above quotations, means the fermented juice of grapes, but it became applied by analogy to all fermented drinks and eventually to any alcohol (Ali, 1983: 86).

In contrast to these prohibitions, though, Muhammad's vision of the Paradise promised to believers does include wine, as the following extract from the Koran (Dawood, 1974: 124–5; Ali, 1983: 1381–2) uncompromisingly indicates,

> This is the Paradise which the righteous have been promised. There shall flow in it rivers of unpolluted water, and rivers of milk for ever fresh; rivers of delectable wine and rivers of clearest honey. They shall eat therein of every fruit and receive forgiveness from their Lord.
>
> (Sura XLVII. 15)

That viticulture and wine were already well established in parts of the Arabian peninsula prior to the birth of Muhammad in 570 is well attested. The *Periplus of the Erythraean Sea* (1912: 77), written towards the end of the first century AD, thus notes that wine was produced in southern Arabia, particularly in the vicinity of Muza, modern Al Mukha, and that 'Italian and Laodicean wines were imported into Abyssinia, the Somali Coast, East Africa, South Arabia, and India'. Most of this wine taken to India was probably date wine, but some may have included wine made from grapes from Yemen. Strabo (XVI.iv.25) mentioned wine in Arabia, although it is significant that he commented that most of it was made from dates. The best confirmation of the presence of viticulture in the hills and mountains of Arabia prior to the sixth century comes from the Arab poets of the pre-Islamic era (Hyams, 1987). Furthermore the poet al-A'sha of Bakr, a contemporary of Muhammad, describes in detail the vintage at 'Anâfit in Yemen, which al-Idrisi, writing in the eleventh century, also describes as being surrounded by vineyards at the time he was writing.

This survival of viticulture in the very heartland of Islam illustrates the important point that it was the consumption of wine rather than the production of grapes that was prohibited. The ease with which grapes could be turned into wine, nevertheless, meant that at times of particularly strict adherence to the precepts of Islam many vineyards were uprooted and destroyed. Goor (1966), for example, thus relates how at the beginning of the eleventh century the Caliph Al-Hakim ordered that all raisins and grapes in Palestine that could be made into wine were to be destroyed, and that as a result many vineyards fell into ruin. It is equally clear, though, that both viticulture and wine production could be maintained, even in parts of Arabia. Thus in Lorimer's (1908-15: II, b, 1756) account of Oman, he noted that at Sharaijah, the principal village in Jebel Akhdar, 'The greatest attention is given to the cultivation of the grape, and from it the inhabitants make a wine with which they regale themselves in the long winter evenings'. These examples well illustrate the difficulties to be encountered in attempting to identify the precise influence of Islam on wine production in areas where grapes were still grown.

The chronology of the Islamic conquests

The early conquests of Islam were extremely rapid (Glubb, 1963). Palestine and Jordan were invaded in 633-4, with Damascus falling in 635 and Jerusalem in 638. Following the two battles of Yarmuk in 634 and 636, Byzantine control of Syria and Palestine was lost. The invasion of Egypt then began in 640, and the Byzantine forces in north Africa soon abandoned Alexandria in 642. Despite numerous attempts, the Islamic forces were nevertheless unable to capture the Byzantine capital of Constantinople before the fifteenth century. To the north-east, they inflicted major defeats on the Persians at the battles of Qadasiya in 637 and Jalula in 638, and then at the battle of Nehavand in 642 the Persian army was effectively destroyed. Resistance in Persia from the local nobility nevertheless continued for a number of years, and it was not until the end of the seventh century, under the Ummayad Caliphate that Kabul, Bukhara and Samarkand were eventually conquered.

It was also under the Ummayads that Islamic forces spread westwards along the southern shores of the Mediterranean to Spain. Byzantine control of Libya was overturned in the campaigns of 642-5, but it was not until 698 that Carthage fell. Thereafter, the Ummayads crossed the Strait of Gibraltar and entered Europe in 711, defeating the Visigothic army under Roderic at a battle on the 'banks of the river Lakka in the district of Sidonia', which has been identified by Livermore (1958; 1971: 285) as probably having taken place by the Guadalete near Arcos de la Frontera. From here they marched on the Visigoth capital of Toledo, occupied it, and proceeded to annex most of the remainder of Iberia by 713, establishing their main power base in al-Andalus in the south, focused around the new capital of Córdoba. The Islamic defeat of the Visigoths brought them into direct conflict with the Frankish kingdom across the Pyrenees. In 721 the Arabs gained Toulouse, but in 732 they themselves were defeated by the Franks under Charles Martel, the illegitimate son of Pepin II, at the battles of Tours and Poitiers. This victory for the Franks established a period of relative stability between the Christian and Islamic powers in south-west Europe, which were effectively divided by the mountain range of the Pyrenees. In the east the subsequent downfall of the Ummayad Dynasty based at Damascus, and its replacement as the ruling Islamic power by the 'Abbasids then led to the formal establishment of the Emirate of Córdoba in Spain by the last surviving Ummayad prince, 'Abdu'r-Rahman, in 756.

Islamic power in Iberia, though, failed to extend fully to the far northwest. Centred as it was on the southern parts of Spain and Portugal, which had been most heavily influenced by the Romans, Livermore (1973: 22) suggests that the Muslims 'found little to attract them in the more tribal north, deeply impregnated with German traditions'. It was thus here, in the mountains of Asturia, that Christian traditions survived, exemplified by the construction of a shrine at Santiago de Compostela in Galicia, and it was from here that the Christian reconquest of Iberia would eventually begin.

The effects of Islamic rule on viticulture

Although Islam theoretically prohibited the drinking of wine, it is evident that, at least in Iberia and Persia, viticulture continued to survive, and indeed flourish, during the early medieval period. What evidence that survives suggests that wine production was also maintained, albeit often illicitly. Perhaps the most significant effect of Islamic rule was that it played an important role in influencing the types of grapes cultivated in the areas under its sway. Thus, since most vineyards were, in theory at least, maintained for the production of table grapes, selectivity in favour of larger, sweeter and seedless grapes was practised (Hyams, 1987). When such grapes, often lacking in acidity, are used for the production of wine, they tend to produce unbalanced wines with a high sugar content, and much of the wine produced in the Islamic world was probably of this nature. The fondness for sweet drinks made from grapes in the Islamic world is also reflected in the production of grape syrups made by the boiling down of grape juice.

The Arab geographer, Muhammad ibn Muhammad al-Idrisi, who completed his description of the world, the *Nuzhat al-Mustaq*, in 1154, provides firm evidence that vineyards were widely established in Spain and Portugal during the twelfth century (al-Idrisi, 1836–40). While some vineyards had been uprooted following the Islamic conquests of the eighth century, they appear to have been re-established relatively quickly once an uneasy peace had been restored. Subsequently a few rulers, such as the Caliph Ozman in the tenth century, who prohibited wine making and ordered the destruction of two-thirds of the vines in Valencia (Smith, 1966), caused a recession in viticulture, but this does not appear to have been long-lasting. Although al-Idrisi (1836–40) fails to comment whether the grapes were used for wine making, his evidence suggests that vines were relatively widespread in Andalucia, Murcia and Valencia, and were also to be found as far north as Salamanca, Burgos and Coimbra. Moreover, Read (1986: 4) notes that 'eulogies of wine by Moorish poets are so numerous as to suggest that wine-drinking was habitual among all but the most pious of Spanish Moslems'.

It is similarly in the works of Persian poets, such as Attar (1119–1230?), Hafiz (1320–91), Iraqi (d.1289), Khayyam (1048–1131), and Rumi (1207–73) that the clearest evidence of the survival of viticulture at the eastern end of the Islamic world is to be found (Arberry, 1954). Hafiz thus brings together the subjects of wine and love in several works, most notably in his poem *Wine and love* from which the following stanza is taken:

> Fill, fill the cup with sparkling wine,
> Deep let me drink the juice divine,
> To soothe my tortur'd heart;
> For Love who seem'd at first so mild,
> So gently look'd, so gaily smil'd,
> Here deep has plung'd his dart.
>
> (Arberry, 1954: 62)

Similar sentiments are to be found in the *Ruba'iyat* of Omar Khayyam. In the following extract, the author integrates several symbolic themes associated with wine and the vine, emphasising not only its role in providing an escape from the harsh realities of the world, but also in the second stanza integrating this with religious overtones:

> I cannot live without the sparkling vintage,
> Cannot bear the body's burden without wine:
> I am a slave to that last gasp when the wine-server says,
> 'Have another', and I can't.
>
> Tonight I will make a tun of wine,
> Set myself up with two bowls of it;
> First I will divorce absolutely reason and religion,
> Then take to wife the daughter of the vine.
>
> (Khayyam, 1981: 65)

The role of wine in providing access to eternal life, that was so prominent in Dionysiac and also Christian symbolism, is also clearly expressed in Khayyam's poetry:

> Drink wine, this is life eternal,
> This, all that youth will give you:
> It is the season for wine, roses and friends drinking together,
> Be happy for this moment – it is all life is.
>
> (Khayyam, 1981: 80)

Furthermore, the connections between wine and human fertility are also well established in his poetry:

Rise up my love and solve our problem by your beauty,
Bring a jug of wine to clear our heart
So that we may drink together
Before wine-jugs are made of our clay.

(Khayyam, 1981: 82)

There are numerous other mentions of wine within Omar Khayyam's poetry, but as the above examples illustrate they all tend to reinforce an imagery and symbolism closely similar to that of other early religions of south-west Asia and the eastern Mediterranean.

The survival of viticulture in Persia, modern Iran, and also further east in north-west India is further attested at a later date in paintings and manuscripts (de Planhol, 1977; Hyams, 1987). Among the most striking of these is an illumination from the Khamseh of Nizami dating from the sixteenth century, at the time of the Moghul Emperor Akbar, which beautifully depicts vines growing on a trellis system supported on forked stakes (Figure 26). Although the precise location depicted in this illustration is uncertain, it nevertheless provides firm evidence of the continuation of a viticultural tradition in this part of the Islamic world. This is further supported by evidence that Akbar's son Jahangir was an ardent drinker of wine, and in later life sank into alcoholism. The overall effect of Islam was thus not to destroy viticulture and wine production in the parts of the world that it came to dominate. Instead, it seems to have redirected the use of grapes, and also to have restricted the amount of wine made from them. Except under the strictest interpreters of Islam, wine continued to be drunk by those who sought its solace.

THE EXTENT OF VITICULTURE *c.* AD 1000

The influences on viticulture and the wine trade of the political, social and economic changes that transformed the European world following the collapse of the western Roman Empire, the emergence of Christianity and the conquests of Islam can be considered at the global, regional and local scales. Globally, viticulture was more widely dispersed in the year 1000 than it had been five hundred years earlier; regionally, the collapse in the efficiency of transport necessitated the restructuring of certain wine producing areas; and locally, the activities of the nobility, the church and the monasteries created new patterns of production and demand.

155

Figure 26   Vine cultivation in an illustration from the Khamseh of Nizami
(1595)
Source: British Library (OR 12208 f.40.b)

The global extent of viticulture

It is extremely difficult to estimate with certainty the overall distribution of viticulture at a global scale around the year 1000. Within Europe, the evidence for the spread of viticulture is comparatively good, and by 1000 it appears that wine was being produced well to the north and east of its extent under the Roman Empire. The establishment of the Duchy of Normandy in the north of France in the tenth century restored a degree of stability following the Viking invasions of the ninth century, and with it viticulture appears to have spread quite extensively to the north of the Loire and the Seine. Eastwards, viticulture had become well established along the rivers Mosel and Rhine, and particularly in the Rheinpfalz, or Palatinate (Hallgarten, 1951). The southern part of the Rhine and Alsace appear to have been relatively late in developing viticulture, largely due to the problems of navigation on the rapidly flowing river to the south of Strasbourg. There is no evidence of Roman viticulture in Alsace, and it seems likely that it began to be developed here during the seventh or eighth centuries under the Carolingians (Dion, 1959). In this context, Barth (1958) has thus identified a number of localities in Alsace where viticulture was certainly practised by the ninth century, and wine from Alsace is also specifically described in a poem dating from 820 written in honour of King Pepin (Lachiver, 1988). Further east still, wine was also grown beyond the boundaries of the former Roman Empire in northern Hungary, and Halász (1962) argues that when the Magyars invaded the country between 892 and 896 they were already familiar with viticultural practices from their contacts with the Caucasus. In southern Europe, peasant viticulture flourished along the northern shores of the Mediterranean in France, Italy and Greece, but in Iberia, Islamic rule probably restricted extensive wine production, except in the north-west, where trees and arbours were still to be found festooned with vines (Stanislawski, 1970).

Further to the north, the evidence for viticulture across the English Channel prior to the tenth century is inconclusive. It is possible that some small scale viticulture was practised in southern England during the late Roman period (Williams, 1977), but the evidence for Anglo-Saxon viticulture has in the past been accepted too uncritically. Bede writing in 731 noted that within Britain vines were grown in certain (*quibusdam*) localities (Colgrave and Mynors, 1969: 14), but this does not provide any foundation for Barty-King's (1977: 14) exaggerated comment that 'vines grew in many places', or Seward's (1979: 124) conclusion that 'these vines must have been planted by the monks'. More tantalisingly, Bede (1968: 37) also mentioned that vines grew in Ireland, but there is little evidence to support this assertion. It is possible that some vines may have been established in the gardens of Irish monasteries with which Bede may have had contact, but if viticulture did indeed exist there it was probably only on a very small scale. Giraldus Cambrensis (1951: 15), in his *Topography of Ireland*, certainly considered Bede to have been incorrect, and that Ireland 'has not, and never had, vines and their cultivators'. The early Irish law books dating from the seventh and eighth centuries (Kelly, 1988), which provide a wealth of information on early Irish agriculture, unfortunately remain silent on Irish viticulture, although there is some evidence in glosses written by Irish monks and preserved in continental manuscripts that that they at least knew something about pruning vines.

The next mention of vines in the English documentary evidence is found in the Laws of King Alfred. On the basis of this Hyams (1987: 187) has argued that 'By the ninth century vineyards had become sufficiently commonplace in Wessex for them to make an appearance in the law-code of Alfred the Great', and this is the generally accepted argument in the wider literature on the subject (Ordish, 1953; Barty-King, 1977; Darby, 1977). The relevant statement in King Alfred's Laws, however, is as follows: 'If any one injure another man's vineyard, or his fields, or aught of his lands; let him make "bot" as it may be valued' (Thorpe, 1840: 23). The critical feature to note about this quotation is that it is found in the preliminary section of Alfred's Laws, where he is quoting from the Book of Exodus (XXII) in the Bible. No mention of vines is to be found in Alfred's own new laws, and indeed most recent commentators have argued that this first section has 'no bearing on Anglo-Saxon law' (Attenborough, 1922: 35). By the tenth century, though, there is firm evidence of at least two vineyards in England, the first mentioned in a Latin confirmation of a vineyard and land at Panborough in Somerset, granted by King Eadwig to St Mary's Abbey, Glastonbury, and dated AD 956 (Sawyer, 1968: 215, No. 626), and the second noted in a grant by King Edgar to Abingdon Abbey of a vineyard near Watchet, together with its vine-growers and the land belonging to it (Finberg, 1964: 141, No. (99)).

By the eleventh century Domesday Book provides clear evidence for the practice of viticulture in southern England (Round, 1900; Darby, 1977). Forty-two unambiguous entries mentioning vineyards may be found in Domesday Book, and in addition under the heading of Holborn in Middlesex, William the Chamberlain was recorded as rendering 6 shillings yearly to the king for land on which his vineyard was found, although this need not actually have been in Holborn. Two other entries also relate to viticulture, one noting a wine render at Lomer in Hampshire, and the other mentioning a certain Walter *vinitor* who held land at Wandsworth in Surrey. Ten of the vineyards were new or as yet only partially yielding, suggesting that the Norman conquest in 1066 may have acted as a considerable impetus to the development of viticulture. Furthermore, it is interesting to note that most of the vineyards were in secular hands, and many belonged to nobles beneath the rank of baron.

This English evidence is important since it reflects the expansion of viticulture beyond its traditionally accepted northern bounds (Dion, 1959), and while some small monastic vineyards may have survived during the Anglo-Saxon period, the main impetus to English viticulture would seem to have come from increased contacts with northern France during the eleventh century.

Outside Europe, viticulture had spread eastwards through Persia to north-west India, and along the line of the Great Silk Road to China, but, according to Laufer (1919), it seems to have still remained something of a curiosity here, with various Emperors regularly having to re-introduce the grape vine. Vines from Persia seem to have first been introduced to China soon after 128 BC, when the envoy General Chan K'ien apparently sent vine seeds from Fergana to the Emperor Wu, but Hyams (1987) argues that little wine was ever made in China, with the few vines that were grown being cultivated mainly for their grapes. Sumptuary laws dating from the Han Dynasty suggest that grape wine was reserved strictly for imperial use (Cooper, 1973), but Chinese poems dating from all parts of the first millenium AD nevertheless indicate that grape wine was more common than has been suggested. Fermented rice wine was the most usual alcoholic drink in China, sometimes being described as white or green wine, but wine from grapes was not unknown (Birrell, 1986: 333). Thus in the *Yü-t'ai hsin-yung* or *New Songs from a Jade Terrace*, an anthology of Chinese love poetry dating from between the second century BC and the sixth century AD, there are several mentions of grape vines, as the following two examples illustrate:

Night breezes puff firefly sparks,
Dawn gleams glint on lichen.
Long faded sweet herb leaves,
Fallen in vain grapevine blooms.
(From *The one I love* by Wang Seng-ju, AD 465–522; Birrell, 1986:
169)

Isolated manor grapevines
Isolated manor grapevines pregnant with dangling fruit,
Chiangnan cardamom boughs grown to an embrace.
Insensate, impassive, they are always the same,
Yet he who loves or hates vainly suffers isolation!
(Hsiao Kang, AD 503–51; Birrell, 1986: 256)

Typical of the many general mentions of wine to be found in *New Songs from a Jade Terrace* is the poem by Chang Shuai (AD 475–527 entitled *Wine*, which begins as follows:

From wine I get real pleasure,
And this wine is the finest!
Wine clear as blossom is valuable,
This wine, sweet as milk, is invaluable!

(Birrell, 1986: 171)

Other specific types of wine mentioned in the anthology include pomegranate wine, cassia wine and mulled wine. Significantly, as in other
parts of the world, wine was frequently used in the symbolic representation of love. As Birrell (1986: 334) has argued, "First wine" meant the
first ritual sharing of wine by bride and bridegroom at their marriage,
and came to mean marriage itself'. She also notes that 'Lovers pledged
with wine, the male or master offering a cup of wine from which he had
drunk the first half' (Birrell, 1986: 334). From a slightly later date, the
poems of Li Po (701–62) and Tu Fu (712–70) also mention wine, with the
following extract from the poem entitled *Coming down from Chung-
Nan mountain by Hu-Szu's hermitage, he gave me rest for the night and
set out the wine*, apparently suggesting the drinking of grape wine:

To green bamboos and a hidden path
With vines to brush the travellers' clothes;
And I rejoiced at a place to rest
And good wine, too, to pour out with you.

(Cooper, 1973: 151)

To the west of Europe, across the Atlantic Ocean, the enigmatic
descriptions of Vínland in the Norse sagas provide further evidence
concerning the distribution of vines at the end of the first millenium AD
(Jones, 1964). Thus in the *Grœnlendinga Saga* written *c.* 1190, and
referring to Leif Eiriksson's exploration of Vínland, Tyrkir the South-
erner reported that

'I have some news. I found vines and grapes.'
'Is that true, foster-father?' asked Leif.
'Of course it is true,' he replied. 'Where I was born there
were plenty of vines and grapes'.

(Magnusson and Pálsson, 1965, 57)

There is, though, no mention in the saga that the vines were cultivated,
or that wine was made from the grapes, either by the indigenous
inhabitants of Vínland or by the Norse settlers. Following Leif Eiriksson's exploration, a further expedition to Vínland was undertaken by
Thorfinn Karlsfeni *c.* 1010. In the description of this expedition in *Eiríks
Saga*, written *c.* 1260, Thorhall the Hunter specifically claims that no
wine had touched his lips while he was in Vínland (Magnusson and
Pálsson, 1965: 97), further suggesting that the inhabitants of Vínland
were not at that time making wine. Adam of Bremen, however, writing in

160

the eleventh century, noted in his *Descriptio insularum aquilonis* (XXXVIIII) which forms the fourth book of his *Gesta Hammaburgensis ecclesiae pontificum* (Adam of Bremen, 1917: 275) that Svein Estridsson, king of the Danes, had told him about Vínland, which was so named 'from the circumstance that vines grow there of their own accord, and produce the most excellent wine' (Jones, 1964: 85). In the more favourable climatic conditions that prevailed at the time of the Norse voyages it is possible that vines could have grown as far north as Newfoundland (Jones, 1964), and that people of Germanic origin, such as Tyrkir mentioned in the *Grœnlengdina Saga*, with experience of wine making may well have attempted to turn the indigenous grapes of North America into wine. Norse settlement in Vínland was, however, short lived, as a result of the constant threat of attack by the Skrælings, the indigenous inhabitants of the country, and following Karlsfeni's attempt at settlement there appear to have been no further successful voyages to Vínland. Indeed, by the twelfth century its exact location had been forgotten (Magnusson and Pálsson, 1965: 28).

At the global scale, viticulture *c.* AD 1000 remained mainly an expression and manifestation of European and Mediterranean societies. In the south, the emergence of Islam had reduced the amount of wine consumed, but to the north vines were being grown beyond the limits of Roman viticulture. Similarly, knowledge of viticulture and wine made from grapes had spread through Asia to China, and if Adam of Bremen is to be believed it is even possible that a small amount of wine had been produced from the indigenous vines of north America.

Viticulture at the regional scale

Within Europe, the changed economic and social conditions following the collapse of the western Empire necessitated a restructuring of the organisation of viticulture at the regional scale. By AD 1000 European viticulture could be characterised by three key features: in the south it formed part of the ubiquitous peasant subsistence economy; in the north vines were cultivated mainly in close proximity to rivers; and vineyards were also to be found surviving in scattered locations beyond their extent both in the Roman and the post-medieval periods.

The importance of riverside locations for vineyards was well exemplified by those belonging to the abbey of Saint-Germain-des-Prés (Durliat, 1968), but it was a widespread characteristic of all northern European vineyards. This is particularly apparent for the vineyards of the Mosel and Rhine in the period before 1050 (Figure 27). Based on the work of Weiter-Matysiak (1985) it is evident that the vast majority of the early vineyards of Germany were located in close proximity to the main rivers of the region, and that the vineyards established over the next two centuries represented both an infilling of this pattern and also a spread to areas further afield. Within France, the dominance of the vineyards of the Paris Basin, along the river Seine and its tributaries such as the Yonne and the Marne, was also noticeable (Dion, 1959; Lachiver, 1988). The main reason for the riverine location of these vineyards was, as it had been in the Roman era, the differential costs of transport by road and by river. By the eleventh century the re-emergence of urban demand for wine in northern Europe provided a market for commercially oriented vineyards. However, the high cost of overland transport gave vineyards situated with easy access to the main fluvial transport routes a distinct competitive advantage.

The cost of overland transport was also, paradoxically, the main reason why viticulture was able to survive in a number of isolated locations. In many parts of Europe far from the great navigable waterways, there was a strong demand for wine from obscure monasteries, from members of the nobility who sought to produce wines on their own estates, and also as a result of the sacramental requirements of Christianity as it spread into the northern parts of Europe. Where such demand existed, the cost of transporting wine long distances by road or track made it prohibitively expensive, and this enabled viticulture and wine making to survive in such isolated places. Once communications improved during the later medieval period, and costs of transporting wine began to decline, it became easier and cheaper for such people to purchase wines produced elsewhere, and it seems that there was consequently a contraction of viticulture to the regions where it was environmentally best suited.

There is some evidence that these economic considerations were also influenced by changes in the climate which enabled vines to be cultivated successfully in the more northern parts of Europe but such evidence is difficult to interpret. Lamb (1965) has, for example, argued that the presence of vineyards in England and the north of France during the period 1000–1200 was a result of the notably warmer climate then prevailing, and that this was then followed by a decline in temperature associated with a disappearance of viticulture in these areas. However, while a range of other evidence does suggest a subsequent cooling of temperature (Lamb, 1965), there is also considerable evidence that the disappearance of viticulture from England had as much to do with the development of English interests in the vineyards of Gascony and Aquitaine, as it had to do with any climatic change. Moreover, even Lamb (1965: 13, 34) suggests that the differences in the average temperature between the years around 1200 and the so-called Little Ice Age around 1600 were only of the order of 1.2–1.4°C. While England is certainly at the edge of the climatic limits of successful viticulture, and wine producers cannot always have guaranteed a successful vintage every year, these temperature differences were probably insufficient by

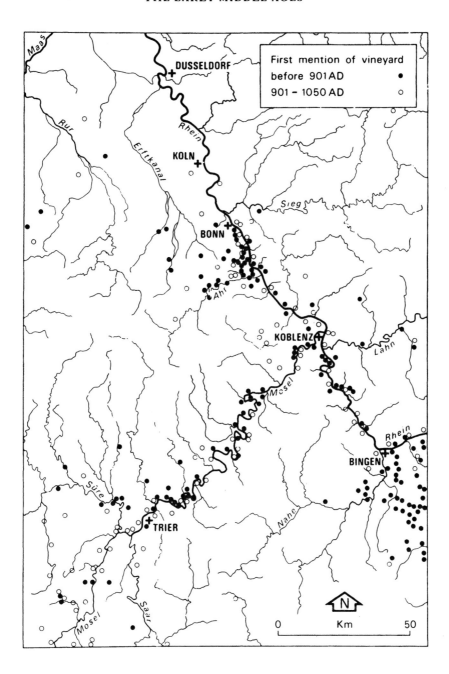

Figure 27   The distribution of vineyards along the rivers Rhine and Mosel
before 1050
Source: derived from Weiter-Matysiak (1985)

themselves to have led to the abandonment of English viticulture. Lamb (1965: 32) supports his arguments for significant climatic change by observing that '20th century temperatures in the best of the former English vine districts appear to fall short of the lowest values required' for successful viticulture. The recent expansion of English viticulture in the 1970s and 1980s, however, indicates that viticulture is indeed possible under these climatic conditions, and suggests that some degree of caution should be used in interpreting the evidence connecting climatic change with the emergence and decline of viticulture.

Viticulture at the local scale

The distribution of vineyards at a more local scale varied between northern and southern Europe. In the north it closely reflected the important role played by the nobility and ecclesiastics in developing a demand for wine. In southern Europe, the great villa estates that had dominated the Roman landscape, and provided wine for the urban markets had been replaced by a system of agriculture oriented much more towards subsistence needs, with peasant holdings continuing a viticultural tradition developed in conjuntion with the cultivation of wheat and olives. In the north of Europe most authorities argue that viticulture was concentrated on the estates of the nobility, the Church and the monasteries, with vines also being found clustered around castles and towns (Dion, 1959; Lachiver, 1988; Hyams, 1987; Younger, 1966). While there is some truth in this view, reflecting the demand for wine generated by the great magnates, there is increasing evidence to suggest that by the beginning of the eleventh century vines were in reality cultivated much more widely. Most towns, castles and monasteries to the south of a line joining Nantes, Paris and Köln probably had their own vineyards, but vines were also to be found scattered elsewhere on the slopes of the river valleys of northern France and Germany. This is certainly the impression to be gained from the distribution of vineyards along the Rhine and the Mosel (Figure 27) (Weiter-Matysiak, 1985), as well as from the peasant holdings of the abbey of Saint-Germain-des-Prés in the Paris Basin (Durliat, 1968). Much of the wine produced on the estates of the great magnates was destined for the consumption of their own households, and there is little evidence in the tenth century of any particular regional specialisation in viticulture. However, the growing urban demand for wine in northern Europe encouraged lesser landowners and the peasantry to produce wine over and above their own needs. It therefore seems probable that vineyards were to be found quite widely distributed in the river valleys of northern Europe which had good access to the urban centres of demand, and not just around the walls of the more important castles and towns.

WINE IN A CHRISTIAN WORLD

Between the third and the eleventh centuries viticulture in Europe had undergone a complete transformation, as the focus of political power shifted from Rome to northern France and the economic structure of the Roman Empire disintegrated. The great estates serving the markets of the Roman Empire had vanished. In the Mediterranean region they were replaced by small scale peasant viticulture, and in the north by the largely self-sufficient estates of the ecclesiastical and secular nobility. However, over this period the limits of viticulture continued to be pushed further towards their maximum extent. By the eleventh century vineyards were apparently well established as far north as Ely in England, and to the east they were an unusual but not entirely unknown element in the Chinese landscape. However, wine, the product of the grapevine, remained essentially an element of the Christian world. Viticulture, although known in China and northern India, did not become widespread there, and in the Islamic world wine was distinctly unfavoured. The importance of ideology and religion in determining the global distribution of viticulture should therefore not be underestimated.

The nature of wine had also undergone a transformation since the great vintage wines produced at the height of the Roman Empire. No longer was wine stored in airtight containers, in which it could mature over several years and produce an enhanced profit for the producer or wine merchant. Instead wine was made for the immediate market, being transported and stored in wooden barrels. The lack of capital investment in the storage and ageing of wine is a marked feature of the medieval wine trade. By the eleventh century, as commercial activity in Italy and northern Europe became revitalised, and as urban communities began to assert their own identities, a new set of economic factors came into play which were eventually to lead to the development of commercially oriented vineyards and the great wine fleets which were such a characteristic feature of medieval trade.

6

MEDIEVAL VITICULTURE AND THE WINE TRADE

Peu d'événements, depuis la conquête romaine, ont autant contribué à promouvoir dans le monde la production vinicole de la
France que le rapide et brillant épanouissement des villes, à partir
du XIᵉ siècle, dans les pays de la mer du Nord, et particulièrement en
Flandre. Les principaux habitants de ces villes, ceux qui animaient
le commerce et régentaient les métiers, avaient acquis, avec la
richesse, un orgueil de classe qui trouvait l'une de ses satisfactions
dans la consommation ostentatoire du vin. [Few events in the world
since the Roman conquest contributed as much to promote French
wine production as did the rapid and brilliant blossoming of towns
from the eleventh century in the countries of the North Sea, and
particularly in Flanders. The chief inhabitants of these towns, those
who animated commerce and dominated the crafts, had acquired,
with the wealth, a class pride which found one of its gratifications
in the ostentatious consumption of wine.]

(Dion, 1959: 201)

Three main images are to be found in the literature concerning
viticulture and the wine trade during the medieval period. The first is the
contrast that is widely thought to have existed between Mediterranean
and northern viticulture, with the former being seen as largely
subsistence based and the latter as being directed much more towards
export trade and commercial profit (Dion, 1959; Postan, 1966; Lachiver,
1988). The second image concerns the organisation and structure of that
trade, which is usually envisaged as being dominated, virtually to the
exclusion of all others, by the regular wine fleets that sailed between
England and Gascony (James, 1971), and the third general conception is
that viticulture during the Middle Ages was static and experienced little if
any technical change (de Blij, 1983). While there is some truth in all of
these generalisations, they hide the many local variations that existed in
medieval viticulture, and the main purpose of this chapter is therefore to
deconstruct these popular images. In so doing it examines the changing

166

tenurial structure of different parts of Europe during the medieval period, together with the broad spectrum of political, economic and social factors that influenced the context of medieval trade. What it reveals is that peasant viticulture as part of a general subsistence economy was to be encountered throughout much of France and Germany during the medieval period, that much wine production within the Mediterranean region was in practice directed towards the supply of wine to the main urban centres of the region, and that some improvements in viticultural practice were indeed implemented.

VITICULTURE AND WINE PRODUCTION 1000-1450

The Mediterranean

The image of a peasant agriculture in the Mediterranean region, based on the subsistence production of wheat, olives and vines, together with some livestock production, is widespread and pervasive (Braudel, 1972-3; Delano Smith, 1979), but it obscures the important differences that existed within the region in terms of tenurial structure, rental agreements, and the purposes for which the crops were grown. For most of the medieval period the idea of an independent Mediterranean peasantry, producing its own subsistence needs on its own land, is largely a myth. Peasant society was highly subdivided in terms of access to land and resources, with the peasantry being subject to a range of surplus extraction demands by the landholding classes. One key insight to the agrarian changes that took place is provided by the balance between communal and individual land tenures. Jones (1966) has thus shown how in medieval Italy an individualistic agrarian system gradually came to dominate an earlier customary system. At the same time, the period 1050-1300 also saw the disintegration in Italy, and particularly in the north of the country, of the manorial system. Moreover, the social and economic position of the peasantry varied appreciably in different parts of the country. In Jones' (1966: 398-9) words,

> No formula, indeed, could express the confusion of rights, part patrimonial, part seigneurial, which by the end of the Carolingian period landlords all over Italy had come to exercise in differing degrees over tenants, 'vassals', *homines* and *fideles* . . . Rural society remained obstinately heterogeneous. Not all peasants became dependent, not all were even tenants.

By the twelfth century, Pounds (1973: 276) has argued, 'the pattern of land-holding in southern France and much of Italy had come to be dominated by small compact holdings held by free peasants for a money rent'. Throughout southern Europe, though, it appears that during the

167

twelfth and thirteenth centuries, and influenced in part by the spread of feudalism (Delano Smith, 1979), the peasantry were increasingly being subjected to greater seigneurial extractions. In Castile, for example, evidence recorded in the *Beccero*, or *Celebrated Book of the Behetrías of Castile*, which lists the dues required from the peasantry in over 600 villages, suggests that by the fourteenth century the full economic rent of the peasants' land was absorbed by seigneurial rents and dues (Smith, 1966). Partly as a result of the 'reconquest' of Spain from the Moors, most of the centre and south of the country eventually came to be dominated by great landed estates, in which 'Slaves, serfs, tenant farmers, and wage earners formed the backbone of agricultural labour' (Smith, 1966: 434).

Amid all this diversity, two generalisations concerning viticulture are apparent: first, vines were widely grown by the peasantry, and secondly, the most common form of rental arrangement associated with them was some kind of share-cropping. The precise nature of rental and share-croppping arrangements, however, varied markedly. Based again on the *Beccero*, it seems that the most common fourteenth century rents in Castile were in the form of wheat, barley and wine (Smith, 1966). In medieval Spain the rents of vineyards varied from one-tenth to one-half of the crop, with figures between one-fifth and one-third being most common. In contrast, in Italy, the percentage of the crop taken by the landowner under the usual share-cropping arrangements for vines and fruit trees was a half or occasionally even more (Jones, 1966). These generalisations have two important implications: on the one hand, they imply that the idea of a free, independent and purely subsistence based peasantry was probably the exception rather than the rule, and, on the other, they suggest that landlords were able to obtain considerable rents of wine in kind which they could then use to supply the urban markets with wine.

Roman viticulture in the Mediterranean region was technically, relatively advanced, and Jones (1966: 371) has argued that 'in medieval Italy this traditional lore was faithfully reproduced and, as leases, show, applied'. This is well reflected in the *Liber Commodorum Ruralium* . . . of Petrus de Crescentiis (1471), completed between 1304 and 1309, which provides a wealth of information concerning the ways in which the classical traditions of viticulture were being interpreted in Italy during the medieval period (Sòriga, 1933). In the fourth book of this work, which quotes liberally from Columella and Cato, Petrus de Crescentiis thus provides considerable detail about the different types of vine being cultivated, their environmental requirements, the seasons for performing different vineyard tasks, and methods of vinification. Stimulated by a growing urban demand, particularly in the northern Italian towns of Venice, Genoa, Florence and Milan (Braudel, 1972–3), production of all of the basic crops, including vines, was both intensified and extended further into more marginal land during the period between 1000 and 1350. New vines were also being introduced as a result of import and selection, and based on a survey of Italian agricultural works of the late Middle Ages, such as Andrea Bacci's treatise on wines, *De naturali vinorum historia, de vinis Italiœ* . . . dating from 1596, Jones (1966) has noted that of the 50 noble vines known in Italy in the sixteenth century, medieval introductions included among many others the Trebbiano of Tuscany and the Marche, the Vernaccia of Liguria and the Schiava of the Po valley. Bacci's (1596) treatise itself, which is based heavily on the works of earlier authors, provides much detail concerning viticultural practice throughout Europe, and in particular in its description of many of the wines of the Mediterranean and Spain as *generoso*, it provides confirmation of the distinction that then existed between these noble wines and the lesser wines of other parts of Europe, many of which were dryer and lower in alcohol (Marescalchi and Dalmasso, 1931–7).

The northern vineyards

In northern Europe, where environmental requirements for successful vine cultivation were less easily satisfied, experiments with vine selection were also widespread. Until the thirteenth century most of the vineyards of northern Europe were planted with white varieties of vine, which were more resistant to the damp and cold of the northern climate (Lachiver, 1988), but gradually new varieties were introduced, and by the end of the fourteenth century the red Pinot Noir was well established in Burgundy, as was the red Gamay against which Philippe le Hardi (Philip the Bold) attempted to legislate in 1395 (Berlow, 1982).

Various different methods of training vines continued to be used in northern Europe throughout the medieval period, but it is often difficult to tell precisely which methods were used in any particular place. Some of the best illustrations of medieval vineyards are to be found in the relatively late illustrated manuscripts, such as *Les Très Riches Heures de Jean de France, Duc de Berry*, dating from *c.* 1416, but earlier manuscripts also provide illustrations of methods of vine cultivation and training. Figure 28, taken from an illustration for the month of September in *Les Très Riches Heures* . . . shows in great detail what must have been a typical vintage scene from Anjou in the late fourteenth and early fifteenth centuries. The peasant labourers are bending down to pick the grapes from relatively low, well pruned vines set in a vineyard laid out below the castle walls, in a scene very different from the polyculture to be found in parts of Mediterranean Europe, where vines might be grown on trellises around fields, or up specially pruned trees. The illustration for March in *Les Très Riches Heures* . . . indicates how such vines were pruned, with each being cut back to three spurs above a trunk, which had

Figure 28 Illumination for the month of September, showing the vintage at the
Château de Saumur, from the *Très Riches Heures du Duc de Berry, c.* 1416 (f.9.v)
Source: Chantilly, Musée Condé; Giraudon (PEC 4906)

been allowed to grow approximately to knee height. A range of other methods of vine training, though, were still to be found in Europe during the fifteenth century. Figure 29, taken from a German woodcut of Virgil and Maecenas published in 1500 (Baird, 1979), indicates four main ways in which vines were trained: up a tree, individually up stakes, over an arched arbour, and over a regular trellis. Although this scene illustrates the artist's impression of the ancient city of Rome, in the guise of a medieval European town, these methods of vine training were almost certainly familiar to him from Germany, and it seems likely that they were to be found throughout medieval Europe, each reflecting different local traditions and environmental conditions. Other evidence concerning vineyard maintenance is to be found in numerous manorial records scattered throughout Europe, and among the most northern of such records is a rent roll from 1235 pertaining to the Archbishop of Canterbury's English vineyard of Northfleet. This mentions the peeling of vines, their layering, digging, hoeing and propping them with trellises, and further information is also provided which indicates that a wine press was used and that the wine was stored in casks (Sutcliffe, 1934).

Turning to the methods of vinification in northern Europe during the medieval period, it is widely argued (Dion, 1959: 192; Johnson, 1989) that the cost of presses meant that they were the exclusive preserve of the nobility and monasteries. This need not imply, though, that there was no peasant wine production, and there are numerous medieval illustrations of grapes being trodden in tubs similar to those shown on the cart in the vintage scene in Figure 28. Typical of these is an early fourteenth century illumination in Queen Mary's Psalter (Figure 30) illustrating the treading of grapes in a wooden vat, and other such illustrations indicate that small scale vinification was undoubtedly practised. Many of the tenants or villeins on great estates would, however, also have been expected to use their lord's press on payment of a fraction of the wine produced. It is evident that the services due from the peasantry on such estates often included not only the maintenance of the vineyard, but also labour services involved in vinification and transporting the wine (Sutcliffe, 1934; Durliat, 1968).

A typical vintage scene is illustrated in Figure 31, which is taken from a Flemish Book of Hours dating from around 1500. In the background, grapes are being picked from vines trained high up on trees. They are then being brought down to the courtyard beside the lord's mansion, where the must is extracted in a screw press. This is subsequently being poured carefully into wooden barrels for fermentation. The artist, anxious to catch every moment of the process, has then combined a number of different stages into the single scene, and he therefore also shows some wine being sampled from the same barrels and offered to a friend or potential purchaser. The key observation to be made about

Figure 29 Woodcut of Virgil and Maecenas, showing different types of vine training, from Hieronymus Brunschwig, *Liber de Arte Distillandi de Simplicibus*, Strasbourg: Johann Grüninger, 1500
Source: Baird (1979)

vinification in northern Europe was that while red wine could easily be made by being trodden, it was still necessary to press the skins to extract the remaining juice, and, more importantly, good quality white wines could only be made by pressing the grapes first in order to separate the juice from the skins before fermentation.

Whereas Dion (1959) argued that French viticulture in the medieval period was dominated by the nobility, the church and the monasteries,

Figure 30 Treading the vintage from the early fourteenth century manuscript of Queen Mary's Psalter
Source: British Library (Royal MS 2.B. vii f.79.v; K2660)

Figure 31   Vintage scene from a Flemish Book of Hours of the Blessed Virgin
(*c.* 1500)
Source: British Library (Add. 24098 f.27.v; K108838)

Durliat (1968) has shown that as early as the ninth century peasant viticulture was flourishing in the Paris Basin. In his words, 'la culture de la vigne n'était pas un passe-temps d'aristocrate, mais une occupation courante dans toutes les couches du monde paysan' [the cultivation of the vine was not a pastime of the aristocrat, but a normal occupation in all the layers of the peasant world]. During the eleventh and twelfth centuries, though, viticulture spread further to the north and east, and in the more hostile environmental conditions pertaining in these regions its success was less assured. Al-Idrisi (1836–40), writing in the first half of the twelfth century, thus noted that vineyards were to be found in Flanders around Ghent (VI.ii.210v) and Bruges (VI.ii.210v), further north still in the vicinity of Utrecht (VI.ii.211r) and even Bremen (VI.ii.212v), and to the east, he also referred to vineyards in Poland (VI.iv.217v), particularly around Kraków (VI.iii.215r). While it is possible that he might have been using the word 'vineyard' to describe places that were reputed to be fertile and prosperous, there is sufficient other evidence, particularly from Flanders (Halkin, 1895), and Poland (Carter, 1987) to indicate that viticulture was well established in the twelfth century substantially to the north and east of its modern limits. Although climatic conditions may have been more favourable to the vine at this period (Lamb, 1965, 1982; Le Roy Ladurie, 1972; Le Roy Ladurie and Baulant, 1980), there were also significant economic and social reasons why vines were grown in these regions where a regular annual yield could not be guaranteed.

THE MARKET FOR WINE

Two critical processes lay behind the spread of viticulture and the shift towards a greater emphasis on the monoculture of vines in northern Europe: the revival and expansion of urban life, and the elevation of wine to a prominent position as a symbol of social status among the burgesses of these towns and cities (Gracia, 1976). Ever since Dion's (1959) classic account of the historical development of viticulture in France, it has been widely accepted that the three driving forces behind the expansion of vineyards were the nobility, the church and the monasteries (Lachiver, 1988). However, in practice it was the growth of towns, and the transformation of the urban economy of Europe during the twelfth and thirteenth centuries, that led to the development of large commercially oriented vineyards on the south-east facing slopes of the major river valleys of northern Europe.

The urban revival

Most Roman cities within Gaul survived the collapse of the western Roman Empire 'though not, in many instances, without a hiatus of greater or lesser duration' (Pounds, 1973: 265). With the construction of castles, the creation of bishoprics, and the establishment of monasteries in the early Middle Ages, a new impetus had been given to urban settlements, but it was in the three centuries between 1000 and 1300 that fundamental changes in the medieval economy transformed the urban face of Europe. Between 1100 and 1300 Pounds (1973) has estimated that the population of Europe approximately doubled, from about 30 million to between 50 and 60 million. In order to feed this ever increasing population there was an expansion in the amount of agricultural land and the development of a market economy in which towns played an important role as centres of exchange. Moreover, industry, and above all the cloth industry, became increasingly urban based.

In 1100, while towns were to be found scattered throughout much of Europe, the most significant urban regions were in northern Italy and Flanders. The densest and most vigorous urban economies were to be found in northern Italy, in Lombardy, and to a lesser extent Tuscany. Here, much of the Roman urban fabric survived, and the towns of Pavia, Milan, Pisa, Lucca and Genoa thrived on the resurgence of commercial activity that began in the eleventh century. In northern France and Flanders, the development of the cloth industry played a major part in leading to the growth of towns such as Saint-Omer, Arras, Douai, and later Bruges and Ghent, most of which developed from a central core in the form of an earlier castle, monastery or cathedral. By the fourteenth century, while these two regions still dominated the urban structure of Europe, new cities such as Prague had emerged, the overall density of urban settlement had become far higher, and there had been a shift in urban emphasis away from Flanders to the towns of Germany. The largest cities in 1300 were Paris, Milan, Venice and Florence, and these were then followed in size by such cities as Ghent, Bruges, Brussels, London, Köln, Nürnberg, Naples, Sevilla and Barcelona. All of these towns were great commercial and manufacturing centres, dependent for their food and other requirements mainly on long distance trade.

In southern Europe, where vines had always been grown and wine was the basic alcoholic beverage of the mass of the people, urbanisation simply necessitated a reorganisation of the production and distribution of the wine that was already being produced. The towns of northern Italy, for example, could at least in their early stages of growth be readily supplied with Italian wine. However, in Flanders, England and northern Germany, where wine had not traditionally been widely consumed, the expansion of viticulture and the development of a long distance wine trade were dependent to a large extent on the elevation of wine into a symbol of high social status.

Urban society, symbolism and northern viticulture

The merchants and burgesses of the towns of northern Europe consumed large quantities of wine, and expended considerable energy on its provision. This is well attested by the creation of vineyards in the vicinity of most major towns and in the significant position that wine held in medieval trade. Moreover, vineyards were also found within the walls of numerous towns in Italy, southern France and Germany, often in the grounds of abbeys and monasteries. Typical of these were the regularly laid out vineyards belonging to the monasteries of St Pantaleon and St Ursula in Köln, which are well illustrated in sixteenth and seventeenth century maps of the city (Schreiber, 1980: 15, 81, 129). The mass of the population of northern Europe, however, drank beer as their basic beverage other than water (d'Haenens, 1984). Most grains were used for malting, but oats, the least valuable, were the most common (Pounds, 1973: 205). Throughout northern Europe many households brewed their own beer, and then sold any surplus. In England, Hilton (1982) has, for example, noted how ale could be sold provided a sign was displayed outside the brewer's house. Officials then had to taste it to guarantee its quality, and it could then be sold 'outside the house, on a level step, from sealed and licensed measures of one gallon or half a gallon' (Hilton, 1982: 14). The activities of the ale-wives and hucksters, supplying provender for the urban poor, were frequently criticised, with many court records containing accounts of how beer was sold contrary to the assize, but it was these petty traders who supplied most of the food and drink consumed in the towns of medieval Europe. That beer was the drink of the poor is well exemplified by William Langland's fourteenth century account of Rose the regrator in his poem Piers the Ploughman, where her husband, Covetousness, describes her as follows:

> Then I bought her some barley-malt, and she took to brewing beer for retail. She would mix a little good ale with a lot of small beer, and put this brew on one side for poor labourers and common folk. But the best she always hid away in the parlour or in my bedroom.
>
> (Langland, 1966: 67)

In contrast, as Dion's (1959) quotation at the beginning of this chapter emphasises, wine was the drink of the merchants and burgesses, the social elite of the medieval town. Wine drinking accompanied all of the great urban celebrations and festivals. It was consumed to welcome royalty, and to entertain guests; it was one of the basic symbols which set aside the burgess from the urban poor. A Flemish text dating from 1256 thus distinguishes peasants and craftsmen from burgesses, because the latter gave wine to their guests to drink (Pirenne, 1933: 235). This symbolic value was derived in part from the cost of wine and its scarcity in the north of Europe, but it was also a consequence of the religious associations that still clung to it. Furthermore, the urban authorities of many medieval towns in Europe were eager to benefit from the wine trade through restrictive trading practices, well illustrated, for example, by the imposition of taxes and the monopoly on wine sales enforced by the authorities of such German towns as Göttingen (Neitzert, 1987), which in some cases could provide over half of a town's annual revenues.

Wine and religion in medieval Europe

As economic motives for the development of vineyards to supply the northern European markets with wine increased in importance, the relative significance of Christian beliefs in the promotion of viticulture diminished. Even in the eleventh century the amount of wine actually consumed in religious ceremonies formed but a very small percentage of that produced in Europe as a whole. Demand for wine by the ecclesiastical population nevertheless continued to remain high, particularly among the bishops and prelates. This is well illustrated by the influence of the Avignon Papacy on the production of wine in eastern and southern France. Until the fourteenth century the wines of Burgundy, lacking good river communications with the markets of Paris and northern Europe, were relatively unknown outside their immediate area of production. However, the transfer of the Papacy to Avignon in 1309, where it remained for much of the fourteenth century, and the extravagence of the Papal court (Knowles and Obolensky, 1969), provided a new source of demand and a way of popularising the wines. Dion (1955) thus notes the particular affection of the Papal court for the wines of Beaune, with one fourteenth century poet describing the town as the queen of wines. While wine was widely produced in southern France, and provided the staple drink of the peasantry, the Avignon Papacy also led to the resurgence of quality viticulture along the Rhône, particularly in such locations as Châteauneuf-du-Pape (Lachiver, 1988).

More significant than the demand for wine by the prelates during the medieval period, though, was the bequeathing of vineyards to the various monastic orders. Within Burgundy, for example, by 1275 the Benedictine abbey of Cluny had come to own all of the vineyards around Gevrey. Likewise, the Benedictine abbey of Saint Vivant on the Mont de Vergy was granted vineyards in Vosne by the Duchess of Burgundy in 1232, and between 1110 and 1336 the Cistercian abbey of Cîteaux gradually acquired vineyards in Vougeot through bequests and purchases, eventually surrounding them with a wall and creating the largest single vineyard in Burgundy, the Clos de Vougeot (Seward, 1979). Along the Rhine, the Archbishop of Mainz in 1130 granted to the Benedictine priory of St Alban in Mainz the hill above Winkel where the monks built a priory and planted the vineyard later known as Schloss Johannisberg, and at the beginning of the twelfth century the Archbishop of Mainz also granted the vineyard of Steinberg near Hattenheim to the Cistercians of Kloster Eberbach (Seward, 1979). Throughout Europe there was a proliferation of vineyards belonging to the various religious orders, ranging from the tenth century vineyards of the Benedictine abbey of San Cugat del Vallés near Barcelona to those of the Knights Templars and Hospitallers in the islands of the eastern Mediterranean.

The particular importance of the vineyards of the monastic orders was threefold: they remained the property of a single owner for a considerable time, thus permitting continuity of cultivation and the opportunity for experimentation; the network of daughter houses and regular communication between monasteries of the same order enabled the rapid and widespread dissemination of any new methods of viticulture and vinification; and the monks were among the most educated people of the age, keen to experiment and develop their vineyards.

Wine as medicine

A further source of demand for wine came from its use in medicine, particularly by monks and the Knights Hospitallers. From its earliest days, wine had been used for medicinal purposes, and Hippocrates of Cos (c.460–370 BC), widely regarded as the most celebrated physician of antiquity, 'prescribed it as a wound dressing, as a nourishing dietary beverage, as a cooling agent for fevers, as a purgative, and as a diuretic' (Lucia, 1963: 36). Claudius Galenus, known as Galen, (c. AD 130–201), likewise made extensive use of medicated wines, and it was through the propagation of his works in the Byzantine period that the medicinal use of wine survived the collapse of the western Roman Empire. Galen's recommendations that wine was to be used for the dressing of wounds, for fevers and for debility came to be widely adopted in medieval Europe. It was, though, the *Liber de Vinis*, written by Arnaldus de Villanova (c.1235–1311) that 'firmly established the use of wine as a recognised system of therapy during the late Middle Ages' (Lucia, 1963: 101; Arnaldus de Villanova, 1943). Among an extensive list of recommended uses of wine, he saw it as being particularly useful as an antiseptic, a restorative, and for the preparation of poultices. Throughout the medieval period wine was one of the few liquids that, as a result of its alcohol content, could dissolve and disguise the taste of the substances that physicians considered to be beneficial. Theriacs, or medicated wines, thus came to be used extensively as cures for a wide range of ills.

THE MEDIEVAL WINE TRADE

Within the Mediterranean world, the increased urban demand for wine pushed the limits of viticulture well up into the mountains and hills, whereas the growing demand from northern Europe led to the spread of vineyards into environmentally favoured areas in northern latitudes. However, the period between 1000 and 1350 saw considerable changes in the organisation of medieval trade, which were to have important implications both for the production and the consumption of wine.

The context of medieval trade

In any discussion of medieval trade it is useful to distinguish both between the long-distance wholesale trade and local retail trade, and also between the great annual fairs and the regular weekly markets that were held in countless villages and towns throughout Europe (Postan and Miller, 1987). Across these distinctions, though, lay the division of Europe into two great commercial networks, on the one hand that of the North Sea and the Baltic and on the other that of northern Italy and the Mediterranean. According to Postan (1987: 169), 'the trade of northern Europe was almost exclusively devoted to the necessities of life'. Apart from furs, the main commodities of this northern trade were bulky and relatively cheap, such as grain, timber, dairy produce, fish, and wool. Already by the eleventh century regions like Flanders, which had food deficits, and the great cities, such as Paris, were being supplied with foodstuffs from regions with a surplus, such as the Rhineland and the lower Seine valley (Pounds, 1973). With the exception of grain and fish, however, 'no other comestible product was more indispensible to medieval diet, or was caried in larger quantities than wine' (Postan, 1987: 172). Such long-distance trade was undertaken almost exclusively by river, and Pounds (1973: 303) has argued that 'the availability of water was, in fact, a condition of the large-scale production of certain bulky commodities'. The trade of southern Europe, in contrast, reflected the commercial revolution that began in northern Italy during the tenth century (Lopez, 1987). It was a trade dominated by Genoese and Venetian ships travelling the sea-lanes of the Mediterranean, concentrating on high value, luxury goods, such as spices, perfumes and silks (Postan, 1987). The economic success of the northern Italian towns nevertheless required large imports of food and wool, and the sweet wines of the eastern Mediterranean were also regularly to be found in the cargoes of Italian galleys.

These two great networks were brought together by the needs of the Italian cloth merchants for high quality cloth and wool, and by the demands of the urban elites of northern Europe for the luxury products of the south. Between them lay the mountains of the Alps, and it was the opening up and improvement of the Mont-Cenis and the Grand St Bernard passes that enabled commercial links between north and south to be developed. The focus of this commercial linkage lay in the fairs of Champagne, which in the words of Bautier (1970: 42)

> constituted a basic economic pivot for the Western world: they were the periodic assizes for long-distance trade, an international center of exchange and credit, a meeting place of northern and southern civilizations, and a melting pot of new commercial techniques and new legislation.

These fairs at Troyes, Provins, Bar-sur-Aube and Lagny originated as local and regional markets, but by the end of the twelfth century they had become the hub of commercial exchange between the merchants of Italy and those of Flanders. Their evolution into international cloth and financial centres was, according to Bautier (1970), largely due to the activities of counts, such as Theobald the Great and Henry the Liberal, who established a range of trading regulations, and in so doing concentrated their efforts and attention on these four towns. Crucially important in their success were the free conducts issued by the counts to merchants on the way to the fairs. By the mid-thirteenth century, however, the role of the Champagne fairs shifted in emphasis, with financial transactions rather than the cloth trade becoming of most importance. As the nature of medieval trade changed in the fourteenth century, with the increasing use being made by Italian merchants of permanent agents in the towns of northern Europe, with the economic emergence of German towns such as Frankfurt and Köln, and the opening of the eastern passes across the Alps, the fairs of Champagne declined, to be replaced by trading practices which could provide for the exchange of goods that were increasingly in regular and continuing demand (Barth, 1958; Pounds, 1973).

While the great annual and periodic fairs of medieval Europe catered for much of the long distance trade of the twelfth and thirteenth centuries, the redistribution of merchandise and the gathering together of local produce, took place in the thousands of small weekly markets of the continent. In England and Wales, for example, grants were issued for the right to hold more than 2,800 markets between 1199 and 1483 (Britnell, 1981). Many of these markets were in small villages, and provided the opportunity for the peasantry to obtain goods, such as salt, cloth and iron which they were unable to provide for themselves. Such markets, though, also enabled the peasantry to sell what agricultural surpluses they had, in order to obtain the money for the taxes and dues which they owed to their lords and which were increasingly being demanded in cash rather than kind (Maddicott, 1975; Unwin, 1981, 1990). Despite the importance of such markets, though, not all trade was conducted through them, and

Donkin (1957), for example, has shown how many English Cistercian monasteries had long standing contracts with Italian merchants for the export of their wool, which they often had to take to the staple ports or other points convenient to the buyer.

Routes and rhythms of the medieval wine trade

Set against this background of markets and fairs, Postan has emphasised that 'The significance of wine in international trade was not only in the quantities in which it was drunk, but also in the conditions under which it came to be produced'. Moreover, he goes on to emphasise the crucial point that 'In the course of the centuries the commercial production of wine, once widespread over the face of Europe, gradually concentrated in regions of highly specialised viticulture' (Postan, 1987: 172). The emergence of such specialist regions reflected not only the environmental requirements of the vine, but also the patterns of demand and the availability of sufficiently cheap transport to which allusion has already been made. While in southern Europe there was a more or less constant flow of wine from the countryside to the town, and while by the thirteenth century the burgesses of northern Europe could obtain a wide range of wines from different origins, three great trade routes came to dominate the structure of the medieval wine trade (Figure 32). These were the westward flow of sweet wines produced in the eastern Mediterranean, the northward flow of wine down the Rhine, and the coastal transport of wine from northern Spain and western France to England and Flanders. From these three main areas of origin, wine found its way in varying amounts to almost all of the ports of northern Europe, although the balance in importance between the different routes varied considerably at different times between the eleventh and fifteenth centuries. Moreover, more than most types of commerce, the wine trade was dominated by a seasonal rhythm, based on the annual cycle of the vine and the inability of the light wines of northern Europe to survive for more than a year in barrel.

One indication of the quality and popularity of different wines in the early thirteenth century is provided by Henri d'Andeli in a poem known as the *Bataille des Vins* probably dating from *c.* 1223 (Lachiver, 1988), or according to Galtier (1968) from around 1240. This poem describes an imaginery review of wines placed before the French king Philippe Auguste, and their assessment by an English priest. Lachiver (1988: 103) suggests that this choice of the judge might in part reflect the view that the English were both good customers for wine as well as being good connoisseurs. The highest quality wines were deemed to be the sweet wines of Cyprus, but those from a number of other regions were also praised. Most of the wines competing for the accolade were from northern

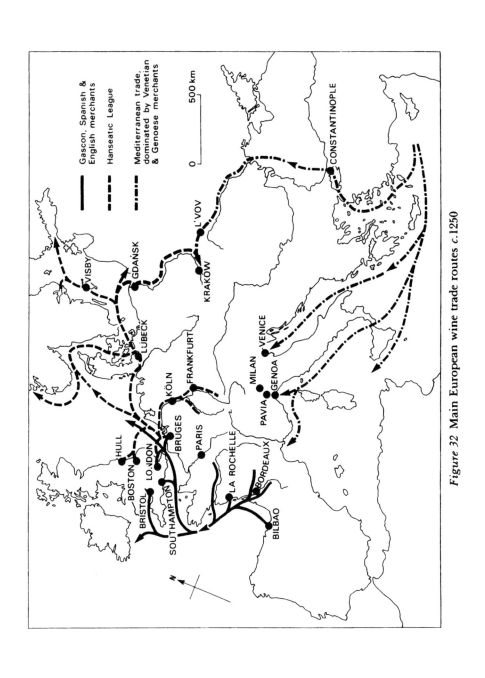

Figure 32 Main European wine trade routes *c*.1250

and western France, indicating the importance of these regions as suppliers of Paris and the French court, but some wines from the south, such as those of Montpellier and Narbonne were also mentioned. Among the best wines were deemed to be those of the Ile-de-France, from the Champagne region around Epernay and Hautvillers, and from Aunis in the west. Several wines, however, mainly from the north-west of France, were considered to be worthy of excommunication, and among these were those of Beauvais, Etampes, Châlons, Tours, Le Mans, Rennes, Chambilly and Argences, the last five of which were criticised because of their acidity. The poem also emphasises the importance of La Rochelle as an exporter of wines, with it being described as supplying wine for the English, Bretons, Normans, Flemings, Welsh, Scots, Irish, Norwegians and Danes.

England, Flanders and the Baltic countries provided the bulk of the demand for wine entering international trade during the medieval period, with much of the English demand being met by the wines of western France. The establishment of formal English links with Gascony dates from Eleanor of Aquitaine's divorce from Louis VII of France, and her marriage in 1152 to Henry, Duke of Normandy and Count of Anjou, Maine and Touraine, who in 1154 became King of England. As part of her dowry, Eleanor brought with her the Duchy of Aquitaine, which included Poitou, Guyenne and Gascony, and rather than buying wines from the annual fair at Rouen as had been the previous practice (Renouard, 1970; Francis, 1972), the English thereafter turned to the ports of Nantes, La Rochelle and Bordeaux for their wine. It is also from this period that wine production in England declined, and it seems that the more regular, reliable and easier production of wine in south-west France gradually made it increasingly uneconomic for English wine producers to compete (Carus-Wilson, 1954). Until 1224 when Louis VIII captured Poitou from the English, it was wines exported from La Rochelle, rather than from Bordeaux, that dominated the trade routes to the north (Figure 32). Renouard (1970) suggests that one of the reasons for this English preference for the wine of Poitou may have been that Eleanor and her son Richard both came from Poitiers, and therefore preferred the wines from there to the more southerly wines of Gascony. However, from 1224 onwards the wine trade between England and Gascony increased apace so that by the beginning of the fourteenth century England was receiving about half Bordeaux's total wine exports of approximately 80,000 tuns a year (Higounet, 1962–72; Francis, 1972). Most of the remainder of Bordeaux's exports were directed to other parts of France, Flanders, and northern Germany, although Renouard (1970) notes that wines from Bordeaux were also exported to Spain and were consumed by the Castilian armies during their campaigns of reconquest against the Moors. Furthermore, with the establishment of trading links between

Gascony and Castile, Castilian wine found its way into the northward flow of trade to England and Flanders (Huetz de Lemps, 1968). Childs (1978: 126) thus notes that the first mention of Spanish wines in England dates from 1227–8, and she argues that during the thirteenth century most of the red and white Spanish wine imported into England was probably from Castile and the Basque country (Figure 32).

The wines of Germany were dominated by those produced from the vineyards along the Rhine and the Mosel (Bassermann-Jordan, 1907; Barth, 1958; Higounet, 1968; Weiter-Matysiak, 1985; Enjalbert and Enjalbert, 1987), and during the thirteenth century several hundred tuns a year of such Rhenish wine were regularly imported into England (Francis, 1972). The term Rhenish was also used to describe the wines from Alsace, and at least by the fourteenth century these had gained a reputation as being the most prestigious of all German wines (Sittler, 1949). Situated along the banks of this great trade artery, the Rhenish vineyards had easy access to the heart of the commercial world of northern Europe. Moreover, as the wider commercial balance shifted away from the fairs of Champagne to the towns of northern Germany, the vineyards of the Rhine and the Mosel were well placed to benefit. Unfortunately, 'Lack of data makes it impossible to trace the evolution of the traffic in Rhenish wine' (Dollinger, 1970: 226), but it is nevertheless evident that it was dominated by the two towns of Köln and Frankfurt (Figure 32) (Barth, 1958; d'Haenens, 1984). Of these, the merchants of Köln were the most important, buying up wine on the spot and then exporting it throughout the Hanseatic network to Flanders, England, Hamburg and the Baltic. Frankfurt appears to have been most important for the wines of Alsace, which were then exported in large quantities northwards and eastwards to Lüneburg, Lübeck and thence to the Baltic (Dollinger, 1970). Wines from western Europe reached as far east as Poland, and Carter (1987), for example, has noted that by the end of the thirteenth century Kraków was obtaining wine from Hungary, Moldavia, and the Rhine. Sweet wines from the eastern Mediterranean also reached Kraków via the Black Sea, the river Dniester and Lvov (Figure 32).

In contrast to the generally dry, light wines of Gascony and Germany, the wines of the Mediterranean, were sweeter, heavier and higher in alcohol, with the result that they travelled better, lasted longer and were thus more valuable. The best known of these wines were malvoisie or malmsey, and romney or rumney. The former was initially produced in the Pelopponese and exported from the Byzantine town of Monemvasia, but by the thirteenth century it seems to have been produced much more widely, and was mainly exported from Crete. The latter also originated in what is now mainland Greece, but by the thirteenth century it was produced principally in the Ionian islands (Allen, 1961; Francis, 1972). During the thirteenth century small amounts of such wines reached northern Europe, either brought back from the crusades or shipped to southern France and thence transported northwards. However, the Islamic domination of Iberia and the difficulties of the long sea voyage meant that it was not until the end of the thirteenth century, following the Christian reconquest of Spain and Portugal, that the sea route from Italy to northern Europe through the Strait of Gibraltar was established. Although a Genoese vessel had reached La Rochelle in 1232 (Ruddock, 1951), it was not until the last quarter of the century that Genoese oared galleys began to appear in any number in northern Europe in the prospect of returning with cloth from Flanders.

The traditional Mediterranean galley was poorly equipped for navigation in the Atlantic unlike the smaller Gascon and English ships, but in the second decade of the fourteenth century the Venetians introduced convoys of larger galleys which regularly undertook the five month voyage to Flanders and England (Ruddock, 1951; Francis, 1972). Such convoys consisted of four or five galleys, each with a capacity of about 140 tuns, and although the early cargoes were mainly of alum, they soon came to carry a range of high-value goods such as spices and silks. Wine appears only to have been carried in order to complete the cargo, but, as Francis (1972: 17) points out, the sweet Mediterranean wines, which travelled better than the light wines of Gascony, could also bear the increased freight costs. During the fourteenth century, with the establishment of a regular trade route around Spain and Portugal, the malmsey and romney wines of the eastern Mediterranean islands became increasingly popular in England. Moreover, the firm establishment of viticulture and wine making in Iberia following the expulsion of the Moors soon led to the development there of similar sweet wine production. This tendency was reinforced by the expansion of the Ottoman Empire in the eastern Mediterranean, which through the fourteenth and fifteenth centuries gradually led to the Islamic conquest of mainland Greece. This had the effect of restricting wine exports to Christian Europe from this region, but it is unclear whether this was the result of vineyard destruction, or rather a reorientation in the trade. Some wine production and export certainly continued from the islands of the eastern Mediterranean, until the conquest of Rhodes in 1522, Cyprus in 1571 and then Crete in 1669, eventually curtailed this source of sweet wines for the northern European market. The overall effect of the expulsion of the Moors from Spain, and the expansion of the Ottoman Empire was to lead to a shift in the source of supply of such sweet, high alcohol wines away from the eastern Mediterranean and towards the Iberian peninsula. By the end of the fourteenth century, Spanish wines were popular in England (Childs, 1978), and Chaucer's account in the Pardoner's Tale, indicates not only the strength of the wine from Lepe near Huelva in southern Spain, but also the common practice of its adulteration:

Keep clear of wine, I tell you, white or red,
Especially Spanish wines which they provide
And have on sale in Fish Street and Cheapside.
That wine mysteriously finds its way
To mix itself with others - shall we say
Spontaneously! - that grow in neighbouring regions.
Out of the mixture fumes arise in legions,
So when a man has had a drink or two
Though he may think he is at home with you
In Cheapside, I assure you he's in Spain
Where it was made, at Lepé I maintain.

(Chaucer, 1977: 265)

Although both red and white wines were exported from Spain, it was the white, sometimes known as Rubidege, that predominated. Much of the wine that entered the export trade came from Galicia and the Basque country in northern Spain, and most of this was probably quite dry and light, but it was the sweet Spanish wines, known as romney, malmsey and bastard, that eventually proved to be most popular (Huetz de Lemps, 1968; Childs, 1978).

As well as being spatially concentrated, the medieval wine trade was also highly seasonal, with its own very specific annual rhythms. These were largely determined by the inability of wine producers to make a product which remained drinkable for more than a single year, but also by the dangers of navigation during the winter months. The key date in the wine merchant's year was the vintage, which in France and Germany usually took place towards the end of September or the beginning of October. However, its precise date was very variable, and in the most comprehensive analysis of vintages from north-east France, French Switzerland and the south Rhineland Le Roy Ladurie and Baulant (1980) have illustrated that for the later period between 1484 and 1879, for which detailed information survives, the average annual date at the beginning of the grape harvest varied from 1st September in 1556 to 25 October in 1816. The new wines were usually available on the market within a few weeks of the vintage, and annual wine fairs were to be found throughout all of the major wine producing regions during October or early November. Typical among these was the fair at Compiègne 'at which the wines of the Paris basin were sold to merchants of the Low Countries' (Pounds, 1973: 404). In her analysis of the Anglo-Gascon wine trade, however, James (1971: 124) has illustrated that, although during the early fourteenth century 'Bordeaux customs registers show that ships habitually arrived any time from late August up to April in the following year' in order to collect wines, two periods dominated the dates of arrival of the bulk of the ships (James, 1971: 124). Most arrived in October or early November, soon after the harvest, in order to take the new vintage wines back to northern Europe in time for Christmas, but there was also a second season in March, when ships arrived to collect the *reek* wines that had already been racked and taken off their lees. A similar seasonal pattern of trade was encountered with the German wines, although it appears that this was less marked than with the wines of Gascony. The Italian fleets, bringing with them the sweet wines of the eastern Mediterranean and Iberia, usually arrived in northern Europe in July and August (Ruddock, 1951), by which time much of the dryer and lighter wine, if it had not already been consumed, had begun to go sour. While wine prices were closely regulated, these seasonal fluctuations in availabilty of wine were nevertheless important influences on the prices that merchants could command for their products, and if good wines were available during the summer months when supplies were normally low, they could often be sold at considerable profit.

Apart from the costs of production and the balance between supply and demand, wine prices were also significantly influenced by trading costs and levels of taxation. James (1971) has classified the costs involved for wine importers into four categories: the freight charges, cooperage expenses, cellarage costs, and labour costs. Of these the freight charges formed the most significant element, varying significantly depending on the season and the level of warfare prevailing on the high seas. During the early fourteenth century, at a time of relative peace, the average wholesale price of wine in England was about £3 a tun, whereas with the collapse of supplies during the One Hundred Years War and the need to protect the wine fleet against piracy and seizure these rose to £8 a tun during the 1370s. Freight, pilotage and carriage charges all varied with the time of the year and the safety of the seas. The greatest element of transport costs was accounted for by the freight charges, and for the journey between Gascony and England during the early fourteenth century these were usually in the order of 6s. to 10s. a tun, representing between one-tenth and one-sixth of the wholesale price (James, 1971). Pilotage charges were an additional element to be paid, and for the journey from the Gironde to the southern ports of England at the beginning of the fourteenth century these amounted to approximately £1 for each ship. Such charges were levied at each port of call, and this therefore added considerably to the costs of coastal shipping. The cost of a pilot to navigate a ship from Bordeaux out into the Gironde was thus between 2s. and 3s., and once a ship entered the Thames the pilotage charge up to London amounted to between 7s. and 8s. (James, 1971: 140). Furthermore, charges for stowage on board ship cost around 4d. a tun, and where ships could not unload directly onto the quay a batellage charge was levied to cover the lading of the wine onto small boats which then took it to the shore, and in London this came to about 2d. a tun.

188

Once ashore the ravages of the sea journey had to be amended, with damaged barrels being repaired by skilled coopers and carpenters. The vulnerability of wine in transit, as described by James (1971: 139), made it a costly product:

> The transport of wine even apart from freight and carriage charges was always expensive, for it was both bulky and heavy, valuable and very fragile; leakage and evaporation were always likely, especially if the journey was rough or the weather too warm, and the vessels had always to be filled up to their proper level at the end of the journey.

As in Roman times, the cost of road transport was far higher than that by sea, and by the fourteenth century in Britain some attempt had been made at regulating the amount that retail wine prices could be increased in order to reflect the distance of sale from a port. Thus in 1330 retailers could increase the price of wine by ½d. a gallon for every thirty miles of carriage and by 1d. for every 54 miles. Twenty-four years later in 1354 these prices were raised to ½d. a gallon for twenty-five miles and 1d. for every fifty miles (James, 1971).

Wine merchants and commercial organisation

The nationalities of merchants involved in the wine trade reflected both the areas of wine supply and its demand (Figure 32). While most wine in the twelfth and thirteenth centuries was transported by merchants from the producing areas of Gascony, Germany and Italy, the balance shifted during the fourteenth century so that increasingly, English and Flemish merchants were to be found buying wine overseas and bringing it back to their home countries. This growing domination of commerce by English merchants was also to be seen in their increasing importance in the wool and cloth trade (Simon, 1906–9; Gray, 1933; Carus-Wilson, 1954).

The opportunities that the wine trade provided for taxation, have led to the survival of a considerable amount of information not only concerning the volumes of wine traded during the medieval period, but also about the merchants involved in that trade. While the early history of customs duties is unclear, the records surviving from English ports from the early fourteenth century onwards provide a wealth of detail about all aspects of the trade. The earliest medieval taxes on wine, such as the *modiato* of Rouen dating from 1055, and the later *prisa* of England, were essentially 'instruments for the supply of the sovereign's table' (Gras, 1918: 14), but such taxes gradually emerged as a more general method of increasing the income of the Exchequer, through the commutation of a seizure of wine into a money payment for the right to import wine. By the middle of the twelfth century a general wine custom seems to have been imposed on wines imported into English ports. From the thirteenth century this varied in amount from 2*d.* to 6*d.* a tun, mainly reflecting local influences (Gras, 1918). Of more importance was the prisage or wine prise, which by the end of the twelfth century had developed in England into a formal right, known as the *recta prisa.* This required from each ship the 'seizure by the crown of two tuns from a cargo of twenty tuns or over, on the payment of 20*s.* a tun to the owners' (Gras, 1918: 41). One tun was to be taken from behind the mast, where the better wine was stored, and the other tun from in front of the mast. This enabled the king to buy the best quality wine at less than its market price, although if the cargo was between ten and twenty tuns only one tun would be seized, and if the cargo was less than ten tuns no wine was to be taken. In the exceptional case of the port of Bristol the sum to be paid was only 15*s.* a tun.

Within northern Europe during the twelfth and thirteenth centuries the wine trade was dominated by the Gascon and Hanseatic merchants. The former maintained a virtual monopoly of the English market, and the latter mainly supplied northern Germany and the Baltic. Typically, the activities of all foreign or alien merchants were restricted. Thus the Gascon merchants were limited to a stay in England of only forty days until 1280 when this period was extended to three months (James, 1971: 71). These Gascon merchant vintners were men of varying position. Some were individual vintners, who took their wines for sale to England following the vintage and then returned rapidly to prepare their vineyards for the next season's vintage , but more usually a single Gascon merchant would represent the interests of many vine growers and wine producers. Individual Gascons usually focused their attention on a particular region of England, arriving at the great seasonal fairs in order to sell their new wines as rapidly as possible, and in years with an early vintage the first merchants might arrive by the end of September (James, 1971: 73). Among the most important fairs for the Gascons was that of Boston, to which vintners and taverners from all of the major towns of eastern England would gather annually in order to purchase the new wines. However, Gascon merchants would also travel widely themselves throughout England, from Cornwall in the south to Northumberland in the north, seeking out markets where the price was good. The largest Gascon community in England, though, was in London, where they generally hired cellars in the Vintry for the duration of their sojourn.

The second great group of wine merchants were those of the Hansa, based initially at Lübeck and Visby (Figure 32). Their power emerged with the foundation and growing strength of the new towns of northern Germany during the second half of the twelfth century (Dollinger, 1970; Lindquist, 1985; Schildhauer, 1985). The main commodities carried by the Hanseatic merchants were salt, amber and furs, and later, following the opening up of new lands in eastern Germany, they maintained a virtual monopoly of the grain trade in northern Europe. An outstanding feature of Hanseatic trade was the variety of products it handled, and wine as the drink of the urban elite, formed an important part of this trade. As Lerner (1984: 136) has argued, 'Everywhere in their own towns, but also in the herring markets of Schonen, Bergen, Königsberg, in the Baltic countries and even in Russia, the Hanse merchants did a wholesale trade in wine and, where they had acquired the necessary privileges, sold retail as well'. Within northern Germany it was the merchants of Köln and Frankfurt that came to dominate the wine trade, and even after the decline of Hanseatic involvement in the Rhine trade in the fifteenth century they managed to maintain their position (Higounet, 1968). They were active in Brabant and Flanders, and in 1157 Henry II granted the merchants of Köln the privilege of selling wine in London (Dollinger, 1984: 223). Here in 1281 the German merchants from Köln, Westphalia and the Western Baltic united to form a single federation or Hansa based on the Steelyard or Stalhof (Schildhauer, 1985), which formed the base of their operations in England until Elizabeth expelled the Hanseatic merchants and ordered its closure in 1598 (Lewis, 1984: 245). Hanseatic merchants, though, not only dealt in the wines of the Rhine, and during the fourteenth and fifteenth centuries they were to be found in France and Spain, particularly at La Rochelle, where wine was taken on board 'as a complementary product for the salt fleets which did not always return fully laden' (Abraham-Thisse, 1984: 235).

From the end of the thirteenth century, the arrival of Italian merchants by sea in northern Europe added a third element to the merchant groups trading in wine. Although their main interest was the acquisition of large supplies of good quality wool, the Italians brought with them the luxury products of the Mediterranean, including wine, in exchange. The trading policies and practices of the different Italian cities varied appreciably over the centuries, with the early individualistic Genoese merchants being replaced in importance during the fourteenth century by the Venetian state fleets which sailed to Europe under the command of a single captain. These Venetian convoys would split into two in the English Channel, half sailing on to Flanders and the others going to London (Ruddock, 1951; Francis, 1972), although later in the fourteenth century the goods for London were usually landed at Sandwich or Southampton. As Italian control of the wool and cloth trade passed increasingly into English and Flemish hands during the later fourteenth and fifteenth centuries, they paid greater attention to the wine trade, and from 1434 the town of Southampton became the main gathering point for the Italian fleets. For sixty years thereafter Southampton virtually became the staple port for Mediterranean wines such as malmsey and romney, as well as the main port for the vernage or vernacchia wines produced around Florence (Simon, 1906–9; Ruddock, 1951; Francis, 1972).

POLITICAL ALLIANCE AND THE MEDIEVAL WINE TRADE

One of the key features of medieval trade was the interaction of economic and political structures in determining its fortunes. This is nowhere better to be seen than in the fluctuations of the Anglo-Gascon wine trade in the fourteenth and fifteenth centuries, for which detailed customs accounts from both Bordeaux and the English ports survive (Simon, 1906–9; Higounet, 1968; Renouard, 1970; James, 1971; Lachiver, 1988). While it is usual to argue that it was political alliances that determined the economic fortunes of the wine trade, the evidence for this is by no means conclusive, and it is equally apparent that economic considerations themselves underlay much political manoeuvering during the medieval period. This is well exemplified by the introduction of a range of legislation in England associated with the *Carta Mercatoria* of 1303.

Alien merchants in England and the *Carta Mercatoria* of 1303

Based on an analysis of Chapter 41 of *Magna Carta*, dating from 1215, concerning the rights of foreign or alien merchants trading in England, Lloyd (1982: 9) has argued that

> The welcome extended by early medieval Englishmen to alien merchants visiting their shores was decidedly mixed, but if there was a guiding principle it was probably that strangers should be encouraged to come here, although their activities should be strictly controlled once they arrived.

From the end of the thirteenth century a new royal policy emerged and this began to favour aliens more than denizens, particularly with respect to commercial activity in the city of London. In July 1285 Edward I, partly in an attempt to overcome the restrictive trading practices of the citizens of London, replaced the mayor with a royal warden who was charged with the task of ensuring that alien merchants were able to trade more easily within the city. There was considerable opposition by the merchants of London to such action, though, and in 1298 the king was obliged to restore the mayoralty. In this struggle, the London merchants were particularly wary of the activities of the Hansa and the Gascons, whose economic power had been growing throughout the thirteenth century.

The struggle with the Gascon merchant vintners culminated in the grant to them by the crown of the so-called Gascon Charter in 1302 (Hall, 1896: 1060-4). This allowed the Gascons safe conduct throughout the king's realm, permitted them to trade wholesale, and allowed them to dwell where they wished and to keep their own hostels. Moreover, on the arrival of the new vintage, all existing wine stocks were to be tested by juries and any which had deteriorated were to be destroyed. The king also released the Gascons from the *recta prisa*, and in exchange they agreed to the payment of a new duty of 2s. on every tun of wine which they imported (Lloyd, 1982). Five months later in 1303 these privileges were extended to the whole alien community in the *Carta Mercatoria* which also introduced a number of new clauses. In particular the *Carta Mercatoria* added a clause which stated that alien merchants should be free of murage, pontage and pavage throughout the king's realm, and another clause which introduced the king's weights and measures to all towns in the country. In exchange a new custom on all merchandise imported or exported by aliens was introduced. For wine this was the 2s. a tun mentioned in the Gascon Charter, which later became known as butlerage, but for wool and hides a new tax of 3s. 4d. on each sack of wool or 300 woolfells, and 6s. 8d. on each last of hides was to be paid. For cloths of assize duty was to be paid at the rates of 2s. for scarlet cloth or cloth dyed in grain, 1s. 6d. for cloth partly dyed in grain, and 1s. for those without grain. For wax the custom was to be 12d. per quintal, and for all other goods a general custom of 3d. in the pound was to be levied (Gras, 1918; Lloyd, 1982). The immediate fortunes of the *Carta Mercatoria* fluctuated, with petitions that it caused an increase in prices leading to its suspension apart from the clauses concerned with wool and cloth in 1309. Within a year, though, it was restored until 1311 when the New Custom was abrogated by the Lords Ordainers. This change came at a time of considerable bad feeling towards the aliens, and it is from this date that their position in English trade can generally be seen to have worsened. Although the new custom was reconfirmed in 1322, by the middle of the fourteenth century English merchants were increasingly gaining in power (Carus-Wilson, 1954). Most of the clauses of the New Custom eventually ceased to be applied by the seventeenth century, but the butlerage on wines only came to an end in 1809.

English and Gascon merchant vintners

By the 1330s the activities of Gascon merchant vintners in England had diminished to such an extent that the king's butler was called to account for the decline in revenue from the New Custom on wine. In explaining this downturn he argued that the Gascon merchants were increasingly turning to Normandy, Picardy and Flanders, rather than to England in order to sell their wines. While it seems that English merchants accounted for no more than a quarter of wine imports from Gascony at the beginning of the fourteenth century, James (1971: 81) has noted that by the end of the 1330s 'the citizens of London alone were importing almost as much as the entire group of Gascons each year, and the King's butler stated that denizens as a whole were importing twice as much as the Gascons'. Judged by the large number of petitions by Gascons against Londoners who regularly seized their cargoes and used physical violence against them, it is evident that the denizens were resorting to all possible measures to overturn the advantages that aliens held as a result of the Gascon Charter and the *Carta Mercatoria*. By 1327 the London merchants gained exemption from the royal prise on wines, and from this date they had the upper hand over the Gascons. While in the political sphere the crown continued to reaffirm the privileges granted to the Gascons, in the economic context they were encountering increasing difficulties in their commercial activities. Eventually, as James (1971: 83) concludes, 'the Gascons came to doubt whether England was any longer as desirable a field for their activities as it had been formerly'.

In their place English merchant vintners came to dominate both the wholesale import trade and the local retail trade in wine (James, 1957). Few were men of the highest eminence, and for most wine formed only part of their commercial enterprise. As the region around Bordeaux came to specialise increasingly in wine production, it provided a ready market for the cloth and grain produced in northern Europe, and English merchants were well positioned to supply such merchandise in exchange for wine. The grain was often purchased from the Baltic, but tin, fish and cloth supplies were readily available in England. Indeed, James (1971: 168) has argued that Gascony was 'at first the greatest overseas market for the expanding native cloth industry in the fourteenth century'. Typical of the London merchants dealing in wine imports at the end of the fourteenth century was Gilbert Maghfeld. Based on the survival of his account books, James (1971) has shown how this ironmonger, who concentrated mainly on importing iron from Bayonne and Bilbao, also dealt in a wide range of other merchandise such as wine. Thus, during the 1390s Maghfeld primarily exported grain and cloth to Bayonne, and on the homeward voyage his cargoes consisted of wine, beaver, saffron, licorice and other local products. Maghfeld's interests were, however, much wider than this would suggest. James (1971: 200) thus notes that

> The known range of his imports included wax, linen, copper, millstones, small quantities of green ginger, woad from Genoa and even asses from Spain; exports of cloth were exchanged against herring in the great mart of Skania, and in partnership with John Hill, a fishmonger of London, he shipped his imports of herring down from Scarborough to Sandwich. Supplies of wine, normally

obtained in the course of his Bayonne trade, were also supplemented
by purchases of the white wines of La Rochelle . . .

While much of the wholesale wine trade was in the hands of a variety of
different types of merchant, it was the vintners who dominated the retail
trade. The origins of the Vintners' Company, which received its charter of
incorporation only in 1437, are obscure, but by the early thirteenth century
vintners played an important role in the administration of the city of
London (Crawford, 1977). During the fourteenth century they gradually
acquired greater control of the London wine trade, and in 1364 Edward III
granted to those enfranchised in the Mistery of Vintry a monopoly to buy
and sell wine from Gascony, which formed by far the largest part of the
English wine import trade. While aliens could still participate in the
wholesale trade, and by the following year the Mistery's purchase
monopoly was broken, the retail trade in Gascon wines thereafter came
under the exclusive control of the London vintners (Simon, 1906-9;
Crawford, 1977). Moreover, all London's wine imports were to be landed
on the west side of London Bridge and in the Vintry, thus reinforcing the
control of the London vintners over the wine trade as a whole.

The retail wine trade in northern Europe in the fourteenth century
could be classified into three broad categories: the supply of the great
royal and noble households, the general provision of lesser households,
and the sale of wine in taverns. The English evidence provides good
examples of all three types of trade. In many instances the great nobles
and ecclesiatics, as well as the crown, purchased their wines overseas,
with such imports being free of custom or subsidy (James, 1971: 178).
Such direct importation seems to have grown during the fifteenth
century, reflecting the increased importance of English merchants in the
wine trade at that time. The crown in addition purchased considerable
quantities of wine at the major ports of London, Hull, Southampton and
Bristol. It was also from the cellars of the merchant vintners at these ports
that the butlers of noble and ecclesiastical households made most of their
purchases. The wealthiest members of medieval society were thus able to
purchase wine in the more distant markets, where it could most profitably be found, and in so doing they were able to avoid the additional costs
involved in purchasing from intermediary retail merchants. Wine however travelled throughout the country, continually being bought and
resold against other merchandise. Where possible the wine was
transported by sea, with the east coast trade being particularly significant
(Carus-Wilson, 1962-3). As James (1971: 189) has noted, given the
cheapness of transit by water, 'the distribution of wines by sea to the
eastern half of England could have cost little more than the overland
distribution to the counties immediately around London'. From these
smaller ports, and the major markets of the interior regions, the lesser

households purchased their wines in bulk. At the bottom of the retail hierarchy were the taverners, selling wine in gallons, pottles, quarts or pints from the tun or barrel. Given the opportunity of adulterating such wine, in order to make it go further, and to hide the taste once it had begun to go sour, the retail trade was closely regulated in all of its aspects. The price of wine was fixed by the assize courts, which regularly determined the retail price at which wine could be sold depending on its general scarcity or abundance. Moreover, taverners were not allowed to sell their wines until they had been assayed. Within London,

> The assayers were eight or twelve men chosen by the mayor and aldermen from among those vintners of greatest experience, and it was their duty to assay the wine in all the London taverns and have each tun marked at its value according to the current price fixed for its type. The mark had to be at the front of the tun so that the buyer could see it, and each tavern customer had the right to see his own wine drawn.
>
> (Crawford, 1977: 28)

The court records of most towns and cities indicate that although such regulations existed they were widely flouted, and given the vast number of taverns this is scarcely surprising; in 1309 London had 354 taverns. Adulteration was treated as the most serious offence, and a proclamation against counterfeit romney wine dating from 1419 issued by the mayor and aldermen of the city of London fully encapsulates the problem.

> for some time both English and aliens, to the common harm of all and the scandal of the city, have been putting wine of Spain and Rochelle and other remnants of broken, sodden, reboiled and unthrifty wine of other countries, which are enfeebled in colour and nothing in value, in various butts and other vessels that are scraped to make resin adhere, gummed with pitch, cobblers' wax and other horrible and unwholesome things in order to reduce and bring again a pleasant colour and a likely manner drinking of Romeneye to smell and taste, to the deceit of all. And not only to put an end to this practice, but also to remedy the great multitude of such wine deceptively counterfeited and medeled with abroad and brought here to sell, called Romeneye, but counterfeited of Spanish and Rochelle, the mayor and aldermen now ordain that no-one within the city or liberties is to sell a gallon of such counterfeited wine on pain of having all his wine and vessels forfeited, and no-one is to meddle with wine in any way, no white with red, old with new, whole with broken or corrupt, Rhenish with Rochelle, or any different kind of growths together on pain of the pillory.
>
> (Sharpe, 1894–1912, *Letter Book I*: 227)

The Anglo-Gascon wine trade and the politics of war

The importance of the producton of wine in Gascony in the fourteenth and fifteenth centuries, and the potential it generated for taxation revenue for the English crown, created a wealth of documentary evidence which has provided the basis for considerable research on the fluctuating fortunes of the Anglo-Gascon wine trade (Higounet, 1962–72; Renouard, 1970; James, 1971; Pijassou, 1980; Lachiver, 1988). Carus-Wilson (1954: 269–70) has well described the ferment of viticultural activity prevailing in Gascony:

> Here in the late twelfth and thirteenth centuries vineyards multiplied, spreading over more and more of the countryside. Forests were cleared to make way for them. The "palus" were for the first time brought under cultivation that they might be planted with them. Vineyards came close up to the very walls of Bordeaux, extending even into the city itself. Everyone, high and low, whatever his rank or profession, possessed vines (the great abbeys of St Seurin and of St Croix planted their lands with little else), and for all, from the archbishop downwards, the sale of their wines to the English became the dominant business of life.

From the middle of the thirteenth century the region around Bordeaux was divided into two districts: the archdiocese of Bordeaux which formed the Bordelais proper, and the *Haut Pays*, or high country to the east of a line joining St-Macaire on the Garonne and Castillon on the Dordogne. The importance of this distinction for the wine trade was that it formed the basis for privileges granted by the English crown to the merchants of Bordeaux. The heaviest tax on wine exports was the *Grande Coutume*, or Great Custom, of Bordeaux, which had to be paid on wines produced in the Bordelais by all exporters other than the burgesses and merchants of Bordeaux. Most towns and districts in the Haut Pays were nevertheless privileged to export their wines at a reduced rate of custom. Although there are considerable problems in interpreting the evidence of customs documents, Table 4 illustrates the broad trend of wine exports from Gascony in the fourteenth century. The customs evidence makes it possible to differentiate between the wines of Bordeaux exported by burgesses and non-burgesses, and between the privileged and non-privileged wines of the Haut Pays and Bordelais. The merchants of Bordeaux were in a particularly advantageous position because, in Renouard's (1970: 75) words,

the high country wines could not be brought to Bordeaux before a date varying between November 11 and December 25; this meant that these wines could not compete with the low country wines produced for the most part on the lands of the Bordeaux merchants, and ordinarily were not sold until the low country supply was exhausted.

By the beginning of the fourteenth century, at which time the export of wines from Gascony was at its peak (Table 4), the privileges granted to the burgesses of Bordeaux had led to them establishing vineyards throughout the archdiocese at the expense of grain production, generating the necessity of grain imports and thus a flourishing trade between Gascony and northern Europe.

The vineyards of the archbishop of Bordeaux at Lormont and Pessac were probably among the best cultivated in the region and produced the highest quality wines for the archbishop's own table (Renouard, 1970). However, the most widely known wines were often those from monastic vineyards in the *Haut Pays*, from places such as Gaillac, Moisac, Bergerac and Cahors. As Table 4 indicates, over half of the wines exported from Gascony at the beginning of the fourteenth century thus came from the *Haut Pays* and 10–20 per cent came from the vineyards of the burgesses of Bordeaux. Although the levels of exports can be ascertained with some accuracy it is not possible to distinguish the destinations of these wines with any precision. England formed the main market, but wine was also exported to northern France and Flanders, whence some of it was re-exported to the Baltic or even across the Channel to England.

The fortunes of this trade partially reflected the annual variability in the quantity of the vintage, but with the onset of the Hundred Years War between England and France the influence of political events both on wine production and the prices of wine in England becomes markedly apparent. The first ten years of the fourteenth century were prosperous ones for the wine merchants of Bordeaux. Despite the problems associated with the interpretation of the customs data, it appears that exports were in the region of 90,000 to 100,000 tuns a year, and the wholesale price of wine in England was usually about £3 a tun. Retail prices ranged from between 3*d.* and 4*d.* a gallon depending on the season and availability of wine. However, in the second decade of the fourteenth century, a series of bad harvests led to a halving of exports in 1310 and 1311, and although there was a recovery in the 1312 vintage, Gascony suffered further from the poor weather, famine and disease that was encountered throughout north-west Europe between 1315 and 1317 (James, 1971; Kershaw, 1973). This was reflected in the prices paid by the king's butler for wine in England, which rose to £5 6*s.* 8*d.* in April 1311 (James, 1971: 11) before falling to £3 again in 1312. Retail prices were somewhat less volatile, although the best quality wines were fixed by the

Table 4 Wine exports from Gascon ports during the fourteenth century

Year Michaelmas to Michaelmas	Wines of Bordeaux Burgesses, nobles, ecclesiastics	Bordeaux non-burgesses[1]	non-privileged wines of the *Haut Pays* and Bordelais	privileged wines of the *Haut Pays*	Total[2] (tuns)
1305-6	13,958		17,956	57,934	97,848
1306-7	13,886		16,034	53,591	93,452
1308-9	12,260		30,947	38,812	102,724
1310-11					51,351
1323-24			6,234	32,305	
1328-29					69,175
1329-30					93,556
1335-36	7,958		14,136	46,901	74,053
1336-37	5,447		2,979	4,645	16,557
1348-49	867		4,586	470	5,923
1349-50					13,427
1350-51	7,282				
1352-53	10,927		8,702	0	19,629
1353-54	8,627		7,659	42	16,328
1355-56	6,698		7,713	0	14,411
1356-57	8,900		11,159	141	20,200
1357-58	10,506		15,559	1,773	27,838
1363-64					18,280
1364-5					43,869
1365-66					36,207
1366-67					37,103
1368-69					28,264
1369-70					8,945
1372-73	5,535	20	8,720	98	14,373
1373-74		605	7,099	76	
1374-75	3,080	323	4,527	0	7,930
1375-76	2,437	1,769	3,522	625	8,656
1376-77	9,636	3,138	10,761	110	23,920
1377-78	6,679	668	5,109	0	12,456
1378-79	7,597	525	5,500	0	13,622
1379-80	2,973	356	2,805	126	6,643
1380-81	4,584	614	3,474	107	9,041
1387-88		397	6,988	120	
1388-89		648	9,205	50	
1389-90		605	5,082	130	

Notes: [1]Non-burgess wines of Bordeaux were those levied at the rate of 30s. (Bordeaux money) a tun.
[2]The total amount includes wines laded at Libourne, Bourg, Blaye and other ports of the Gironde below Bordeaux. The accounts of the *Grande Coutume* are found in the Public Record Office, London, series E101.
Source: derived from James, M. K., *Studies in the Medieval Wine Trade*, edited by E. M. Veale, Oxford: Clarendon Press, 1971: 32-33.

London assize at 5*d.* a gallon in 1311, and in 1316 they rose to 6*d.* a gallon. These fluctuations brought about mainly by the weather were short lived, however, and by 1320 the retail prices of wine in London had fallen back to 3d. a gallon for the best quality and 2d. for the less good quality.

The outbreak of war between England and France in 1324, when much of the *Haut Pays* was in rebellion against the English crown, led to a further decline in exports to England, but with the onset of peace in 1327 the trade apparently recovered, and by 1329-30 Gascon wine exports had risen again to 93,556 tuns (Table 4). The resumption of hostilities in 1337, at the outbreak of the Hundred Years War, was however far more serious, and total exports for that year were less than a quarter of the level in 1335-6. As Table 4 illustrates, the effects were felt most seriously in the *Haut Pays*, whose exports were only one-tenth of their previous levels. Not only were production levels decimated as vineyards were destroyed, but the wine fleets were also liable to increased harassment. In James's (1971: 16) words, 'As the dangers of the sea became increasingly manifest England was forced to organize the defence of her wine-fleets lest her sea-borne trade with Gascony should become paralysed'. A convoy system was introduced, and this had the effect of limiting the periodic shipments of wine that had previously entered England outside the usual arrivals of the vintage and *reek* wines. Prices in England reflected these changes, with those at which the king bought his wine rising to £5 a tun, and retail prices reflecting a similar rise. Over the next century Bordeaux's exports and the wine prices in England closely mirrored the fortunes of the war. During the truce between 1340 and 1345 export levels again increased with a comensurate fall in retail prices to 4d. or 5d. a gallon obtaining in London (James, 1971).

In 1348, with the advent of plague in Bordeaux, a new crisis of exceptional severity hit the wine producers and merchants of the Bordelais. Wine exports in 1348-9 were only 5,923 tuns, representing less than 6 per cent of their level in 1308-9 (Table 4). Prices in England rose to new levels, with wholesale prices as high as £8 a tun, and even during periods of peace, such as that following the Peace of Brétigny from 1360-69, they were never to fall again to the levels that had pertained around 1300. During the 1360s retail prices in England were thus usually in the range of 8d. to 12d. a gallon (James, 1971). Renewed warfare from 1369 led to further destruction of the vineyards. Not only were the English pushed back to the immediate vicinity of Bordeaux, but plague and famine during the 1370s further reduced both the production and export of wine. While the 1350s and 1360s had seen wine exports in the region of 20,000-30,000 tuns a year, during the 1370s annual exports were more usually in the order of 10,000 tuns. With the return to peace at the end of the century there is some evidence of reconstruction, with vineyards being restored and exports increasing again, but both the volume of exports from Bordeaux and the level of imports into England could not compare with those at the beginning of the century.

Table 5 Wine exports from Bordeaux during the fifteenth century

Year	Wines of Bordeaux: Burgesses, nobles, ecclesiastics	Wines from outside Bordeaux: Burgesses	Wines from *Haut Pays* under English rule:		Other wines[1]	Total[2] (tuns)
			Burgess	non-burgess		
1402-3						10,067
1409-10	4,840	3,533	3,618	1,223	56	13,270
1412-13	5,171	2,810	4,246	557	264	13,158
1418-19		1,086	1,195	102	232	
1422-23	6,107	2,886	5,822	1,388	105	16,258
1427-18	3,796	2,225	2,410	705	28	9,074
1428-29	3,747	2,817	2,769	1,004	228	10,765
1429-30	5,333	3,055	3,527	1,197	10	13,222
1430-31	5,528	2,882	3,934	1,244	46	13,634
1431-32	2,929	2,608	2,307	782	0	8,626
1433-34		2,663	2,994	604	0	
1435-36	5,586	2,616	3,466	911	14	12,603
1436-37	4,748	2,316	3,170	669	0	10,903
1437-38	2,333	712	1,430	268	18	4,761
1438-39	2,638	257	871	286	0	4,052
1443-44	4,502	2,141	1,549	735	0	8,827
1448-49		5,638	2,339	905	124	
1452-53	5,337	2,712	1,448	422	0	9,919

Notes: [1]Other wines includes wines of the Agenais and wines on which custom was compounded at one franc or one noble a ton.
[2]Totals do not include returns of customs collected at Libourne.
Source: derived from James, M. K., *Studies in the Medieval Wine Trade*, edited by Veale, E. M., Oxford: Clarendon Press, 1971: 55-56

Until the culmination of the Hundred Years War in the middle of the fifteenth century, there was much greater stability in the level of Gascon wine exports to England (Table 5). These continued at around 10,000 tuns a year, with retail prices remaining at about 6*d.* a gallon. The short-term fluctuations that did occur were usually the result of poor harvests due to adverse weather conditions. In 1438, though, with the renewed French assaults on the Bordelais coinciding with a period of plague, wine exports fell to just under 5,000 tuns (Table 5). Eventually in 1451 Bordeaux surrendered to the French, but English and Gascon merchants not wishing to ally themselves with the victors were given six months to leave with their goods and ships, thus enabling most of the 1451 vintage to be exported (Renouard, 1970). The temporary restoration of English rule in the following year was brought to a conclusive end in 1453 at the battle of Castillon near St Emilion, after which conditions became very much more difficult for trade. In 1455 Charles VII imposed a tax of 25*s.* (tournois) on every tun of wine exported from Bordeaux, but chose not to prohibit entirely the wine trade with England, seeking instead to benefit from it through taxation. By the last quarter of the fifteenth century there had been some recovery, with Bordeaux's wine exports generally approaching 7,000 tuns a year (James, 1971), but this was but a fraction of the level of trade when it had been at its peak in the early fourteenth century. Louis XI recognised that the basis of Bordeaux's wealth was its trade with England, and rather than impoverish the city's economic basis, he agreed with his Counsellor Regnault Girard, that the English should again be allowed to trade in the city (Carus-Wilson, 1954; Renouard, 1970). The effects of the loss of Gascony were, though, reflected in the price of wine in England, where from the mid-1450s prices increased permanently to at least 8d. a gallon, more than twice their level at the beginning of the fourteenth century.

THE FOURTEENTH CENTURY: CRISIS AND CHANGE

The above example illustrates the importance of the interaction between political and economic factors in determining the fortunes of the Anglo-Gascon wine trade during the fourteenth and fifteenth centuries. However, set against these structural changes, and scarcely commented upon by authorities such as Renouard (1970), must be considered the effects of the demographic decline of the mid-fourteenth century, and the broader social and economic changes associated with the so-called crisis of feudalism (Brenner, 1976, 1982; Hilton, 1978; Postan and Hatcher, 1978; Aston and Philpin, 1985).

Table 4 clearly illustrates that, although the various stages of the war between England and France did influence the levels of wine exports from Bordeaux, it was the coincidence of renewed warfare in the mid-fourteenth century with the onset of plague in 1348 that led to the permanent reduction in such exports from their peak at the beginning of the century. Although vineyard planting, wine production, and wine exports from Gascony experienced some resurgence at times of relative peace, such as the 1360s, this never again reached the levels that they had earlier in the century, suggesting that the reduction of population by perhaps one third as a direct result of plague may have been at least as significant as warfare in determining the fortunes of viticulture in the region. The cultivation of vineyards was labour intensive. Renouard (1970: 77) thus describes the usual practice on the estates of the archbishop of Bordeaux as follows:

> the soil was plowed four times a year, carefully avoiding the roots; after the spring pruning the branches left on the plant were tied to stakes with strips of wicker, which were also produced in the

vineyard. Finally, in summer, the women thinned the vines in order to give the grapes more sunlight.

Under such conditions, any reduction in labour would be expected to lead to a reduction in both the yields of the vines and the quality of the wine. In some places where viticulture was marginal, it may also have led to the abandonment of inferior vineyards. The effects of plague throughout Europe in the mid-fourteenth century not only dramatically reduced the labour supply, but also led to a relative improvement in the bargaining power of the peasantry, who were to some extent able to press for better returns for their labour power. In monetary terms, rising labour rates and difficulties of supply, would have expressed themselves in an increase in the cost of wine, even without the effects of war.

However, it can also be argued that as grain became cheaper in relative terms during the second half of the fourteenth century and through into the fifteenth century, there would have been more money available for people to spend on wine, which in northern Europe was a luxury item. Wine demand therefore probably declined somewhat less than the amount by which the population as a whole, and thus the total demand in the economy, declined. Although it is extremely difficult to identify precise levels of wine imports into England during the fourteenth century, what evidence there is generally supports the above arguments. While alien wine imports to England usually ranged between 3,000 and 4,000 tuns between 1334 and 1348, and they fell to between 1,000 and 2,000 tuns for most of the remainder of the fourteenth century (James, 1971: 34), this decline needs to be weighed against the general increase in importance of English merchants dealing in wine at that time. Even by the 1330s alien imports of wine were well below the 8,636 tuns imported in 1322–3 and the 7,704 tuns imported in 1323–4 (James, 1971: 34). As the supply of wine from Gascony decreased, the English increasingly turned elsewhere, to countries such as Spain (Childs, 1978), for their imports. Indeed, the upward pressure on wine prices in England in the second half of the fourteenth century, caused largely by the contraction in supply, provided a ready incentive for merchants to import whatever wine they could.

Another effect of the mid-fourteenth century population decline can be seen in Burgundy, where one of the reasons for the expansion of Gamay production during the second half of the fourteenth century may well have been that its greater yield enabled producers to offset the losses resulting from the decline of labour inputs. The Gamay can yield up to three or four times more than the Pinot Noir, and given the reduction of labour inputs following the plague of 1348 it made sense to convert to the higher yielding, but lower quality, vines (Berlow, 1982).

Population changes during the fourteenth century were, though, associated with far more than the reorganisation of vineyards and the wine trade. They formed part of the complex series of economic and social transformations that led to the eventual disintegration of the feudal mode of production (Brenner, 1976; Postan and Hatcher, 1978). The responses of different states to the social problems engendered by a falling population, which expressed themselves most forcibly in the growing power of the peasantry and the falling price of grain, were to lay the foundations for the rise to dominance of new economic forces and the states with the power to control them.

7

WINE IN THE AGE OF DISCOVERY

Outside Europe, wine followed in the wake of Europeans. Great feats were accomplished in acclimatizing the vine in Mexico, Peru, Chile (reached in 1541) and in Argentina, after the second foundation of Buenos Aires in 1580. In Peru vineyards rapidly prospered in the hot and fever-ridden valleys because of their proximity to Lima, an exceptionally wealthy town. They prospered still more in Chile where the soil and climate were propitious; vines were already growing among the *cuadras*, the blocks of the first houses of the growing town of Santiago.

(Braudel, 1981: 232)

In a little over a century, between the Portuguese expedition to Madeira in 1419 and Magellan's circumnavigation of the world in 1519-21, the entire scale of European life changed. New horizons were opened up, and new possibilities were created for the individual appropriation of profit. The Portuguese and Spanish explorations of the fifteenth and sixteenth centuries revealed vast 'new' lands across the Atlantic, as well as new routes to the east, bypassing the traditional Islamic monopoly of the vital luxury trade with India. In the wake of these voyages of discovery, the practice of viticulture and the consumption of wine were eventually taken to the furthest corners of the world.

However, the period between 1500 and 1750 witnessed more than just the discovery of new lands by the Europeans; it also saw a range of technical and social innovations which were to transform the structure of the economy. Four key changes can be identified. For the first time in history, population in the eighteenth century began to increase apparently unconstrained by disease and famine. In contrast to the plagues and demographic declines of the sixth and fourteenth centuries, new economic and social structures emerged during the sixteenth and seventeenth centuries which laid the way for subsequent rapid population increases. Secondly, there was a considerable expansion of the productive economy, with changes in agrarian practice and organisation enabling higher population levels to be sustained. New relations of production had been introduced in both agriculture and industry, which provided the basis upon which the economy was able to expand at an unprecedented rate. Thirdly, an economy which had been essentially European in the fifteenth century had, by the end of the eighteenth century, become global in scale, leading to a transformation in both the patterns and also the mechanisms of trade. Fourthly, as commercial negotiations became more sophisticated, this in turn enabled the global economy to become increasingly spatially differentiated and specialised. All of these changes manifested themselves in a reorganisation of viticulture and the wine trade. Not only were new regions planted with vineyards, but new methods of viticulture and vinification were introduced, and new types of wine created.

THE CREATION OF A NEW WORLD, 1500-1750

From feudalism to capitalism

Since the publication of the first volume of Marx's *Das Kapital* in 1867, a vast literature has arisen on the so-called transition from feudalism to capitalism between the fifteenth and the seventeenth centuries (Wallerstein, 1974, 1979, 1980; Dunford and Perrons, 1983; Aston and Philpin, 1985; Dodgshon, 1987; Glennie, 1987). However, the terms *feudalism* and *capitalism* as used to describe social formations are recent and were introduced long after the periods to which they refer. Bloch (1965: xvii) thus notes that *feudalism* was first used to 'designate a state of society' only in 1727 by the Comte de Boulanvilliers, and Braudel (1982: 237) observes that, although the word *capitalism* apparently gained its current meaning in 1850, it was 'not until the beginning of this century that it fully burst upon political debate as the natural opposite of socialism'. Although Marx never used the word 'capitalism', it has achieved widespread use to describe the third epoch in his sequence of modes of production from the ancient or slave, through the feudal, to what he called the modern bourgeois mode (Marx, 1971).

In many ways, the use of such broad terms as feudalism and capitalism to define stages of economic development has restricted understanding of the underlying processes of social and economic change. In particular such terms can convey the impression that feudalism and capitalism were, and are, uniform and monolithic in their expression. In practice, however, while there were common elements to feudalism, such as, the holding of land from a lord by a vassal in return for military service, there were many variations in its precise manifestation in both space and time. As Bloch (1965: 444) has emphasised, it is important 'to note that feudal Europe was not feudalized in the same degree or according to the same rhythm and, above all, that it was nowhere feudalized completely'. Moreover, the acceptance of such a stage model presents a fundamental conceptual problem, since the period of transition between what might be called high feudalism, which ended in Britain in the fourteenth century, and mature capitalism, which began in the eighteenth century (Dunford and Perrons, 1983), actually lasted longer than either stage. Even if it is argued that the essence of Marx's model was that one mode of production would emerge from structural contradictions in the previous mode in a process of dialectical change (Dodgshon, 1987), the length of this transition period suggests that it might be more profitable to place greater emphasis on the actual processes of economic and social change, and rather less on the definitions of an abstract feudalism or capitalism.

Nevertheless, certain elements of Marx's description of the feudal and modern bourgeois or capitalist modes of production are particularly useful in an interpretation of the changes in viticulture and the wine trade which occurred during the period between the fifteenth and eighteenth centuries. These are specifically concerned with the creation of profit, the definition of capital, and the social relations of production. Apart from in the most primitive tribal societies, all social structures are predicated upon the expropriation of a surplus by one class from another. This enables one class to undertake activities other than the direct production of food. The essence of the feudal mode of production was thus that an unfree peasantry performed labour services on the lord's land and owed rent and other feudal dues to the lord, which themselves defined the unfree relationship between peasant and lord. In return the peasants were permitted to cultivate land which provided the means for their own subsistence and reproduction. The lords were thus able to extract a profit both from the labour power of the peasantry and also through feudal restrictions on the freedom of that peasantry (Kosminsky, 1956; Hilton, 1965, 1978).

For Marx (1976: 875) the economic structure of capitalist society grew out of the structural contradictions of the economic structure of feudal society. The essence of his view of capitalism was built on four attributes: the production of commodities, wage labour, acquisitiveness and rational organisation (Howard and King, 1985). Central to his arguments was the view that, although capital, or money bent upon the accretion of money, existed in pre-capitalist societies, it hardly ever entered the sphere of production in such societies. Mandel (1976: 32) notes, therefore, that 'A basic difference between the capitalist and pre-capitalist modes of production is that under capitalism capital not only appropriates surplus-value; it produces surplus-value'. This concept of surplus value, though, can only be understood in terms of Marx's relational concept of commodities. For Marx, commodities differ from articles in that they are products of private individuals who only come into contact with each other when they exchange their products. 'Thus the relation between the commodities (things) is simultaneously a relation between people (commodity producers)' (Howard and King, 1985: 45). In distinguishing between *use value*, the basic value of the use of a thing conditioned by its physical properties, and *exchange value*, the quantitative relationship between commodities expressed as the proportion in which use values of one kind can be exchanged for use values of another kind, Marx was then able to explain the creation and appropriation of surplus value by capitalists. In brief, he argued that 'There exists one commodity, to wit labour-power, whose use-value for the capitalist is its ability to produce new value larger than its own exchange value' (Mandel, 1976: 33). The secret for the capitalist is therefore to expropriate this surplus value by paying the labourer sufficient only for the exchange value of his labour power. Moreover, through his control over the means and processes of production vested in land, the capitalist is able to force the labourer to sell his labour power to the capitalist as a commodity.

For capitalism to develop out of feudalism, two main conditions therefore had to be satisfied: a free peasantry able to sell its labour power had to be created, and it had to be restricted from gaining access to its own means of production and reproduction. Marx used the term primitive accumulation to describe this process, and he argued that the conditions under which it was able take place were as follows:

> In themselves, money and commodities are no more capital than the means of production and subsistence are. They need to be transformed into capital. But this transformation can itself only take place under particular circumstances, which meet together at this point: the confrontation of, and the contact between, two very different kinds of commodity owners; on the one hand, the owners of money, means of production, means of subsistence, who are eager to valorize the sum of values they have appropriated by buying the labour-power of others; on the other hand, free workers, the sellers of their own labour-power, and therefore the sellers of labour . . .
> The process, therefore, which creates the capital-relation can be nothing other than the process which divorces the worker from the ownership of the conditions of his own labour; it is a process which operates two transformations, whereby the social means of subsistence and production are turned into capital, and the immediate producers are turned into wage-labourers.
>
> (Marx, 1976: 874).

The unfree feudal peasantry, previously tied to the land, thus had to gain its freedom and engage in a wage contract with the owners of capital.

The implications of these changes for viticulture and the wine trade reflect both changes in the methods of production and also a fundamental reorientation of the motives of wine producers and merchants. During the medieval period most wine, other than that required for the subsistence needs of the peasantry, was produced either through the labour services of an unfree peasantry performed on the lords' vineyards, or as rent in kind from the vines of peasants. As the peasantry gained its freedom, in varying degrees and rates in different parts of Europe (Aston and Philpin, 1985), these production methods changed. Gradually, a new awareness of the potential profit to be gained from the wine trade became apparent among the mercantile community. However, this awareness was not immediately reflected in the reinvestment of such profits in the production of wine.

Wallerstein, and the creation of a modern world system

One of the most comprehensive and influential (Butlin, 1987) neo-Marxist analyses of the expansion of European interests overseas during the late fifteenth and sixteenth centuries has been that of Wallerstein (1974, 1979, 1980, 1983). In its insistence on the integral connection between social, economic and political change, Wallerstein's analysis is in marked contrast to the earlier neo-classical work of North and Thomas (1973), with its central focus on the nature of organisations and nation states. According to North and Thomas, the driving force behind the rise of the Western world was population growth, which forced changes to be made in the organisation of the European economy. They suggest that a potential population crisis in the seventeenth century was averted due to three main factors: the disintegration of feudal social relations; improvements in shipping, which made possible overseas discovery and settlement; and changes in institutional arrangements, particularly property rights, which made it worth while to undertake socially productive activity. Underlying all other changes, though, according to North and Thomas, was the form of the evolving nation state induced by the expanding market economy. Thus in France and Spain they suggest that the power of the monarchy gave rise to high levels of taxation, which stifled innovation. In contrast, in Holland, dominated by an oligarchy of merchants, and in England, where the power of parliament was all important, the creation of a hospitable environment for the evolution of property rights, involving fee-simple absolute landownership, free labour, the protection of privately owned goods, and the development of patent laws, all generated a situation favourable to sustained economic growth. North and Thomas's (1973) emphasis on the role of property rights is, however, exaggerated (Dodgshon, 1987), and suffers further from its consideration of population growth as being exogenous.

In contrast, Wallerstein (1974, 1980) separates the economic from the political realm, arguing that

The distinctive feature of a capitalist world-economy is that economic decisions are oriented primarily to the arena of the world-
economy, while political decisions are oriented primarily to the
smaller structures that have legal control, the states (nation-states,
city-states, empires) within the world economy.

(Wallerstein, 1974: 67)

Wallerstein's (1974: 67) central tenet is that 'It was in the sixteenth
century that there came to be a European world-economy based upon the
capitalist mode of production'. This was forged from the reorganisation
of the earlier separate medieval world economies of northern Italy and
Flanders. Three processes were essential for the establishment of this new
world system: an expansion of the geographical size of the world in
question; the development of variegated methods of labour control for
differing products and in different zones of the world economy; and the
creation of relatively strong state machineries in the emerging core states
of the capitalist world economy. Crucial to these arguments is his
suggestion that different types of labour control were utilised by
capitalists in different parts of the world. Thus, tenancy and wage labour
came to dominate in the core states, sharecropping and serfdom in the
semi-periphery, and slavery in the periphery. While the actual locations
of these zones changed through time, Wallerstein suggests that the geo-
economically peripheral areas were dominated by primary activities,
including mining and agriculture, and the core regions, with higher
population densities, witnessed the intensification of agriculture and the
emergence of a hierarchical division of labour. The semi-periphery then
represents a mid-point on the continuum running from core to periphery. By 1640, he sees Holland, England and to some extent northern
France as being the core states, with slavery dominating in the periphery
of Latin America, and Iberia and eastern Europe forming the semi-
periphery. Critical to these arguments is the way in which groups of
people in the strong core states came to dominate activities in the
peripheral areas, and in this context Wallerstein (1974: 356) argues that
two tipping mechanisms operated at the level of the creation of state
machinery in which strength in the core areas created more strength and
weakness in the periphery led to greater weakness. For Corbridge (1986,
35), 'it is this which cranks the handle of Wallerstein's model of surplus
transfer'. The existence and functioning of these mechanisms is, however,
poorly elucidated, and Wallerstein's analysis has been subjected to a wide
range of critiques (Dodgshon, 1987; Brenner, 1977, 1982; Laclau, 1979;
Wolf, 1982; Corbridge, 1986; Butlin, 1987).

Three particular problems with Wallerstein's thesis can be
highlighted. Firstly, he argues that a number of different 'European
capitalisms' existed within the sixteenth century (1974: 77), each with its

own particular organisation of labour. These emerged because 'each mode of labor control is best suited for particular types of production', and this European division of labour assured 'the kind of flow of surplus which enabled the capitalist system to come into existence' (Wallerstein, 1974: 87). Such an argument is incompatible with definitions of capitalism derived from Marx which suggest that a central feature of the capitalist mode of production is the relationship between wage labour and the owners of the means of production. If this definition is accepted, only certain parts of Europe can be considered to have been capitalist during the sixteenth century. Moreover, at an empirical level, Wallerstein's (1974: 87–95) suggestion that different types of agriculture are best suited to different types of labour control is also open to some doubt. A second problem with Wallerstein's very wide definition of capitalism is related to the investment of profit. As Corbridge (1986) has pointed out, a basic feature of later capitalism was the investment of profit in the productive process, and this was significantly unusual in the sixteenth century. Brenner (1977) thus suggests that while the profits of plunder in Latin America were frequently used for conspicuous consumption and reinvestment in further mercantile adventures, they were only very rarely invested in new methods of production. A third criticism of Wallerstein's model is that it is notably Eurocentric in its formulation, and largely ignores the changes taking place within other political and economic systems. In particular, the Ottoman and Chinese empires were both highly organised and effective, playing a major role in the global economy during the sixteenth and seventeenth centuries, and to relegate them merely to side-shows is inappropriate.

Despite these difficulties with Wallerstein's proposals, they do emphasise the importance of a consideration of labour relations in an examination of the expansion of European interests overseas. Moreover, vines were one of the few crops that came to be grown throughout the lands that increasingly came to be dominated by the interests of the north-west European states, from Hungary (Zimányi, 1987) in Wallerstein's semi-periphery, to his periphery in Latin America. An examination of the methods of viticulture in these different regions can therefore provide important insights into the changing economic structure of the period.

Trade, profit and ideology

The concentration of much critical attention on Wallerstein's arguments concerning different systems of labour control has tended to overshadow the importance of a second crucial element necessary for the transformation of the economy of sixteenth and seventeenth century Europe. This concerns the origins of profit, and changes in mercantile financial institutions which facilitated the development of new credit arrangements (Spufford, 1988). In this context, Braudel (1982: 245) has argued that 'Quite clearly, the economies of the past produced a considerable quantity of gross capital, but in some sectors this gross capital just melted away'. The emergence of capitalism is fundamentally related to the way in which this melting away was transformed; how money came to be used to make more money. While levels of capital investment remained relatively low during the sixteenth and seventeenth centuries, Brenner (1977) and Corbridge (1986) place insufficient emphasis on the importance of capital in the changing structure of international trade during this period. Braudel (1982: 374) has, thus, emphasised that rather than being central to the *production* process, 'Capital was most at home in the sphere of circulation'. It was in the new credit and banking systems that capitalism found its true home.

Three main phases can be identified in the development of credit. The first emerged in Florence during the late thirteenth and early fourteenth centuries, with the activities of the Lucchese, Fresobaldi, Bardi and Peruzzi families who at various times all became the bankers of the kings of England. In Braudel's (1982: 395) terms, 'The triumph of the Florentines lay not only in holding the purse strings of the kings of England, but also in controlling sales of English wool which was vital to continental workshops and in particular to the *Arte della Lana* of Florence'. The Bardi and Peruzzi used money from the English Crown to pay for the wool, and in return provided the king with the credit he needed abroad. However, these families loaned large sums of money from their depositors, well in excess of their own capital, and when Edward III of England defaulted on his debts in 1345 the banking edifice of the Bardi and Peruzzi collapsed. Following the wider mid-fourteenth century economic crisis, Florentine banking did revive somewhat under the Medici in the fifteenth century, but by then it only formed a part of a much broader base to the city's economy. In Goldthwaite's (1987: 17) words

> The very structure of the renaissance economy of Florence, with its wide-ranging and dispersed activities and lack of central focus, precluded dependency of its various parts on any one function, so that it was virtually impossible for any one operator to hold the key to the whole system.

The second banking development identified by Braudel (1982) was associated with the increased use of paper bills of exchange by the Genoese during the sixteenth century. These worked on the principle that the seller of a bill in any market could obtain money immediately, with the receiver being repaid at another market at some later date according to the rate of exchange then prevailing. The receiver therefore had to calculate the potential profit that could be gained, and thus the relative risk of the transaction. During the second half of the sixteenth century, the Genoese provided loans to Spain, and were repaid in silver bars and pieces of eight deriving from the Spanish exploitation of Latin America. Genoa thus became the leading silver market in Europe, and through the use of bills of exchange the Genoese also came to control the circulation of gold. However, bills of exchange could only gain value if there were different discount rates in the money-markets where they circulated, and at times when much money was in circulation the system would grind to a halt. At the end of the sixteenth century Braudel (1982) notes that the financial markets of Europe were inundated with money, and he suggests that it was probably for this reason, rather than any dramatic decline in shipments of silver from America, that the Genoese paper mountain collapsed.

The third short-lived banking experiment took place in Amsterdam during the eighteenth century, and was also based on paper credit. Again quoting Braudel (1982: 395), 'Here various types of paper credit came to assume an unprecedented and extraordinary role. The entire commodity trade of Europe was governed, remote-controlled so to speak, by the rapid movement of credit and discounting'. However, the large profits that this engendered meant that the Dutch bank had excessive amouts of money on its hands, and began lending to European states, with the result that the bankruptcy of France in 1789 led rapidly once again to the collapse of a banking system based on paper.

The importance of this discussion of banking developments, and the introduction of new credit systems, is that it underlay and facilitated changes in the organisation of trade. Quite simply, long-distance trade was unable to operate successfully without the existence of a complex network of credit. Moreover, distance itself, at a time of costly and difficult transport, provided the opportunity for profiteering. It was the activities of the great merchants, keen to appropriate increasing amounts of profit from their activities, that were at the forefront of economic change. Money was being invested in the expectation of increased profits; capital was being used to create more capital, but it was not yet being diverted into the productive economy. It was only when this system began to break down during the eighteenth century, that attention turned to the use of capital in the processes of production. In Braudel's (1982: 400-1) words, the mechanisms of profit appropriation by the merchants of the sixteenth and seventeenth centuries were thus as follows:

> the great merchants, although few in number, had acquired the keys to long-distance trade, the strategic position *par excellence*; that they had the inestimable advantage of a good communications network at a time when news travelled very slowly and at great cost; that they normally benefited from the acquiescence of state and

society, and were thus able regularly, quite naturally and without any qualms, to bend the rules of the market economy.

The remainder of this chapter explores the significance of the arguments of both Braudel and Wallerstein for changes that took place in viticulture and the wine trade between the fifteenth and eighteenth centuries. The possibilities for increased profit from long distance trade in wine, the opportunities opened up by new markets, the option of vine cultivation in newly conquered lands, and the eventual investment of capital in the creation of new types of wine, all played a part in the economic transformation that has become known as the transition from feudalism to capitalism.

DISCOVERY AND CONQUEST: THE VINE IN LATIN AMERICA

The Great Discoveries

The voyages of exploration and discovery that left the shores of Portugal and Spain in the fifteenth and sixteenth centuries provided the agency through which European *Vitis vinifera* vines were taken across the Atlantic, and for the first time planted in the southern hemisphere. However, no one motive can be used to explain the extraordinary explosion of energy and vitality represented by these voyages, and the purposes and indeed characteristics of the explorations, discoveries and exploitation that occurred varied considerably over both space and time.

The earliest voyages of discovery were undertaken by the Portuguese and concentrated initially on the exploration of the southern sea-route around Africa to India (Chilcote, 1967; Boxer, 1969; Bell, 1974). Their first foray overseas was the capture of Ceuta, in what is now Morocco, in 1415. Soon afterwards contact was established with the Atlantic islands of Madeira in 1419, and the Azores in 1427. During the 1430s and 1440s various expeditions followed the coast of Africa southwards, with Cape Bojador being passed by Gil Eanes in 1434 and Cape Verde being reached in 1444. The Cape of Good Hope was eventually passed in 1488 by Bartolomeu Dias, and then in July 1497 Vasco da Gama set out on the voyage that at last reached Calicut in India, having also called at Mozambique, Mombassa and Malindi (Camões, 1981). Meanwhile, Castilian interests were being established in the Caribbean and Latin America through the voyages of the Genoese navigator Christopher Columbus, who first arrived in San Salvador, Cuba and Haiti in 1492 (Andrews, 1978). Competition between Portugal and Castile over their newly discovered lands was settled by the Treaty of Tordesillas in 1494, which split the world into two zones of influence along the meridian line 370 leagues to the west of the Cape Verde Islands. Lands to the east of this line were to belong to Portugal, and those to the west to Spain. The choice of this particular line, which meant that much of Brazil became Portuguese territory, suggests that the Portuguese had some knowledge of South America before the treaty, and thus before Cabral's formal 'discovery' of Brazil in 1500. From the beginning of the sixteenth century further exploration and conquest continued apace, with Magellan circumnavigating the globe during the years 1519-21. The Portuguese reached the Pacific in 1511, China in 1513 and Japan in the 1540s, while the Spanish led by Cortés conquered the Aztec empire in Mexico between 1519 and 1521, and under the command of Pizarro subjugated the Incas between 1531 and 1534.

Until the sixteenth century, Portuguese interests overseas were mainly concerned with the opening up of new lands for trade and commercial enterprise, but when Francisco de Almeida was appointed Viceroy of India in 1505 they embarked on a new strategy involving the building of key fortresses in Africa and Asia. This policy was developed under Almeida's successor, Afonso de Albuquerque, who conquered Goa in 1510 and thereafter made it the base of Portuguese activities in India. This shift from trading, albeit often on an unequal basis, to the imposition of alien rule over a subjugated people heralded the rise of the colonial policies of the various European powers which were to dominate global political relations for the next three centuries.

The motives for the fifteenth century discoveries were complex and multifaceted. In economic terms three issues seem to have been paramount: the desire to have free access to trade with India and the east, which had previously been dominated by the Islamic powers of the Middle East; the wish to obtain control of Islamic gold production in western Africa; and later the exploitation of the silver resources of Latin America. However, there were also strong social, political and ideological motives for the voyages of discovery. Politically, they provided one mechanism by which the rulers of Castile and Portugal were able to keep a potentially restive nobility occupied, and the voyages also served to detract attention from increasingly difficult social problems at home in the two countries. However, as Marques (1972: 140) has argued, while these explanations provide a reasonable platform for action 'they omit the colorful wrapper that each man requires to rationalize his own actions and to convince others of a noble and idealistic undertaking: the fight against the infidel and the salvation of souls'. Central to the Portuguese voyages of discovery was the ideological conflict with Islam, and the desire to establish contact with the Christian kingdom of Prester John in what is now Ethiopia (Ley, 1947). Moreover, the importance of ideology in Spain was reflected in the activities of the 'Catholic Monarchs' Ferdinand and Isabella, who in 1492 finally conquered Granada from the Moors and in the same year instituted the Inquisition and promulgated a decree expelling the Jews (Vives, 1970). This Catholic religious fervour was also to play an important part in the Spanish conquest of Latin America, and is well exemplified, for example, in the *Historia Verdadera de la Conquista de la Nueva España* of Bernal Díaz, who served under Cortés (Cohen, 1963). It is this religious motive which is also most usually used to explain the expansion of viticulture in the Americas (Hyams, 1987: Johnson, 1989).

The vine in Latin America

The speed with which vine cultivation spread through Latin America is ample testimony to the esteem in which wine was held by the conquering Spaniards. Spanish wine, much of it produced from the region around Jerez, was taken on most of the voyages of discovery, and considerable quantities were shipped to Mexico during Cortés's campaigns against the Aztecs between 1519 and the destruction of Tenochtitlan in 1521. As Andrews (1978: 57) has noted, with reference to the fleets sailing to the Caribbean during the first two decades of the sixteenth century,

> Andalucía from its own resources and those of Castile could satisfy most of the needs of the Spaniards in the New World because those needs were elementary: flour, wine, oil and miscellaneous stores filled most of the cargo space in outward-bound vessels.

The first indication of vine cultivation in Mexico is Cortés's importation of vines, together with other European crops, to *Nueva España*, New Spain, in the 1520s (Prescott, 1936: 639). Following the conquest of the Aztecs, 'Cortés distributed almost the entire Indian population of central Mexico in *depósito* or encomienda to himself and his conquistador companions' (Gerhard, 1972: 8), and by 1524 virtually all of the old Aztec Empire had been brought under Spanish domination. The *encomanderos* were supposed to ensure that the Indians under their control became Christians and vassals of the Spanish king, and in return they were entitled to services and tributes from their Indian subjects. Cortés also gave Spanish settlers grants of land and Indian slaves to work on it, known as *repartimientos*, and in the municipal ordinances of the new Mexico City, promulgated in 1524, the following regulation specifying the cultivation of vines is found: 'Item, que cualquier vesino que tubiere Indios de repartimiento sea obligado á poner en ellos en cada un año con cada cien Indios de los que tuvieram de repartimiento mil sarmientos, encogiendo la mejor que pudiese hallar' *(Ordenanzas Municipales*, año de 1524, MS, quoted in Prescott, 1936: 639); for each grant of land equivalent to 100 Indians, proprietors of these new estates thus had to plant 1000 vines of the best quality available.

Despite the presence of local indigenous varieties of vine in Mexico at the time of the Spanish conquest, there seems to have been no attempt by the indigenous people to make wine from their grapes. The traditional alcoholic drink was mainly *pulque*, made from the agave, known locally as the *maguey*, but other drinks, such as *tesgüino*, a kind of beer made from the sprouted kernels of maize, and *balché*, a honey mead made from *Lonchocarpus* leaves, were also common (Bruman, 1967). In his *Historia de los Indios de Nueva España* written in 1536, the early Spanish missionary, Fray Toribio de Benavente, known as Motolinía (1950), in particular noted that the indigenous peoples made a kind of wine from the *maguey*. Moreover, Motolinía (1950: 244) goes on to observe that they worshipped a god, Ometochtli, who was 'one of the principal Indian gods and was worshipped as the god of wine and very much feared and respected; for they were in the habit of getting drunk and from their drunkenness came all their vices and sins'. The absence of wine made from the grapevines existing in America prior to the Spanish conquest was possibly due to the greater ease with which alcohol could be made from other plants, such as the agave, but it also suggests that the symbolic link between wine, grapevines and religion was in some way a specifically European phenomenon.

Turning to the actual cultivation of vines, Motolinía (1950: 217) commented that

> In many places in the mountains there are big wild grapevines, and no one knows who planted them. They grow out very long shoots and bear many clusters and produce grapes which are eaten green. Some Spaniards make vinegar out of them, and some have made wine, but only in small quantities.

Such wine made from local grapes apparently did not suit the Spanish palate, and most vineyards were instead planted with the Criolla or Mission grape which the Spaniards introduced. However, Motolinía (1950: 265) notes that, except in particularly favourable conditions, these grapevines brought from Spain suffered from frost, because a cold period often followed a warm spell around Christmas, and this damaged the new shoots which had been encouraged to emerge by the earlier heat. He also points out that the Spanish friars taught the local people how to graft fruit trees (Motolonía, 1950: 219), and it is therefore possible that they experimented with the grafting of European vines on to the indigenous species that they found growing there, although Motolinía himself is silent on this practice. In terms of the labour used to cultivate the vineyards, the Spaniards were soon forced to turn to slavery because of the dramatic population decline among the indigenous population, which may have fallen from around 22 million in 1519 to about three million in

1550 and as little as one million in 1620, as a result of smallpox, measles, typhus and unspecified plagues (Gerhard, 1972). Negro slaves had been brought to America by the earliest conquistadors, and particularly following the outbreak of *cocoliztli* in 1545-8, which was probably measles and which decimated the indigenous population, the flow of slaves from the Antilles and Africa increased dramatically.

From Mexico viticulture spread rapidly throughout Latin America in the wake of the Spanish conquests of the Inca empire. Peru fell to Pizzaro in 1531-4 (Prescott, 1936), Bolivia to Almagro (1535-7), Colombia to Queseda (1535-8), and Chile to Valdivia in 1540-5. In Peru it seems that the Spaniards realised that high altitudes would compensate for the hotter temperatures found in such low latitudes, and while vine cultivation was first actually recorded near Cuzco in 1550 it is probable that vines were introduced soon after Pizzaro's first victories in the 1530s (Figure 33) (de Blij, 1983; Hyams, 1987). Vines were reported to be widely grown in Chile during the second half of the sixteenth century (Guerrero, 1978), and Valdivia himself observed that wine was being made at El Reino de Chile by 1556 (Hyams, 1987: 294). By 1556 vines had also been introduced by a Jesuit priest across the Andes in what was to become Argentina (de Blij, 1983: 59). In a mere thirty years, therefore, viticulture had spread from Mexico down the entire western coast of Latin America as far as Concepción. While the Spanish were conquering Latin America, French Huguenots had reached Florida (Figure 33), and between 1562 and 1564 they also attempted to produce wine from the indigenous Scuppernong grapes (Adams, 1984: 16, 56). However, in 1566 this settlement was destroyed by the Spanish under Menendez, and all of the Huguenot population were massacred. It was only with the subsequent establishment of Catholic missions here that limited amounts of wine were once again produced in Florida (Hyams, 1987: 262).

' The traditional explanation for the extent and rapidity of the introduction of European vines to central and southern America was that wine formed an essential part of the religious rituals of the Catholic church. This has been well expressed by Hyams (1987: 255) in the following terms:

> One of the problems which faced the Conquistadors in their conquests and colonizations in America was that of providing a supply of wine for the Mass. It should not be thought of as a minor problem; to the Spaniard of the sixteenth century it was of the very first importance.

Such arguments, though, considerably overstate the specific religious importance of wine. Undoubtedly it did form an essential part of the ritual of the Mass for the clergy, but wine did not at that time form part of

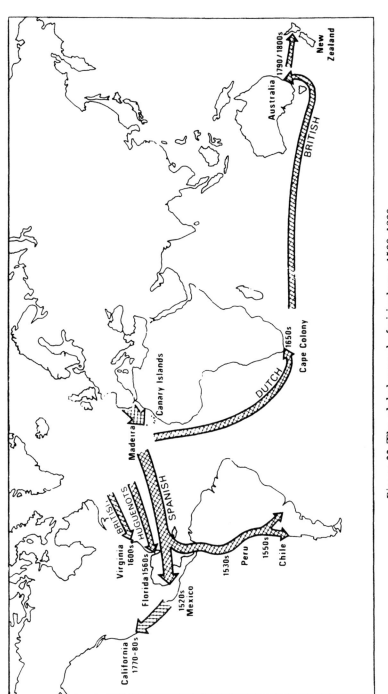

Figure 33 The global spread of viticulture, 1500–1800

the rite for the secular population. The total religious demand for wine was not therefore great, and could readily be supplied by the fleets from Spain and the Atlantic islands. Of much more importance in explaining the spread of viticulture were social and economic reasons. The Spanish leaders of the conquest were eager to create a new Spain in Mexico, and in symbolically recreating their homeland they introduced as many European crops as possible, including the vine. Wine was central to the lifestyle of the Spanish conquerors. As Motolinía (1950: 148) noted, guests at a wedding of eight Spaniards in Telzoco in 1526 brought the couples gifts of 'good jewels and also much wine – the jewel which gave most joy to all of them'. Moreover, wine formed an essential part of the diet of the conquistadors, and its consumption provided one way of, at least temporarily, overcoming the hardships of campaigning in the inhospitable alien environments that they encountered. It was also expensive to transport wine to New Spain, and the high costs involved provided a ready incentive to begin the planting of vineyards and the production of wine in these newly conquered lands. During the sixteenth century the largest vineyards were cultivated on secular estates, although the Catholic missions did maintain a small number of vines, and this again reinforces the view that it was economic rather than ideological factors that were of most importance in explaining the spread of viticulture in the Americas. By the end of the sixteenth century the repercussions of this expansion for Spanish wine producers were keenly felt, and, in a remarkable parallel to Domitian's edict 1500 years previously, Philip II in 1595 promulgated a decree forbidding any further planting of vineyards in Spain's American possessions (de Blij, 1983: 57; Adams, 1984: 514–15). This does not appear to have had much success, despite its attempted reinforcement by successive viceroys over the next two centuries. However, because Catholic religious orders were exempt from the decree, it does seem to have resulted in a relative shift in balance from secular to religious wine producers in the Spanish American possessions during the seventeenth century.

In summary, the development of viticulture in Latin America was essentially based on the use of slave labour, which accords well with Wallerstein's arguments concerning the peripheral regions of his modern world system. However, the wine produced in New Spain was destined for consumption in the colony itself and there is little evidence of any wine from Latin America being exported back across the Atlantic to Europe. Instead, the early stages of the Spanish conquests in America generated a considerable demand for wine exports from Spain to the new colonies. By the end of the sixteenth century this trade had declined, with the increased production of wine in the Americas in turn giving rise to complaints from Spanish wine producers seeking to protect their own positions.

EUROPEAN VITICULTURE IN THE
FIFTEENTH AND SIXTEENTH CENTURIES

Spain and Hungary

The Spanish conquests of Latin America during the first half of the sixteenth century coincided with the integration of Spain into the Habsburg Empire, which under Charles V (1519-56) incorporated Spain, Naples, Sicily, Burgundy, the Netherlands, Austria and Hungary. Following Charles's abdication, Philip II received the Spanish inheritance, and in opposition to France and England attempted to enforce Spanish hegemony in Europe. Spain's Golden Age (Ortiz, 1971), however, was short lived, and the gradual emergence of the independent United Provinces through the eighty years of the Dutch War of Independence (1568-1648) can be seen as symbolising the shift in the balance of economic and political power from Iberia to the countries of northern Europe. By the end of the seventeenth century Wallerstein (1974, 1979) thus sees Spain and Hungary as being relegated to the semi-periphery of his modern world system, while Holland and England had emerged as the new core.

These shifting balances of political power were intimately related to the economic strengths of the different states of Europe, and had important implications for wine production and trade. In particular the market for Spanish wine was considerably enhanced in the sixteenth century by increased demand from the Netherlands and England, as well as the growing urban demand within Spain itself. Wines from the northern Spanish coast, such as Rubidage (Huetz de Lemps, 1968), had regularly been exported to northern Europe in the fifteenth century (Childs, 1978), but the establishment of Spanish suzerainty over the Low Countries provided the context for a considerable boost to Spanish wine exports. By 1540 Antwerp had thus emerged as a major centre for the redistribution of sweet wines, although many of these, particularly from the eastern Mediterranean, were transported by the Genoese (Francis, 1972).

The loss of Gascony to France in 1453, and the growing preference of the English for the sweet wines of the Mediterranean, led to a substantial shift in the balance of English wine imports in the fifteenth and sixteenth centuries to the benefit of Spain and Portugal (Childs, 1978). While periods of political conflict between Spain and England, such as that culminating in the Great Armada of 1588, led to a recession in trade between the two countries, the sixteenth and seventeenth centuries can generally be seen as a time of increasing wine exports from Iberia to England. The popularity of these new wines can readily be gleaned from contemporary English literature, and particularly from the plays of William Shakespeare written in the second half of the sixteenth century.

Pride of place among the wines exported from Spain to England was *sack*, or *seck*, the forerunner of modern sherry, to which Shakespeare frequently referred in his plays *King Henry the Fourth Parts One* (I.ii.3; II.ii.52; III.iii.50) and *Two* (II.iv.120; II.iv.306), *King Henry the Sixth Part Two* (II.iii.60) and in *The Merry Wives of Windsor* (III.v.3; III.v.22; V.v.171), to give but a few examples. Indeed, Shakespeare's description of sack, given in the words of Falstaff, cannot be bettered:

> A good sherris-sack hath a two-fold operation in it. It ascends me into the brain; dries me there all the foolish and dull and crudy vapours which environ it; makes it apprehensive, quick, forgetive, full of nimble fiery and delectable shapes; which, deliver'd o'er to the voice, the tongue, which is the birth, becomes excellent wit. The second property of your excellent sherris is, the warming of the blood; which, before cold and settled, left the liver white and pale, which is the badge of pusillanimity and cowardice: but the sherris warms it and makes it course from the inwards to the parts extreme. It illumineth the face, which, as a beacon, gives warning to all the rest of this little kingdom, man, to arm; and then the vital commoners and inland petty spirits muster me all to their captain, the heart, who, great and puffed up with this retinue, doth any deed of courage; and this valour comes of sherris. So that skill in the weapon is nothing without sack, for that sets it a-work; and learning, a mere hoard of gold kept by a devil till sack commences it and sets it in act and use. Hereof comes it that Prince Harry is valiant; for the cold blood he did naturally inherit of his father, he hath, like lean, sterile, and bare land, manured, husbanded, and tilled, with excellent endeavour of drinking good and good store of fertile sherris, that he is become very hot and valiant. If I had a thousand sons, the first human principle I would teach them should be, to forswear thin potations and to addict themselves to sack.
>
> (*King Henry the Fourth Part Two*, IV.iii.103-36)

Unlike the usual dry and light wines previously available in England, sack was sweet and strong in alcohol. This meant that it could last longer, a point evidently known to Shakespeare who at one point referred to 'old sack' (*King Henry the Fourth Part One*, I.ii.3). The name sack probably originated at the end of the fifteenth century, and derives from the Spanish word *sacar*, to draw out, signifying that it was a wine for export (Jeffs, 1982). At that time the English had established a small merchant colony at Sanlúcar, at the mouth of the Guadalquivir, and it was from here and nearby Cádiz, that the sacks produced from the region around Jerez de la Frontera were exported. Sack was not, though, the only new wine to appear in England during the sixteenth century, and

Shakespeare also notes other wines, such as charneco (*Henry the Sixth Part Two*, II.iii.63), possibly from the region around Colares in Portugal (Francis, 1972), and wine from the Canaries, described by Mistress Quickly as 'a marvellous searching wine', which 'perfumes the blood ere one can say' (*King Henry the Fourth Part Two*, II.iv.29–30).

Of more importance than the export trade for the development of viticulture in Spain, however, was the growth of the home market, influenced both by the increasing extravagance of the royal court, and also by the rise in urban demand among all classes of the population. Both of these sources of demand coincided in Castile, centred on the court at Valladolid, and the burgeoning industrial towns of Segovia, Salamanca and Burgos. Here on the high continental plateau to the south-west of Valladolid, in the vicinity of Rueda and Medina del Campo, there emerged a specialist area of wine production known as the *Tierra del vino* (Huetz de Lemps, 1968; Enjalbert and Enjalbert, 1987). In this relatively dry area, with cold winters, yields were low, and the wines were of a good quality. However, unlike the wines of Jerez which had good access to the sea and thus foreign markets, those of Castile had to undergo journeys of between 250 and 300 kilometres by mule or cart to the ports of northern Spain. Their prime market was therefore Valladolid, where, by the middle of the sixteenth century, it is estimated that average wine consumption reached 100 litres per person per year (Braudel, 1981: 236). This is but one example of an increasing trend towards quite widespread urban drunkenness throughout Europe during the sixteenth century, which, as Braudel (1981) has argued, did not require high quality wine, and which therefore led to the expansion of coarse types of high-yielding vines.

The importance of viticulture to the economy of Valladolid during the sixteenth century is indicated by a collection of ordinances published by order of Philip II in 1597, and entitled *Ordenanzas con que se ha de Gobernar y guardar la entradda del vino y venta del en la ciudad de Valladolid*. These were primarily designed to ensure that the wine trade was controlled by the urban authorities of Valladolid rather than by people living in the surrounding countryside, but they also provided a range of laws to minimise fraudulent mixing and sale of wines. Some of this legislation is derived from earlier ordinances, such as those of the Emperor Charles V issued in 1549, which sought to forbid such practices as the mixing of red and white wine, and the illegal use of various additives which could be dangerous to health (*Ordenanza* XXVI). The second ordinance of 1597 (*Ordenanza* II), which sets the tenor for much of the remainder of the legislation, is particularly interesting in that it comments on the decline of viticulture and the uprooting of vines in parts of the region. This situation is blamed on the failure to enforce previous laws which restricted the time of year when wine could be brought into the city. As a result, the ordinance argues, people were storing poor quality wine in the countryside and were then bringing it into the town whenever it was required. Much of this wine was probably destined for the taverns (*Ordenanza XV*), fuelling the urban drunkenness noted above by Braudel. As well as lowering the quality of the wine, this was deemed to be prejudicial to royal revenues, and consequently the ordinances (*Ordenanza II*) specified that thereafter all wine and must had to be brought into the city between the end of the vintage and the last day of the following February. Moreover, all vineyards were to be registered (*Ordenanza* III), and wine could only be made within the city by licence (*Ordenanza* V). The presence of the court at Valladolid was therefore crucial to the maintenance of quality wine production in the region. It was the liking by members of the court for white wines, high in alcohol content and aged underground in large oak casks, that determined the fortunes of the wine producers of Medina and Rueda. Indeed, at the beginning of the seventeenth century, before the court returned to Madrid in 1606, documentation survives to suggest that it was the usual practice to mature such wines for several years before they were consumed (Enjalbert and Enjalbert, 1987). Thereafter, however, this tradition was lost, and the wine producers of the *Tierra del vino* reverted to the production of cheap wines for drinking within the year in the urban taverns of northern Spain.

At the eastern end of the Habsburg Empire, in Hungary, a different set of processes was at work, transforming the viticultural landscape. In the medieval period Hungary's gold and silver mines had made it one of the richest countries in Europe, but following the defeat of the royal army by the Turks at the battle of Mohács in 1526 the state disintegrated. The west and north of Hungary became part of the Habsburg Empire, while the remainder survived in the hands of a newly elected Hungarian king, John Zápolyai until his death in 1541 when it became fully integrated into the Ottoman Empire. Despite this national tragedy, Hungary became increasingly integrated into the wider economic development of Europe during the sixteenth century, particularly through the export of agricultural products. Zimányi (1987: 20) has thus argued that 'It was in this century that a basic international division of labour was developing: Western Europe exported industrial goods for mass consumption into the eastern part of the continent, which in turn supplied the West with foodstuff for its growing population'. The rise in grain prices caused an agrarian boom in Hungary and a transformation in the surplus extraction methods used by the landlords. During the fifteenth century, the agrarian structure of Hungary was characterised by a large number of small peasant holdings, but throughout the sixteenth century there was an increase in demesne farming in the western and northern regions of Royal Hungary outside the area of Turkish domination. This was associated with a considerable expansion of serfdom. Thus Zimányi (1987:32) has argued that

> The establishment of demesne farms contributed to the increasing employment of unpaid peasant labour, the *robot*. Before demesnes were set up on a large scale, the landowners did not insist on the weekly one day *robot* that had been stipulated in the laws of 1514, passed after the defeat of the peasant uprising. However, by the middle of the century the development of demesne farming led to an increasing demand of labour. At first it could be satisfied by enforcing the laws on servile *robot* for ploughing, sowing, haymaking, harvesting, as well as of wine and food transportation. . . However, the landlords were unable to extort from their serfs all the labour needed for cultivating their increased demesnes, and had to employ hired labourers as well.

This description of the expansion of serfdom associated with increases in grain production on the semi-periphery of Europe accords well with Wallerstein's (1974, 1979) model of an emerging capitalist modern world system. However, as Brenner (1977) has pointed out, there is a paradox in Wallerstein's argument, because serfdom, if it is identified as a fundamental characteristic of the *feudal* mode of production, can hardly be seen to be at the same time an essential element of a wider European *capitalist* mode of production based on the social relations involved in the links between landlords, capitalist tenant farmers and free labour. For Brenner (1977), the non-capitalist organisation of the economies of eastern Europe meant that they were unable to respond in a capitalist manner to market opportunities. It is therefore more reasonable to argue that while in parts of north-west Europe capitalist relations of production were beginning to come into operation by the seventeenth century, Europe was by no means fully integrated into a capitalist mode of production at this time. Instead, the emergence of capitalism in England and Holland can in part be seen as having been enabled by the enforcement of feudal surplus extraction relations in eastern Europe.

While the above arguments pertain largely to the production of grain in eastern Europe, the changing relationships between lord and peasant with respect to vineyard cultivation and the wine trade (Zimányi, Makkai, Kirilly and Kiss, 1968; Szakály, 1979) shed further doubt on Wallerstein's (1974, 1979) model of the development of a modern world system. In Zimányi's (1987: 35) words, 'While in earlier times the serfs of viticultural regions sold their wine in districts without vineyards, towards the middle of the sixteenth century a large part of this market was taken over by the landlords'. This was mainly achieved through the enforcement of the right by which landlords could purchase the wines of their serfs before any other buyer. However, paradoxically viticulture also provided one means by which some peasants could escape from the oppressive restrictions imposed by servile tenure. Unlike arable land, vineyards in Hungary were outside the system of servile tenure, and their owners only had to pay a vine-tax to the landlord and tithes to the Church. The loss to the Turks of Syrmia, which had been the main wine producing area of medieval Hungary, provided a considerable impetus to wine production in other parts of the country, particularly in the Tokaj district in the north-east and the region around Sopron and Lake Fertö in the west. In these areas three different types of wine producers emerged: independent peasants, profiting from the opportunity to avoid the constraint of 'perpetual serfdom'; landowners, cultivating their vineyards through the use of hired labourers; and urban producers, benefiting from the sale of their wines through the municipal wine cellars (Zimányi, 1987). In all of these instances, though, an essential difference between the cultivation of vineyards and arable farming was that the former was both more labour intensive and also required more highly skilled labour. It was for this reason that landlords preferred to hire skilled labourers for their vineyards, rather than to use the *robot* system. Unlike the remainder of the agrarian economy which was bound up within essentially feudal relations of production, the emerging viticultural sector can therefore be seen to have incorporated certain elements of capitalist production from as early as the sixteenth century.

Coincident with these changes in labour organisation, a new style of wine also began to be made in Hungary. In the last third of the sixteenth century wine producers in Tokaj began to add a small amount of juice from late harvested grapes dried on the vines to their wines. This produced sweet dessert wines, which under the name *Tokay aszu* became famous throughout Europe. During the early seventeenth century wine production appears to have continued to expand, and Zimányi (1987: 89) suggests that 'The market-towns of the wine-growing regions remained the centres of peasant commodity production resisting successfully to [sic] the overwhelming competition of the manorial farmsteads'. Despite the pressures on them from landlords and urban owners of vineyards, the peasantry were still able to sell their wine at a reasonable price. Much of this wine was consumed within Hungary, but there was also a considerable export trade, particularly of the sweet Tokaj wines to Poland (Carter, 1987).

French and German viticulture in the fifteenth and sixteenth centuries

The wine producing areas of France and Germany lay closest to the emerging core of Wallerstein's modern world system. With increasing population levels, and the development of a thriving economy in the fifteenth and sixteenth centuries the population of the Netherlands and northern Germany provided an ever increasing market for wine. Typical of the towns importing large quantities of wine from all over Europe was Bruges, which during the fifteenth century drew its wines from Poitou, La Rochelle, Bordeaux and Orléans, as well as countries further afield such as Spain, Italy and Greece (d'Haenens, 1984; *Van Rank tot Drank*, 1990). Moreover, Bruges also acted as an important entrepôt for wine, with customs registers from the end of the fourteenth century, for example, indicating that consignments of Spanish, French and malmsey wines were regularly imported to Gdansk and other Pomeranian towns from Bruges (Carter, 1987). The activity at the Bruges quayside has been marvellously captured in a miniature illustrating the month of October from a breviary attributed to Simon Bening and dating from about 1508 (Figure 34). In this illustration, the wine barrels, or pipes, are being unloaded by a man-powered crane from lighters, which have brought them into the city from larger ships, probably docked at Ecluse or Damme, the outer harbours of the city. One of the barrels is being filled up to compensate for wastage on the sea journey, and to minimise further oxidation and spoilage of the wine. Another barrel is being sampled, with a merchant offering his wine for a prospective customer to taste, while a third person, possibly a broker, looks on between them. Nearby a carrier waits on horseback to transport the purchased wines, with a pottery vessel nestling on straw at the front of his sledge suggesting that he expects to be paid in kind.

Within Germany, Köln continued to act as the main centre for the wine trade, with its merchants exporting the wines of the Rhine northwards to the markets and fairs of Scandinavia and the Baltic. By the end of the fifteenth century Antwerp was one of the largest clients of Köln's wines, with half of all the wines imported into its port of Escaut between 1488 and 1499 coming from the Rhine (Higounet, 1968). Increasingly, however, through the sixteenth century the nature of this trade changed. Dutch merchants gradually came to dominate the wine trade on the sea lanes of northern Europe, and as merchants from Danzig and Lübeck began to travel more extensively southwards they brought back wines from France and even Spain, which were able to compete favourably in terms of quality and price with those of the Rhine. This can be seen as part of a wider change in European viticulture during the sixteenth century, when many northern vineyards, particularly in Flanders, were abandoned in the face of competition from the wines of southern and western Europe, which were increasingly becoming more readily available in the northern markets (Pounds, 1979; Carter, 1987).

Within northern Germany itself, vineyards were found as far north as Stettin and Königsberg, and they continued to have a considerable economic importance along the rivers Oder and Spree near Berlin as late

Figure 34 Early sixteenth century miniature from a Bréviaire attributed to Simon Bening, illustrating the arrival of wine at Bruges
Source: Munich, Bayerishe Staatsbibliothek, Codex lat. 23638, f.11.v.

as 1700 (Schröder, 1978). However, as communications and trade contacts with the rest of Europe improved, consumers were increasingly able to turn to other sources of origin for their wines. Moreover, during the so-called Little Ice Age of the sixteenth and seventeenth centuries (Lamb, 1982; Grove, 1988) colder conditions will have increased the risk of frost damage to the vines in the spring, and will also have shortened the growing season, thus making successful viticulture less reliable and at the same time producing more acid, less alcoholic wines. At a time when stronger, sweeter wines were particularly fashionable, this will have encouraged the merchants of northern Europe to look further afield for their wines. It was the wine producers of Spain and France, who were to benefit most from these shifting balances of trade.

Within France, the collapse of the English market for wines from Bordeaux in the fifteenth century was in part replaced by the increased demand for the wines of the Bordelais from Flanders and the Baltic countries. However, during the fifteenth and sixteenth centuries, with a resurgence of population growth following the devastating plagues of the fourteenth century, the home market, particularly that of Paris also began to play a significant role in the development of French viticulture. The wines of the west coast, with ready access to the sea continued to supply the bulk of the country's wine exports, while those of the Ile-de-France and Champagne supplied Paris. Orléans, situated on the Loire and also near Paris, was in a fortunate position, able to supply both markets. Indeed, it was during the fifteenth and sixteenth centuries that Orléans rose to a position of importance in the wine trade equivalent to that of Bordeaux in the south-west of the country, with its wines frequently being considered to be of an equivalent quality to those of Beaune (Lachiver, 1988). The wines of the Loire were exported from Nantes, with the bulk going to the English, French and Flemish ports either side of the English Channel, although there was also a considerable flow to northern Spain following the peace signed between the two countries in 1559, and smaller amounts found their way as far afield as Ireland, Scotland and the countries of the Baltic. Both red and white wines were regularly exported from Nantes, but there is good evidence that by the sixteenth century certain regions were beginning to specialise in wines made from particular grape varieties. Thus, in Anjou, renowned for its white wines, there was a distinct preference for the Chenin Blanc, known locally as the Pineau de la Loire, whereas around Tours and Orléans the preference was for the Cabernet Franc, also known as the Breton, which gave the wines their much sought-after flavour of raspberries and violets (Lachiver, 1988).

Here, as elsewhere in France, most red wine production seems to have been based on a relatively short fermentation. This has been most clearly attested for the wines of Bordeaux, known as *vinum clarum*, *bin clar* or *clairet.* Until the middle of the sixteenth century most black grapes grown in the Bordelais underwent a fermentation of about 48 hours, after which time the wine was usually a pale colour, approximating the rosé wines of the twentieth century (Marquette, 1978). These were the wines known as *clairet,* from which the English word claret was derived. However, other wines, called variously *vinum rubeum purum, bin vermelh,* or *pinpin,* were also produced, and Marquette (1978) argues that these were of a much darker red colour, produced by pressing the remaining skins following the first fermentation. In many cases water was then added to this dark wine to make a type of *piquette.*

It was not only in the Loire, though, that some degree of regional specialisation in vine varieties was to be found during the sixteenth century. Thus, in Burgundy the reputation of the wines of the Côte d'Or had been established in the fourteenth century, based on the cultivation of the Pinot Noir, and sales of these wines in the north European market were considerably enhanced in the fifteenth century through the acquisition of Flanders and Brabant by the Dukes of Burgundy. At a somewhat later date, in the sixteenth century the three varieties of Pinot (Noir, Gris and Blanc) are also mentioned at Riquewihr in Alsace. It was not the Pinot, though, that was to achieve greatest fame in Alsace, where the foremost grape variety to emerge in the sixteenth century appears to have been the Muscat, first attested at Wolxheim in 1523, at Colmar in 1552, and at Riquewihr in 1575 (Lachiver, 1988: 151). During this period the most important commercial centre for wine in Alsace was Colmar, and from here the wines were exported along the Rhine southwards to Switzerland, and northwards to Frankfurt, Köln, and the Netherlands (Sittler, 1949).

The prime market for wine in northern France during the sixteenth century was the burgeoning city of Paris. Elsewhere in the north, particularly in Brittany and Normandy, cider, which was cheaper to produce, had begun to replace wine as the main alcoholic drink of the poor, but for the region around Paris, the populace of the city provided an ever increasing market for wine. By the sixteenth century, it was not only the urban rich who drank wine; it had also become the standard drink of the poorer classes (Braudel, 1981). There was, however, an important distinction in the types of wine drunk by the different classes: only the rich could afford wines of greater quality from further afield, while the poor consumed the cheaper local wines, whose prices had not been augmented by the high cost of long-distance transport. This production of low quality wines was centred on the Ile-de-France, but to the east, the valleys of the Oise, Aisne and Marne provided easily navigable routes along which the wines of Laon, Reims and Epernay could be brought to the capital. The close proximity of this region to the market of the Netherlands was also of significance, and Sivéry (1969) has shown how, particularly in the early fifteenth century, the high quality white wines of Laon were greatly prized by the Counts of Hainault.

THE STRUCTURAL REORGANISATION OF VITICULTURE AND THE WINE TRADE

Viticultural and vinification practices between 1400 and 1600 owed much to traditions built up over previous centuries, but they also experienced a structural reorganisation which laid the foundation for yet further change in the seventeenth century. Central to any discussion of this reorganisation is the observation that prior to about 1600 profits from wine production and trade were only very rarely reinvested in the production process. Profits were certainly to be made from the wine trade, but these were largely generated by the demand for wine in areas where it was not produced, or where it was only produced in limited quantities. As mercantile activity increased, and it became easier and cheaper to obtain wine from further afield, the vineyards of northern Europe, where wine production was at best unreliable, gradually disappeared. Moreover, the light dry wines of France, which were low in alcohol content and which travelled badly, also began to encounter growing competition from the heavier, more alcoholic wine of Spain and the Mediterranean, which were able to survive the long distances over which they travelled. It was thus the merchants of northern Europe, with good access to new credit arrangements and with knowledge of the markets, who were best able to take advantage of these circumstances.

Wallerstein's (1974, 1980) arguments, although they draw attention to the expansion of European interests in the New World, and usefully emphasise the need to bring together political, social and economic considerations in any understanding of the changes that took place, fail satisfactorily to account for the fortunes of viticulture and the wine trade over these two centuries. In particular the development of viticulture in Latin America, although it was largely undertaken by slave labour, can only be understood in the context of the imposition of a particular Iberian cultural tradition on the newly conquered lands of Mexico, Peru and Chile. Wine was produced here largely to obviate the need to import European wines, which were not only expensive, but which also suffered badly from the voyage across the Atlantic. There is no evidence to support an argument that slave labour was used to produce wine more cheaply in Latin America than it was in Europe, in order to enable American wine to be exported to Europe at a profit.

Moreover, the evidence from eastern Europe, where wine production was generally undertaken either on demesne estates through the use of relatively skilled wage labour, or by independent peasants as a way of obtaining a livelihood free from the servile implications of the *robot* system, also suggests that Wallerstein's (1974, 1980) generalised arguments concerning the appropriateness of particular methods of labour control for specific types of agriculture need to be refined. Particularly significant in this context is the great diversity of labour relations encountered in European wine production during the fifteenth and sixteenth centuries. Within France, for example, Lachiver (1988: 178) has suggested that there was a major shift in the methods of vineyard cultivation between 1300 and 1600, with the nobility abandoning direct exploitation of their lands in favour of leases. With the loss of population resulting from the plagues of the fourteenth and fifteenth centuries, demand for wine fell, and production became less profitable. The result of this, according to Lachiver (1988), was that by 1600 there were many more small scale independent producers of wine than there had been three centuries previously.

Towards the end of the sixteenth century other changes in the organisation of viticulture and the wine trade were also apparent. Certain regions, such as the Côte d'Or in Burgundy, had begun to develop a reputation for quality wines based on particular named grape varieties. This was closely related to the increasing ability of the nobility and higher levels of urban society to purchase wines from further afield. However, as wine consumption spread down the social hierarchy during the sixteenth century, there emerged a distinct division between quality wines, which retained a symbolic element of social status, and locally produced wines, generally of low quality, which were destined for mass consumption by the urban poor. In general, where producers had the choice in areas close to large towns such as Paris and Valladolid, it seems that they usually opted for the production of low quality wines suitable for the mass market.

In such circumstances, there was little incentive for the production of wines that would last for any length of time. However, in their search for wines that would survive the long sea voyages to England, the Netherlands and the Baltic, the merchants of northern Europe increasingly turned to the sweeter, more alcoholic wines of Iberia and the Mediterranean. These fetched higher prices than the wines of France and Germany, and this price differential therefore provided one opportunity for possible exploitation by merchants wishing to invest capital in the production of wine. Moreover, it also illustrated that greater profits could be made from wines that lasted for longer than a year, thus providing an important incentive for the creation of a whole new range of alcoholic beverages. It was in the seventeenth century that these new forces began to transform viticulture.

8

CAPITAL IN THE SPHERE OF PRODUCTION

Up, and having set my neighbour Mr. Hudson, wine cooper, at
work drawing out a tierce of wine for the sending of some of it to my
wife - I abroad, only taking notice to what a condition it hath
pleased God to bring me, that at this time I have two tierces of claret
- two quarter-cask of canary, and a smaller vessel of sack - a vessel of
tent, another of Malaga, and another of white wine, all in my wine-
cellar together - which I believe none of my friends . . . now alive
ever had of his own at one time.
   (*The Diary of Samuel Pepys*, 7th July 1665, [Pepys, 1972:151])

The dawn of the seventeenth century heralded a turning point in the
history of wine. While there had been a number of changes in viticultural
practice during the previous two centuries, particularly in the selection of
specific grape varieties for cultivation in different parts of Europe, capital
had generally been invested in trade rather than in production. From
around the beginning of the seventeenth century all this was to change;
in Lachiver's (1988: 253) words, 'Avec le XVIIe siècle s'ouvre une ère
nouvelle dans l'histoire du vignoble français.' [With the seventeenth
century there opened a new era in the history of the French vineyard].
One aspect of this change can be seen reflected in the contents of Pepys's
wine cellar, noted in the above quotation. This illustrates that by the
mid-seventeenth century a great diversity of wines was being imported
into England, and in particular it reflects the popularity of wines from
areas where viticulture had previously never existed, such as the Canary
Islands. Enjalbert and Enjalbert (1987: 36) see this development as being
primarily a response to political instability. Thus they argue that the
incorporation of Crete and other eastern Mediterranean wine producing
lands into the Turkish Ottoman Empire in the sixteenth and seventeenth
centuries led to the substitution of Malmsey made in Madeira for that
from Crete. They also suggest that the political instability and destruction of vineyards caused by the Thirty Years War (1618–48) in Germany
led to the replacement of the wines of the Rhine in the wider European

market by those from Jerez and Oporto. However, capital was not only being invested in the development of completeley new vineyards. It was also being used to produce new types of alcohol, the products of distillation, brandy, calvados, vodka, whisky and gin. Moreover, by the late seventeenth century entirely new wines, most notably champagne and later port, were also being created through the investment of capital in the processes of production.

While the chronology of these changes is relatively well established, their genesis is far less clear. In particular, the reasons why capital began to be invested in production rather than in trade, and why patterns of demand for different alcoholic drinks changed, need to be further elucidated. The recognition of the prime agents of these changes, however, provides one insight into their likely causes. It was thus not the wine producers themselves, but rather the foreign merchants, first the Dutch and then the English, who first began to introduce new methods of wine and alcohol production. This does not, though, altogether resolve the argument as to whether these changes occurred in response to new patterns of demand in northern Europe, or whether this demand was in fact created by the enterprise of these merchants.

ALCOHOL AND ALCOHOLISM IN SEVENTEENTH CENTURY EUROPE

Braudel (1981: 227) has drawn attention to the important observation that between the fifteenth and eighteenth centuries 'the rise of alcoholism was continuous in Europe'. Although wine had been widely available in the medieval period it was generally expensive to transport, and the more alcoholic wines of the Mediterranean remained a relative luxury in the towns of northern Europe. However, with the discovery of distillation, a beverage could be produced from grain, wine or sugar, which was not only able to survive the hazards of transportation and storage, but which was also far stronger in its alcohol content. While production costs were more expensive, the transport costs per unit of equivalent alcohol content were dramatically reduced, and the final product, vodka, brandy or rum, could always be diluted at its destination. The urban poor, keen to escape temporarily from their hardship, provided an eager market for these new spirits. In Braudel's (1981: 247) words, 'Alcohol succeeded all too well in Europe, which discovered in it one of its everyday stimulants, a cheap source of calories and certainly an easily accessible luxury, with vicious consequences'.

Beer and cider: the northern boundary

Traditionally, ale made from grain was the main alcoholic beverage produced in the countries of northern Europe beyond the latitudes where it was possible to grow the vine. In medieval times in England, the Netherlands, Germany, eastern Europe and Russia, whereas the rich imported wine from abroad, ale was the normal drink of the poor. Such ale was normally quite sweet and, lacking preservatives, it was designed for drinking soon after it was made. However, with the introduction of hops, which added a bitter taste and acted as a preservative, a new type of beverage, beer, was created. The use of hops is first mentioned in Germany in the twelfth century, in the Netherlands in the early fourteenth century, and in England in the early fifteenth century (Braudel, 1981: 238). Subsequently the area of beer production spread southwards through France and Germany to Spain, and Braudel (1981: 238) has noted that 'There was even a brewery in Seville, which was not only the centre of local wine production, but of the international wine trade, in 1542'. Despite this expansion of beer production, though, the main areas of viticulture established in the medieval period survived. Rather than creating a crisis for wine producers, the new brewers served to supply the poor, particularly in the northern towns, with a cheap alcoholic beverage, and in so doing they led to a general increase in alcohol consumption throughout Europe.

Likewise, the spread of cider production in the north of France appears not to have had a major detrimental influence on viticulture. Cider was established in the Basque country well before the twelfth century (Forbes, 1956), and by the eleventh and twelfth centuries it was also being made in Cotentin, the area around Caen and in the Pays d'Auge (Braudel, 1981). During the thirteenth century it was to be found in parts of south-east England (Sutcliffe, 1934), and by the end of the fifteenth and the beginning of the sixteenth century its production had spread to eastern Normandy, as well as to Brittany, but it remained a drink mainly of the poorer classes. The continuing preference of the people of Brittany for wine during the sixteenth century is well attested by the large quantities of wine imported through the Breton ports (Lachiver, 1988), and rather than undermining wine production cider appears to have competed mainly with beer. In Braudel's (1981: 241) words again, 'The newcomer did not interfere with wine; it competed with beer and met with some success, since beer was made from grain and drinking it sometimes meant going without bread'.

Through the seventeenth century, as a result mainly of the greater ease with which wine could be transported, but also in part because of the increasing unreliability of the grape harvest in northern Europe, both beer and cider did make some headway in northern France and Germany at the expense of wine production. However, this influence should be seen mainly in the context of a wider transformation leading to the availability of a greater choice of alcoholic beverages throughout Europe, and whose main characteristic was the veritable explosion in the production of spirits. Thus for Braudel (1981: 241), 'The great innovation, the revolution in Europe was the appearance of brandy and spirits made from grain - in a word: alcohol. The sixteenth century created it; the seventeenth century consolidated it; the eighteenth century popularized it'.

Burnt wine: the origins of brandy

There is much debate concerning the origins of distillation, and whether it was invented first by the Greeks, the Arabs or by the Chinese (Forbes, 1948, 1956; Rose, 1977; Braudel, 1981; Hyams, 1987; Enjalbert and Enjalbert, 1987; Lachiver, 1988). The derivations of the words alcohol and alembic, from the Arabic words *al-koh'l* and *al-anbiq*, suggest that it was from the Islamic world that the practice of distillation first entered Europe. Nevertheless, the ultimate origin of the word alembic, meaning a still, is derived from the Greek ἀμβικ-/ἀμβιξ, meaning a cup or beaker, and it is possible therefore that primitive stills had been developed by the Greeks and Romans. By the eleventh century distilled alcohol appears to have been known to the alchemists of Salerno, in southern Italy, and it probably entered France in the second half of the thirteenth century through the introduction of the works of Arabic alchemists by Spaniards, such as Raymond Lull (Lully) (c. 1233-1315) and Arnaldus de Villanova (c.1235-1313), working at the University of Montpellier. Until the end of the fifteenth century distilled wine, known as *aqua vitae*, seems to have been used largely as a medicine, but in 1493 a Nuremburg doctor noted that everyone in the city had got into the habit of drinking *aqua vitae*, and three years later the city authorities were obliged to forbid the sale of alcohol on feast days (Forbes, 1956: 144). The general spread of distillation seems to have been relatively slow, inhibited in part by the expense of stills, and the closely guarded secret of successful distillation. Thus, within France, the production of brandy only emerged from the control of doctors and apothecaries when Louis XII granted the privilege of distilling it to the guild of vinegar makers in 1514. It was then further encouraged when Francis I also enabled the victuallers to distil it in 1537 (Braudel, 1981).

Demand for distilled wine was mainly from the northern countries of Europe, which had previously suffered most from the detrimental effects on the wine of the long sea voyages from western France and Spain. Moreover, it is widely argued that the colder climates of northern Europe were more conducive to the consumption of beverages high in alcohol content, which provided the drinker with an immediate sense of warmth and well being (Braudel, 1981; Enjalbert and Enjalbert, 1987). Above all, it was the Dutch who, having gained their independence from Spain, provided the largest market for this new type of alcohol. Indeed it is the Dutch word *brandewijn*, literally meaning burnt wine, that came to be used for the new product now known as brandy. Four main factors lay behind this development of Dutch interests in brandy. First, there was a growing demand for cheap alcohol in the burgeoning towns of the Netherlands, which already meant that the Dutch were the most import-ant wine merchants of the era. Secondly, the reduction of Spanish wine imports during and immediately following the Dutch War of Indepen-dence, forced the Dutch to develop new sources of supply for their alcohol. Thirdly, the ready availability of money in the Netherlands, provided the capital necessary for investment in the new methods of production, and fourthly, the wider trade links developed by the Dutch, enabled them, for example, to purchase the copper from Sweden which was essential for the construction of the stills. During the sixteenth century, the Dutch were importing a range of raw materials for distilla-tion into alcohol, but from an early date the government discouraged the use of grain, since this had a tendency to increase the price of bread for the poor (Faith, 1986). Consequently it was the low quality wine, mainly from western France, which found its way into the growing number of Dutch stills.

The strength of the new distilled spirits, which had eight or nine times the alcohol content of wine, provided a strong incentive for the Dutch to establish stills in the areas of wine production so that transport costs per equivalent unit of alcohol would be considerably reduced. Traditionally, Bordeaux, La Rochelle, and the Loire vineyards furnished most of the French wines exported to the Netherlands, and distilleries therefore came to be established in all of these districts, as well as further inland among the vineyards of Périgord and Béarn (Braudel, 1981). However, one region, that of the Charente, was soon to dominate the production of brandy. The reasons for this are not entirely clear, but the wines of the Loire produced to the east of Ingrandes had to pay a heavy export duty when they passed through that town (Lachiver, 1988), and this will have acted as a disincentive to the Dutch, who appear to have been primarily concerned with the exploitation of a cheap source of alcohol. Likewise, to the south in the vicinity of Bordeaux, developments were taking place which led to the later harvesting of the grapes and thus the production of sweeter wines, which were better able to survive the sea voyages to northern Europe. It was thus particularly to the Charente region, midway between the Loire and Bordeaux, that the Dutch turned for their main source of cheap, low quality wine for distillation into brandy. Moreover, this region also had considerable reserves of woodland and forest which were able to supply the new stills with fuel. Furthermore, during the sixteenth and seventeenth centuries the Charente had already experienced a degree of industrial change, with the development of paper mills, iron works, and textile production (Pinard, 1987). This generated an atmosphere of capital investment, which together with new systems of labour relations, predisposed it well to the subsequent development of the distillation industry.

Flemish merchants had been established at La Rochelle during the fifteenth century, and by the mid-sixteenth century the town was exporting *eau-de-vie* to London (Delafosse, 1963). The dominant vine in the coastal region was the Folle Blanche, known locally as the Gros Plant, and this provided large quantities of low quality wine suitable for distillation. However, the specialisation by the vine growers around the town of Cognac on a quality vine, the Colombard, also attracted the Dutch merchants further inland in their search for good quality wines, which were less harsh and more alcoholic than those produced from the Gros Plant (Lachiver, 1988). Paradoxically, brandy produced from the Colombard was less popular than that of the Gros Plant, but the existence of the two types of wine nevertheless enabled the Dutch to begin exporting both good quality wines and brandy from the same region. Thus in 1625, a Dutch merchant is recorded as exporting 124 barrels of *vin de Cognac*, as well as sixty-one casks of *eau-de-vie* from Tonnay-Charente, indicating that by that date the name Cognac was not yet synonymous with spirits (Lachiver, 1988). With the development of their interests in the vineyards of the Charente, more merchants moved to the port of Tonnay-Charente at the mouth of the river, and in 1624 two Dutch merchants also established a distillery here. These developments were further enhanced by the introduction of a tax on spirits at La Rochelle in 1640 (Enjalbert and Enjalbert, 1987: 37).

The Dutch acted largely as merchant brokers, buying up wine for distillation, and exporting it to Holland and the countries of the Baltic. However, the outbreak of war between France and Holland in 1672, and the subsequent banning of French imports into the Dutch ports, was potentially disastrous for the young brandy industry in the Charente. Although the Dutch had provided much of the original capital, together with the stills and the technology necessary for distillation, local producers had also invested in the production of brandy, and they were therefore forced to seek new markets for their products. Elsewhere throughout France, but particularly in Armagnac, other wine producers had also begun to distil their wines, especially when they were of poor quality. Eventually, as Braudel (1981: 244) argues, 'Brandy gradually came to be made wherever the raw material was available. Inevitably it poured out of the vine-growing lands of the south: Andalusia near Jerez, Catalonia, Languedoc'. Fortunately for the producers in the Charente, a vigorous market in luxury beverages was emerging in the late seventeenth century in England. Following the Restoration, an influx of new drinks, such as tea, coffee and liqueurs, entered the London market, and among these brandy featured prominently (Allen, 1950; Faith, 1986).

In explaining this apparent rise of alcohol consumption in late seventeenth century England, most authorities (Allen, 1961; Younger, 1966; Francis, 1972; Braudel, 1981) suggest that it was market led. Faith (1986: 35), in discussing cognac brandy, thus argues that 'its appearance as the superior form of a routine drink would have been impossible without the existence of a market sufficiently rich and fashion-conscious to pay a premium price'. Moreover, at the bottom end of the market it is also widely argued (Francis, 1972; Braudel, 1981; Enjalbert and Enjalbert, 1987; Lachiver, 1988) that soldiers and sailors throughout Europe became accustomed to these new types of alcohol on campaign, or at sea, where brandy and other spirits survived better than wine, and that on return to their home countries their demand provided an added incentive to the enhancement of the international trade in alcohol. At both ends of the market, price was all important, and it was a critical factor influencing the patterns of trade during the seventeenth and eighteenth centuries, with merchants continually seeking the cheapest available wines and spirits in the international market. Differential taxation rates on imports of wine to the countries of northern Europe, however, distorted these patterns, and reflected the various changing political alliances of the period.

Other social and political factors were also important in leading to an influx of new drinks into England in the late seventeenth century. During the Civil War from 1642 to 1648 the wine trade had been much reduced, as a result of the diminished circumstances of many of the nobility and gentry, as well as the introduction of a heavy excise tax on wine (Francis, 1972). Moreover, the rise of Puritanism had also generated an ethos in which drunkenness and extravagant living were frowned upon, further tending to diminish the demand for all types of alcohol (Clark, 1983). Consequently, with the Restoration in 1660, and the return of the royal court to London, there was a resurgence in demand for the luxuries of life, including wine and the new alcohols, such as brandy. Moreover, Charles II had been educated at the French court, and his entourage included many people who had grown accustomed to a wide range of wines and spirits while in exile in France. This further enhanced the demand for a greater diversity of alcoholic beverages to be imported into England.

This demand-led view of the development of the wine and spirit trade is, however, incomplete as an explanation of the changes which occurred. While the addictive character of alcohol undoubtedly meant that once people had started to consume it, many of them tended to demand greater quantities and to degenerate into alcoholism, it is too simplistic to suggest that merchants and producers were simply profiting from a demand inherent in the population for a narcotic, which enabled them to escape from the harsh realities of their lives. The increasing importance of fashion in determining the types of drinks consumed in the late seventeenth century is thus of particular significance, since it suggests that the market was to an extent being manipulated and developed by those with capital to invest in the production of, and trade in, these new types of beverage.

Among the rich, the relative scarcity of wine compared with beer in northern Europe had imbued it since early medieval times with a particular symbolic value. The introduction by merchants of small quantities of new types of high value beverage from the 1660s onwards, likewise created a new range of social symbols for the aspiring elite, who could readily be exploited by those with access to sufficient supplies of the relevant beverage. This system was further enhanced by important changes in the retail trade which saw the emergence of around 2,000 coffee houses in London between 1660 and 1700 (Lillywhite, 1963). Most provincial towns also had their own coffee houses, and in addition to coffee, they provided a range of other new drinks such as cocoa, tea, and brandy. Moreover, coffee houses developed as important social centres for the upper and middle classes, and in the capital as a meeting place for merchants. All of these changes exemplify the way in which the market for alcoholic beverages was being consciously developed by a profit motivated group of merchants. At the bottom end of the social scale, alehouses introduced spirits to the mass of the poor, and with relatively higher excise duties being imposed on beer and ale, the new spirits enjoyed a considerable price advantage. Here again, then, the importers of foreign spirits, and those investing in the production of grain spirits within England, were able to profit from the development of a new market.

Alcohol from grain: the challenge of the new spirits

Brandy remained a product of the vine, but, compared with wine, it was one in which greater capital had been invested in the production process. Until the discovery and spread of distillation, wine, the product of the grape, had been the strongest source of alcohol available in Europe. However, from the fifteenth century onwards (Forbes, 1956) it became possible to produce drinks high in alcohol content from a range of other fruits and grains through distillation. This enabled the countries of northern and eastern Europe, beyond the limits of viticulture, at last to produce their own sources of strong alcohol, and from the seventeenth century these provided serious competition for wine. The great advantage of alcohols such as vodka, kvass, whisky and gin, which were made from grains, was that they were cheap, and they thus provided a ready source of escape for the poor of Europe beyond the Elbe and in the countries adjoining the Baltic and North Sea. Furthermore, spirit producers also often gained from beneficial legislation. Clark (1983: 240) has thus noted how, in late seventeenth century England anyone could 'set up a distillery on giving 10 days' notice to the excise'. Profits from distillation were enhanced by a rate of excise duty on spirits of only 1d. a gallon, by a ban imposed on imported spirits in 1689, and by falling grain prices. Consequently, it was widely reported apocryphally that one could be drunk on gin for a penny, and dead drunk for twopence, with the effects being well illustrated, for example, by Hogarth's well known image of Gin Lane.

Numerous other types of new alcoholic beverage also emerged in different parts of Europe during the seventeenth century. In particular *apéritifs*, also known as *ratafias*, and roughly equivalent to the liqueurs of the twentieth century, achieved a considerable fashionable status. In medieval times several such drinks had been created in Italy, most notably *rosoglio* or *rossoli*, which was a sweet liqueur prepared from raisins and, so it was claimed, sun dew (*ros solis*) (Forbes, 1956; Braudel, 1981). By the fifteenth century, it was widely known by the apothecaries throughout most of Europe that distilled wine, as well as other types of alcohol, could be used as a solvent for the essences of fruit, spices and herbs. This led to a proliferation of different types of liqueur, many of which were produced by the monastic orders. Thus around 1510 a Venetian monk, Dom Bernardo Vincelli, working at the Abbey of Fécamp in Normandy is reputed to have created a liqueur made from distilled wine, 'sweetened with honey and an infusion of some 27 herbs including balm (melisse), hyssop and angelica' to which he gave the name Benedictine (Seward, 1979: 152). Likewise, in 1605 the Carthusian monks of Paris were presented with a formula for an elixir produced from the infusion of 130 herbs in brandy, which eventually became known as Chartreuse. It was not, though, until about the middle of the seventeenth century that such sweet liqueurs became widely fashionable in the main cities of Europe. Another particularly important alcoholic introduction was rum, distilled from West Indian sugar cane. This became popular throughout Holland, England and the English colonies in America, and during the eighteenth century it became the staple drink of the English navy. From the middle of the seventeenth century it was produced throughout the sugar producing lands of the Caribbean under the name first of *killdevil* and then *rumbullion*, the latter eventually becoming shortened to rum (Watts, 1987). Until the development of rum production, the European settlers in the Caribbean had relied upon expensive and often tainted imports of wine, beer and cider, or local drinks such as *mobbie*, made from fermented sweet potatoes. Rum was soon exported back to Europe as well as to the north American colonies, and it was also widely used by the Dutch slave traders in order to acquire their African slaves. However, there is little evidence that rum production seriously damaged the wine trade during the seventeenth and eighteenth centuries. Rather, it served to fuel the widening consumption of alcohol both in the new world and in the old.

PRODUCTION IN NEW ENVIRONMENTS

As well as capital investment being directed to the production of new types of alcoholic beverage to compete with wine, considerable investment also took place during the seventeenth century in the cultivation of vineyards in areas of the world where viticulture had previously been absent, or at best of limited extent. Three particular factors encouraged this development: the expanding market, the establishment of differential customs duties, and the shifting balance of political treaties. Again, as in the medieval period, economic restrictions on trade, expressed largely through the imposition of customs duties, were closely linked to the exercise of political power. The growing population of wine drinkers in northern Europe, enhanced by the greater fashionability of wine among larger sections of society, provided a demand which the Dutch and British governments sought to supply increasingly from their own colonies.

This is clearly reflected in the history of the taxation of wine imports into England during the seventeenth century (Simon, 1906–09; Dietz, 1964; Francis, 1972; Chandaman, 1975; Hoon, 1986). Since medieval times, customs duties had been used both to provide revenue for the Crown, and also as a method of promoting home industry while generating a favourable balance of payments. For much of the sixteenth century, though, the Customs revenues were farmed out, with individuals being granted the right to collect the Customs in return for an annual fixed payment to the Crown. In essence, English policy with respect to wine was both to encourage the importation of wine from territories belonging to the English Crown, and to hinder the trade with countries with which England was at war, which at different times during the seventeenth century variously meant France, Spain and the Netherlands. During the medieval period this policy had found its expression in the generally higher taxation of sweet wines, which was designed to benefit the non-sweet wines from the English province of Gascony. In 1620 James I had imposed an additional duty of 40s. on every tun of wine imported as a means of financing his support for his son-in-law the Elector Palatine during the Thirty Years War (Simon, 1906–09; Francis, 1972). A year later this was reduced, and then in 1634 cancelled altogether, leaving the overall customs duty on non-sweet wine at 40s. a tun. In 1632 the wholesale prices of wine had been fixed at £32 a tun for canary, muscadel and alicante, £26 a tun for sacks and malaga, £18 a tun for wines from Gascony, and £15 a tun for those from La Rochelle and lesser French wines. Retail prices were likewise fixed at 12d. per quart for canaries, 9d. for sacks and 6d. for the wines of La Rochelle (Simon, 1909: 38; Francis, 1972: 54). However, at the time of the the the Civil War the introduction of a new duty had the effect of raising total duties payable on wine, which were specified in the 1642 Book of Rates (Chandaman, 1975). Then in 1651 a new Navigation Act was passed which prohibited the importation of goods in foreign ships other than products of the nation concerned (Thirsk and Cooper, 1972: 502–5). This effectively precluded the Dutch from bringing French, Spanish or other wines to England.

At the Restoration in 1660 there was a considerable reorganisation of trade policy, and two further new Acts, one granting to the King a Subsidy of Tonnage and Poundage dated 28th July 1660 (12 Car. II, c.4), and the other an Act for the Encouragement of Shipping and Navigation dated 13th September 1660 (12 Car. II, c.18), both raised and simplified the duties on wine (Chandaman, 1975). The payment of tonnage was differentiated in terms of the types of wine imported, the ports at which the wines entered the country, and whether they were carried by subjects of the Crown or aliens. Broadly speaking wines carried by subjects paid less duty than those imported by aliens, and wines entering the port of London paid more duty than those imported elsewhere. Thus, French wines imported into London by subjects paid a duty of £4. 10s. 10d., but if carried by aliens paid £6, and when they were brought into other ports aliens had to pay £4. 10s. 0d. whereas subjects only paid £3. These tonnage subsidy rates also applied to the importation of 'Muscadels, Malmseys, Cuts, Tents, Allicants, Bastards, Sacks, Canaries, Malligoes, Maderaes and other Wines whatsoever, commonly called Sweet Wines, of the Growth of the Levant, Spain, Portugal, or any of them, or of any the Islands or Dominions to them or any of them belonging, or elsewhere' (12 Car. II, c.4), although the rate was specified as being per but or pipe, rather than per tun. Wines from the Rhine were taxed more heavily at a rate of 20s. per awm, equivalent to about £6 a tun. The Act also specified additional duties of £3 a tun for wines from France, Germany, Portugal and Madeira, and £4 for all other wines, to be paid within nine months of the wine being imported. If the wine was then exported within 12 months of its importation this duty was to be returned. Francis (1972: 62) notes that in addition to these duties, all wines had to pay a coinage charge of 10s. 0d. a tun. In effect, therefore, French and Portuguese wines had to pay gross customs charges of around £8 a tun, Spanish and Mediterranean wines £9 a tun, and wines from the Rhine £9. 10s.0d.. Although it was possible to claim various allowances, the result was to favour the importation of wines from France and Portugal as against those of Spain, the Mediterranean and the Rhine. Situating these changes within a broader fiscal policy, Chandaman (1975) has noted that these increases in duties on imported wines were designed in part to balance the financial loss from reductions in the duty on cloth exports. The 1660 Navigation Act (12 Car. II, c.18; Thirsk and Cooper, 1972: 520-24) reinforced the main provisions of the 1651 Navigation Act, but it also required that certain plantation exports from the colonies destined for the European market should thenceforth pass through England and thus be subject to the English Customs system. The net effect of these new Acts was that retail prices rose, and the maximum prices at which wine could be sold were therefore specified in a further Act entitled, 'An Act for the better Ordering the Selling of Wines by Retaile, and for preventing Abuses in the mingling, Corrupting and Vitiating of Wines, and for Setting and Limiting the Prices of the same' (12 Car. II, c.25). As well as forbidding a wide range of abuses, this specified that from 1 September 1661 the maximum prices at which wine could be sold were to be 18*d*. a quart for Canary, Muskadell, Allegant or other Spanish or sweet wines, 12*d*. a quart for Rhenish wine, and 8*d*. a quart for Gascoigne or French wines.

In the latter part of the seventeenth century, war with France was accompanied by a range of policies designed to limit the importation of French goods, particularly wine, into England. An Act of 1678 (29 & 30 Car. II, c.1) prohibited imports of French merchandise (Simon, 1909; Kennedy, 1913; Chandaman, 1975), but on the accession of James II in 1685 this was repealed, and overall Customs duties were raised by £8 a ton for French wines and £12 a ton for other wines (1 Jac. II, c.3). At that date Simon (1909) has calculated that overall duties for French wines were £14. 2*s*. 10*d*. per tun, whereas for Spanish and Portuguese wines they were £17. 3*s*. 3*d*., and for Rhenish wines £19. 17*s*. 3*d*. From then until 1689 French wines were therefore taxed less heavily than other wines, and enjoyed a short-lived boom in England. Francis (1972: 99) has thus commented that 'The four years 1686-9 saw more French wines imported than there had ever been since the separation of Aquitaine or were to be again until the twentieth century'. However, following the so-called Glorious Revolution of 1688 which brought William of Orange to the English throne, and the establishment of the Grand Alliance against France in 1689, duties on French wine were once again raised appreciably. Subsequently, further Acts of Parliament provided for additional customs revenues, together with various deductions and exemptions, making precise generalisations about the rates of customs duties hazardous. The most significant of these, though, was an Act of 1696 which raised the duties on French wines by a further £25 a tun (7 & 8 Gul III, c.20), and Simon (1909) has estimated that at that date the various rates had risen to a total of £47. 2*s*. 10*d*. per tun for French wine, £26. 2*s*. 10*d*. for Rhenish wine, and £21. 12*s*. 5*d*. for

Spanish and Portuguese wines. A year later the duty on all wines was raised yet again (9 & 10 Gul. III, c.23) and Simon (1909) argues that the total duty on French wine was by that time £51 2s. 0d. While Francis (1972, 115) prefers a higher estimate of gross duties on French wines of nearer £58 a tun, with those on Portuguese wines being about £30 a tun, there is little doubt that the effects of these differential duties were in practice to force English wine merchants to turn elsewhere for the bulk of the wine which they had previously obtained from France.

One alternative was direct investment in production overseas, not only in the new colonies, but also in Spain and Portugal. Thus the seventeenth century saw an outburst of vineyard development in the Atlantic islands of Madeira and the Canaries, as well as in Virginia. Both of these developments owed much to the increasing market for wine in the colonies of America. Moreover, the Dutch also began direct investment in vineyards in their Cape Colony in southern Africa, with much of the production being used for brandy for export back to northern Europe.

Viticulture in the Atlantic islands: Madeira and the Canaries

The Atlantic islands of Madeira and the Canaries had been discovered in the fifteenth century, and with the Iberian exploitation of Latin America in the sixteenth century, they provided important ports of call for the victualling of fleets bound for the Americas (Figure 33). However, by the sixteenth century, the Portuguese and Spanish authorities had also begun to realise the potential for cultivating luxury crops for which there was a growing demand in northern Europe on these Atlantic islands. Land was readily available and cheap. Moreover, the plantations could be cultivated by slaves, and merchants who controlled the ships and could buy land were able to make considerable profits. The crops that were grown were first sugar, and then vines, tobacco and bananas, but each in turn suffered the fate of all boom economies, in that, as production increased, prices fell.

As early as 1452 Diogo de Teive had been granted a contract for the establishment of a sugar mill and plantation on Madeira, and within four years the first exports to England were recorded (Marques, 1972: 154). Production by the Portuguese expanded rapidly, and it is estimated that in the mid-sixteenth century there were some 40 sugar mills on Madeira, with over 3,000 slaves working on the plantations. In a similar manner, following the conquest of the indigenous Guanche population of the Canary Islands by the Spaniards in the 1490s, sugar plantations were established in Tenerife (Steckley, 1980). However, by the late sixteenth century increasing quantities of Brazilian and Caribbean sugar began to reach the European markets, at a price well below that of the Madeiran and Canarian producers, and by the 1650s West Indian sugar sold for half the price of that from the Canary Islands (Steckley, 1980). As a result merchants and plantation owners in Madeira (Croft-Cooke, 1961) and Tenerife converted their land to vineyards and began producing wine for export both to the Americas and to the European market.

In an analysis of the wine economy of Tenerife, Steckley (1980) has traced in detail the seventeenth century expansion of wine production in the Canaries, and its subsequent collapse in the eighteenth century. The dominant grape variety in Tenerife was the Malvasia, which had spread westwards from the Mediterranean, and produced the sweet white malmsey wines which were gaining such popularity in northern Europe during the sixteenth century. Two other kinds of wine also appear to have been made, one a greenish dry wine, and the other a purplish sweet wine made from late harvested grapes (Steckley, 1980).

Throughout the seventeenth century the price of Canary wine increased, leading to an expansion in the area of vineyards along the entire northern coast of Tenerife. Indeed, from the 1640s onwards, both arable and pasture lands were being converted to vineyards, and this in turn led to recurring subsistence crises. Gradually, therefore, the merchants alongside their purchase of wines for export were also able to develop a thriving grain import trade to the island.

In the late sixteenth century, there were three main markets for the wines of the Canaries: the Spanish and Portuguese provinces in America, Portuguese Cape Verde, and the northern European markets, most notably France, the Netherlands and England. However, by the middle of the seventeenth century, with Portugal having asserted her independence from Spain, the Canaries lost out to Madeira in terms of the wine provisioning of the Portuguese territories in Brazil and Cape Verde as well as the trade to Portugal itself. The prosperity of the wine producers of Tenerife therefore came to rely increasingly on the north European market, and in particular on that of England, where Canarian malvasia attained considerable popularity. Steckley (1980) thus estimates that in the 1630s approximately half of Tenerife's malvasia was marketed through London, with this share rising to two-thirds by the 1690s. Moreover, 'In 1681 London customs officers taxed enough Canary wine to fill roughly 4.5 million quart bottles' (Steckley, 1980: 343). The extent of this trade attracted a number of English merchants to Tenerife, where they established a small factory. However, there is little evidence of English merchants investing directly in wine production through the purchase of vineyards in Tenerife, and in the early years of the seventeenth century it appears that they mainly sold textiles in Tenerife, and used the resultant profits to purchase the necessary wines. As demand grew for the Canary wines, they found great difficulty in financing their trading activities, since it was only possible to sell a small amount of woollen goods on an island where there was a limited need for warm clothing. Even diversifying their imports to Tenerife, to include grain, fish, oil and paper, did little to solve the balance of trade. As a result some English merchants 'regularly handled Canarian tax payments to Madrid, sending bills of exchange drawn upon English partners in the peninsula and retaining the unique Canarian coinage for wine purchases in Tenerife' (Steckley, 1980: 345). Other merchants picked up Canary wines on their way east, as evidenced by John Evelyn's diary entry for 16 January 1662, where, having been entertained at an East India vessel at Black-Wall in the company of the Duke of York, he noted that 'among other spirituous drinks, as *Punch* &c, they gave us *Canarie* that had been carried to, & brought back from the *Indies*, which was indeede incomparably good' (de Beer, 1955: 313). By the end of the seventeenth century, though, rising wine duties in London reduced the margins of profitability in the Canary wine trade, and with the outbreak of the War of the Spanish Succession in 1701 English merchants were forced to depart from Tenerife. This coincided with an expansion in English imports of Portuguese wines, particularly of port wine, and on the resumption of peace the wine trade with the Canaries never again resumed its earlier levels.

As on the Canary Islands, the development of wine production in Madeira was also closely related to the fortunes of political relationships between England and the mainland Iberian polities. However, whereas Tenerife eventually suffered in this respect, Madeira was able to benefit from the establishment of close political links between Portugal and England during the late seventeenth and eighteenth centuries (Vizetelly, 1880; Croft-Cooke, 1961). Despite the development of viticulture in Madeira, the collapse of the sugar industry during the seventeenth century had led to a substantial decline in the affluence of the island and precipitated considerable emigration abroad, particularly to Brazil (Marques, 1972). Following the English Civil War, when Catholic Spain and Portugal supported the Royalist cause, Portugal signed a new treaty of alliance with England in 1654, which among other items confirmed the privileges of the English merchants in Portugal, who were at that time based largely at Lisbon and Viana de Castelo (Walford, 1940; Delaforce, 1979). Direct trade between England and Madeira was relatively insignificant during the late seventeenth and early eighteenth centuries, and as Fisher (1971: 13) has pointed out

> Madeira's significance for English overseas trade at this time largely lay in its position in the English Atlantic trading system, as a market for Pennsylvanian grain and New England cod and a principal source of the wines drunk in the mainland colonies and the West Indies.

This position was largely derived from interpretations of the 1660

Navigation Act and subsequent legislation in 1663 (15 Car. II, c.7), which enabled wines from Madeira, an island off the coast of Africa, to be exported directly to the Americas without having to pass through an English port. Some wines of Madeira had been sent to the Portuguese possessions in Brazil as early as the sixteenth century, but it was this development of the trade with the English colonies in North America and the West Indies in the late seventeenth century that heralded the real expansion of vineyards and wine production on the island. At this date most Madeira wine was sweet, like the wines of the Canary Islands, but, even though it had a good reputation for keeping, it is interesting to note that early in the eighteenth century English merchants such as William Bolton used to rack the wines in the January after the vintage and ship them in the spring, aiming to dispose of them all within the year (Francis, 1972). Meanwhile, other grape varieties were being introduced into the island, most notably Bual, Sercial and Verdelho, and together with the Malvasia these gave rise to the four main types of Madeira wine still produced to this day. Surplus wines were also being distilled into brandy, and by the middle of the eighteenth century many merchants were adding a small amount of this brandy to their wines, to enable them to survive the long Atlantic crossing more readily. The heat of the equatorial zone, together with the motion of the sea voyage was found to improve the wines considerably, and this eventually led to the practice in the 19th century of deliberately sending Madeira wine on ocean voyages across the equator before releasing it onto the market. Later this was to be replaced by the artifical heating of the wines in *estufas*.

During the eighteenth century the attractions of the Madeira trade brought several English and Scottish merchants to the island, and by the end of the century the wine had achieved considerable popularity in England. This was enhanced by the experiences of British troops in North America during the American War of Independence, who gained a liking for the wine of Madeira that they found there, and also by the expansion of the triangular trade between England, the Iberian peninsula and America. Much Madeiran wine thus found its way to England via the Caribbean, and at the end of the eighteenth century Francis (1972: 257) has observed that 'almost every ship reaching Bristol from the West Indies brought at least a few gallons or a hogshead or two' of Madeiran wine.

Virginian wine

Despite the early Huguenot and Spanish attempts to cultivate vines in America north of Mexico during the sixteenth century, it was not until the early seventeenth century, with the development of English colonies in Virginia that viticulture became firmly established in the north of the continent (Pinney, 1989). Initially, the settlers made wine from the indigenous grapes, and the first such recorded wine dates from Jamestown as early as 1609, only two years after the establishment of the settlement (Adams, 1984). The high costs of transport across the Atlantic, and the poor quality of the wine on arrival provided a firm incentive for the first settlers to develop their own vineyards. However, as Lee and Lee (1987: 12) point out, there was another purpose behind these experiments: 'The Virginia Company, chartered by King James I to exploit the economic opportunities of the New World on the mid-Atlantic Coast, had looked forward to developing a wine industry and thereby reducing England's dependence on the European continent for wine'. Unlike the Spanish development of vineyards in central America therefore, the investment of capital in the vineyards of the north was specifically designed to provide wine for export back to northern Europe.

Unfortunately for the Company, the first attempts to make drinkable wine from the indigenous Scuppernong grapes were unsuccessful, and even the introduction by Lord Delaware of French vines and *vignerons* to cultivate them in 1619 proved of little use, since the vines failed to survive. This was largely a result of diseases and pests, such as phylloxera, which at that time remained unidentified. In part it was also due to the delicate nature of the vines imported from Europe, which were not hardy enough to withstand the harsh winters and hot summers of north America. In ignorance of the reasons for failure, the settlers continued to try to develop vineyards, and in 1623 the Virginia House of Burgesses passed an Act requiring all settlers to set aside land for the cultivation of vines as well as other specified crops. In 1624 this was reinforced by a declaration requiring the planting of twenty vines and four mulberry trees for every male over the age of 20 (Lee and Lee, 1987). In the same year the Company was dissolved, and Virginia made a Crown Colony. The successful expansion of tobacco production, and the importation of spirits, particularly rum, to the colony during the seventeenth century acted as further disincentives to the development of a wine industry, and despite prizes being offered to successful wine producers little headway was made. Elsewhere in north America, Governor John Winthrop of Massachusetts was granted Governor's Island in Boston harbour in 1632 for the planting of vines, Lord Baltimore attempted to cultivate them in Maryland in 1662, and William Penn brought Spanish and French vines to the new colony of Pennsylvania in 1683 (Adams, 1984). All of these attempts, though, failed despite the expenditure of considerable energy and capital, and it was not until the nineteenth century that successful vineyards were established on the eastern seaboard of North America (Pinney, 1989).

The Dutch Cape Colony

In contrast to the ill-fated development of vineyards in the English North American colonies, the Dutch investments in viticulture in their Cape Colony in southern Africa proved to be highly successful, and formed the basis for a thriving wine industry through to the present day (Figure 33). Unlike the Catholic Iberian conquest of central and southern America, where viticulture was taken to the New World largely as part of the cultural, social and ideological assemblage of the *conquistadores*, the introduction of vines to southern Africa was essentially driven by economic motives. During the early seventeenth century the Dutch provided the largest export market for wine and brandy (Enjalbert and Enjalbert, 1987), and as the leading trading nation of an expanding global economy they were eager to produce their own supplies of wine.

During the 1640s the Dutch East India Company, Oost Indische Compagnie, described by Braudel (1982: 446) as 'the first of the great companies to enjoy spectacular and fascinating success', was seeking to establish a victualling station to provision its vessels on the long voyage to the east. Consequently a small group of settlers, under the command of Jan van Riebeeck, was despatched to southern Africa, and on arrival in Table Bay in April 1652 they began to construct a fort and houses, together with farms to supply not only themselves, but also the Dutch ships that were to call at the Cape on their way east. Although there seems to have been little encouragement from the *Heeren Zeventien*, or Lords XVII, who managed the Dutch East India Company, van Riebeeck himself was keen to plant vineyards in the Cape. The reasons for this are not entirely clear, but it is evident that he fully appreciated the cost of importing wine to southern Africa for the Dutch settlers, and as a doctor by trade he also seems to have recognised the value of wine for sailors as a defence against scurvy. Consequently he pressed the Lords XVII to send him vines from Europe, and eventually in July 1655 a ship arrived carrying a container of vines. There is some uncertainty about the origins of the vines which he received, but two, the *Hanepoot*, a Muscat of Alexandria, and the *Steen*, a Chenin Blanc, both probably originating from south-west France, came to be of particular importance (Orffer, 1979; Robinson, 1986; Hyams, 1987). These first vines were planted near the fort, but with the arrival of more vines from Europe in the following years van Riebeeck planted a new vineyard at Boschheuvel, the present Wynberg, and on 2nd February 1659 he was able to write in his diary 'To-day – God be praised – wine pressed for the first time from the Cape grapes' (Leipoldt, 1952: 17). Van Riebeeck, nevertheless, had difficulty in persuading the other Dutch settlers to cultivate vines, largely because of their lack of expertise in viticultural methods and the losses caused by animals and birds attacking the vines. The first wines made in the Cape were sweet and strong in alcohol, reflecting the grape varieties, the climate, and the north European taste for such wines, but they were also generally of poor quality. One remedy for poor wines was to distil them, and in 1672, within twenty years of van Riebeeck's landing, a ship's cook is reputed to have distilled the first brandy made from Cape grapes. Subsequently, considerable quantities of brandy were produced in the Cape, largely for local consumption and for the Dutch fleets that stopped there on their way to and from the east.

A second impetus to wine production in the Cape came in 1679 with the arrival of a new Governor, Simon van der Stel, who began planting a new vineyard at Groot Constantia in 1685 (Leipoldt, 1952). There is some debate concerning the first vines that he cultivated here, but within a few years the main cultivars were Red Muskadel, Muscat de Frontignan, and, perhaps somewhat later, Pontac (Orffer, 1979). By 1705 he had also established vineyards of imported German cultivars. Groot Constantia soon developed a reputation for its sweet red dessert wines, which were of much higher quality than those that had previously been made in the Cape. Van der Stel's efforts to expand viticulture were considerably boosted following Louis XIV's Revocation of the Edict of Nantes in 1685, which led to the flight of approximately half a million Huguenots from France. Many first settled in Holland, but partly encouraged by the Lords XVII, who recognised their potential as *vignerons*, a number then emigrated to the new Cape Colony in 1688, where they were settled in Franschhoek, Paarl and Stellenbosch. By the end of the seventeenth century the Company's policy concerning direct agricultural production changed, and in 1695 the Lords XVII decided to rid themselves of their farms, beginning with the vineyards, and to encourage the colonists to produce their own subsistence needs.

During the first half of the eighteenth century wine production, using slave labour, expanded slowly to serve the local population. This expansion was enhanced by legislation which meant that vines were the only crop upon which tithe payments were not due. The government, nevertheless, maintained a monopoly on the sale of wine, and, as Katzen (1969: 207) has argued, 'There was little incentive to improve quality, because the Company paid low prices for a limited quantity of wine and brandy, and levied dues on volume brought into Cape Town for sale without taking quality into account'. By the middle of the century the Dutch East India Company was seeking to replace the European wines imported to Holland by those produced in the Cape, but with the exception of the wines of Constantia the Cape wines were generally considered to be inferior to those obtainable from elsewhere. This resulted partly from the poor prices paid to the wine producers, but also from a continuing shortage of good quality containers in which the wine could be exported to Europe. The indigenous woods of southern Africa proved to be unsuitable for making barrels, and consequently most wine had to be exported in barrels that had already brought produce, such as arak from the east or salted pork from Holland, to the Cape. Moreover, it was possible to buy French and Spanish wines on the open market in Europe far more cheaply than those of the Cape, and Dutch traders therefore sought to resist the pressures put on them by the Company to buy Cape wines.

The Dutch East India Company nevertheless made considerable profits from its encouragement of viticulture, and its monopoly of wine sales. There was an excise duty payable on any wine made in the Cape, and all retail wine shops were taxed. Moreover, it was the Company, rather than the producer, who benefited most from the sale of Cape wine in Europe, with Hyams (1987: 315) noting that

> For the leaguer of White Constantia which the Company was selling in Europe for between £120 and £190, the grower and maker received £12. For the Red Constantia selling to the merchant at up to £333 in Amsterdam, the grower and maker received £16.

By the end of the eighteenth century, though, the fortunes of the Dutch were on the wane. The British first occupied the Cape between 1795 and 1803, and then following the resumption of the Napoleonic Wars again attacked the colony in 1806 (Davenport, 1969). Although they were not confirmed in their possession of the Cape until 1815, the British rapidly developed the potential of Cape wines as an alternative to those of France. In 1811 an official wine-taster was appointed to ensure the quality of wines exported from the Cape, and by 1815 wine exports to Britain had risen from a mere 60 leaguers in 1811 to 3,647 leaguers. Thereafter, wine production expanded considerably, and by 1864 some 55 million vines had been planted, compared with the 13 million that had existed in 1819 (Leipoldt, 1952).

The examples of the Atlantic islands, of Virginia and of the Cape Colony, therefore, all provide evidence of the introduction and expansion of viticulture into parts of the world where vine cultivation had previously been absent. This development, although not always successful, was fuelled by the growing number of wine consumers in the markets of northern Europe, but it was effected by landowners, merchants and companies seeking to profit from capital investment in both trade and also in production.

THE CREATION OF NEW WINES

As well as the production of new types of alcohol, and the development of viticulture in new places, the seventeenth century also witnessed the creation of entirely new types of wine. The most notable of these were champagne and port, but the changes that took place in France and Portugal were to have profound implications for the future of the wine trade throughout the world. Until the seventeenth century the fundamental problem which had faced all wine merchants was that they were dealing in a commodity which had only a limited life span. Particularly when tossed around in barrels on the high seas, wines rapidly went sour and became unpalatable. Wine, therefore, had almost always to be sold within the year, and it was usually the merchants rather than the producers who took the risk of having old stock on their hands at the time of the new vintage. The dislocation between the main areas of wine production in France, Iberia and Germany and the areas of consumption in northern Europe necessitated the development of a complex series of commercial relationships in the early modern wine trade, many of which lasted well into the twentieth century. In general the vine growers and wine producers sought to sell their wines as soon as possible after the vintage, so that they would be able to pay off debts built up over the previous year, and have cellar space and sufficient barrels for the next year's vintage. For the merchants it was also desirable to purchase new wines as quickly as possible after the vintage in order to replenish their stocks and satisfy the orders of their customers. Some Dutch and English merchants therefore began to establish themselves in the main areas of wine production in Iberia and France in order better to furnish supplies of wine. Typical of these was the Spanish Company established by English merchants at San Lucar de Barrameda, Seville, Puerto de Santa Maria and Cadiz between 1530 and 1585 (Croft, 1973), and the Dutch communities in Bordeaux and Tonnay-Charente. Other foreign merchants relied on purchases from indigenous French, Spanish or German merchants, who bought up supplies of local wine and then sold them for export.

During the sixteenth century there had been some limited experiments to store wine for longer than a single year. In Germany it had been discovered, probably during the fifteenth century, that wines kept in larger barrels lasted longer than those in smaller barrels (Rohr, 1730). Although the precise reasons for this were unknown at the time, larger barrels presented a smaller percentage of their contents to the air, and this therefore reduced the rate at which the *Acetobacter* turned the ethanol in the wines to acetic acid. Providing these large barrels were not racked, and were kept topped up by new wine every time any older wine was withdrawn, the wines within them were able to last for longer than just a year. This effect was considerably enhanced when wines from a particularly hot year, with a high sugar content, were used. One of the earliest recorded such barrels was the Strasburg Tun of 1472, and among the most famous were the Heidelberg Tuns of 1591 and 1663, the latter of which held some 150 hogsheads, equivalent to 37,500 gallons (Allen, 1961). Such barrels, however, were expensive to construct, and the impossibility of moving them meant that the wines could only be aged near their place of origin.

Elsewhere, it had long been recognised that sweet wines with high alcohol levels lasted better than light dry wines, and in the late sixteenth century it appears that Spanish wine producers were also attempting to age their wines. Writing at the end of the sixteenth century, Bacci (1596: 265) thus refers to the addition of cooked wines, *vini cocti*, to those being exported from southern Spain. As a result of evaporation, these cooked wines were particularly sweet, and their addition to normal wines increased their overall sugar content, making them less liable to spoilage. Typical of the wines of southern Spain were those from Jerez, which by the seventeenth century had begun to be known as sherris-sack, and later sherry. There is much debate about whether or not these wines had been fortified with brandy during the sixteenth century (Croft-Cooke, 1955; Younger, 1966; González Gordon, 1972; Jeffs, 1982), but even without this fortification their strength ensured a higher degree of stability than that of most other wines. Moreover, the development of *flor* on some sherries, which was certainly recognised by 1616 (Younger, 1966), further helped to prevent their deterioration. *Flor* consists of yeasts of the species *Saccharomyces bayanus*, *Saccharomyces capensis* and *Saccharomyces fermentati* (Amerine *et al.*, 1980: 172), and occurs naturally in the fermentation of wines in both Jerez and the Jura region of France. Its effects are to form a film of yeast on the surface of the wines, which protects it from the activities of *Acetobacter*, and it is found to develop most successfully when the temperature of the wine is between 15° and 20°C, with the alcoholic strength being between 13.5° and 17.5°, preferably in the more limited range of 15°–15.5° (Fornachon, 1953). *Flor* will only develop on dry white wines, and it may well be that some such wines, with a sufficiently high alcohol content may have developed naturally in Jerez prior to the late sixteenth century, to provide a dry white wine which would last for a considerable time in barrel. Furthermore, elsewhere in Spain, in Valladolid, in the early seventeenth century there is also evidence that wines were being aged in the city's bodegas for several years before being released for sale (Enjalbert and Enjalbert, 1987). Despite these early beginnings, though, it was not until the discovery later in the seventeenth century that wine kept in bottles, sealed with a cork, would last for much longer than a single year, that the new vintage age of wines was born.

Glass bottles and vintage wines

It is widely argued (Allen, 1961; Younger, 1966) that it was the technical advances of the seventeenth century that were central to the development of vintage wines, with the two innovations of most importance being the introduction of glass bottles and shaped corks. However, glass had been made in Venice probably continually since Roman times, and by the eleventh century a small number of glass bottles were apparently already being produced in northern Europe (Charleston and Angus-Butterworth, 1957). Glass bottles, though rare, were therefore nothing new in the seventeenth century. For vintage wines to be created there also had to develop an awareness of their potential as a means for producing a new kind of wine. This was an acquisitive capitalist awareness, keen not only to appropriate surplus value, but also to produce it. The opportunity for this was enhanced by the Venetian development of a method of producing clear crystalline glass during the fifteenth century, and its subsequent dissemination throughout Europe during the sixteenth century, which led to an expansion in the use of glass bottles for serving wine.

The subsequent shift from the serving of wine in bottles to its maturation therein took place during the course of the seventeenth century, in association with the development of a number of new types of wine, most notably the new wines of Bordeaux, champagne and port. However, changes were also taking place elsewhere, and Rohr (1730), writing about German viticulture and wine making at the beginning of the eighteenth century, commented that wines were best when stored in bottles tightly sealed with corks and placed in sand in a dry cellar. The critical characteristic of these new wines was that they had the ability to age, and could be sold for more than the other types of wine then available. Thus, capital investment in their production was rewarded by a higher return than had previously been possible. This not only required merchants with the foresight to begin experimenting by putting wine into bottles, but it was also essential for this investment to be realised through the subsequent sale of the wine. Many factors influenced the changing tastes of wine consumers in northern Europe during the late seventeenth century, but central to any explanation of those that occurred was the desire by the ruling classes to preserve a symbol of their position in society. At a time when other types of alcoholic beverage, such as gin and cheap wines, were becoming increasingly available to the mass of the population, the existence of a new and expensive type of wine provided just such a symbol.

The characteristics of these new wines, and the methods of their production, however, played an important role in the subsequent restructuring of the wine trade. The fragility of glass bottles meant that wine was still generally transported in barrels, and only bottled on its arrival in the northern markets. Consequently, the bottling was undertaken by the merchants or eventual purchasers rather than by the wine producers, and this therefore shifted the relative balance of power in favour of the former. It was the merchants and consumers who invested the capital in bottling the wines, and it was they who would thus benefit from any additional profits. Moreover, the storage of wines for more than a year was beyond the scope of most small wine producers. If producers were to store their wines in barrels, they also had to purchase additional ones for the next year's wine production, and any storage of wine also involved the tying up of capital which could not be realised until the wines were sold. Most wine producers were unable to afford either of these expenses. Again, therefore, it was only the merchants and some large scale producers, who were able to invest the necessary capital in the storage of wine, and who would consequently reap the greater profits to be gained from its sale.

The new wines of Bordeaux

On 10th April 1663, Samuel Pepys wrote in his diary,

> Off the Exchange with Sir J. Cutler and Mr. Grant to the Royall Oake Taverne in Lumbard-street, where <Alexander> Broome the poet was, a merry and witty man I believe, if he be not a little conceited. And here drank a sort of French wine called *Ho Bryan*, that hath a good and most perticular taste that I never met with.
>
> (Pepys, 1971: 100)

This wine was none other than that produced on the property of Haut-Brion in the Graves, by Arnaud de Pontac, first President of the Bordeaux Parlement. Three years later in 1666, Arnaud de Pontac sent his son to London, where he opened a restaurant, grocers and tavern named the Sign of Pontac's Head, and here he introduced his wines to the discerning elite of London society, including Daniel Defoe, Jonathan Swift and John Locke (Pijassou, 1980; Penning-Rowsell, 1985; Enjalbert and Enjalbert, 1987). Unlike most other wines imported into England, the wine of Haut-Brion was produced on a single vineyard of 38 hectares, with its particular quality deriving from its soil, the westward facing slope on which the vines were planted, and the age of the vines (Lachiver, 1988). Pontac had grasped, as would other merchants and landowners around Bordeaux in the ensuing decades, that the future of the vineyards of the region lay in the production of high quality wines, which could be sold at a suitably high price to warrant the additional investment involved in their production (Pijassou, 1974).

Following the collapse of trade with England in the middle of the fifteenth century, the wine merchants of Bordeaux increasingly turned to the Netherlands as a market for their wines during the sixteenth and seventeenth centuries. Dutch merchants had also begun to establish themselves in the Chartrons quarter of Bordeaux, from where, with the complicity of local *négociants*, they exported a range of wines and in particular those of Gaillac and Cahors, which were stronger and more popular than those of Bordeaux. Moreover, during the seventeenth century a number of Dutch merchants settled permanently there, becoming naturalised citizens of Bordeaux, so that they would benefit from the tax exemptions granted to the burgesses of the city (Enjalbert, 1953). In addition to the strong wines of the Haut-Pays, the Dutch merchants were also interested in exporting sweet white wines and *eaux-de-vie*. As a result, during the seventeenth century numerous stills were constructed throughout the Garonne valley (Enjalbert and Enjalbert, 1987), and the date of the vintage was pushed back so that the grapes had a higher sugar content and the resultant wines were sweeter. By 1666 this last practice was well established in the Sauternes, and by the end of the century it was also known further east at Monbazillac (Lachiver, 1988). Eventually, under the influence of the Dutch, it was recognised that if the Sémillon grapes in these particular areas were left until they were attacked by a *pourriture noble*, later identified as *Botrytis cinerea*, they would attain à sweetness unmatched elsewhere, and for which a high price could be demanded.

However, as Enjalbert (1953: 325) has argued, 'Les Hollandes, qui avaient si largement contribué à la prospérité agricole et viticole de l'Aquitaine, furent les agents conscients de sa ruine' [The Dutch, who had contributed so largely to the agricultural and viticultural prosperity of the region, were the conscious agents of its destruction]. Following the restoration of peace between the Netherlands and Spain in 1648, and the subsequent outbreak of war with France in 1672, the Dutch turned to the sweet wines of Portugal and southern Spain, in particular those of Setúbal, Jerez and Malaga, rather than those of France for their supplies of wine. Moreover, although the wines of the Graves had reached London society through the activities of such enterprising men as Arnaud de Pontac in the 1660s, the outbreak of war between France and England, the 1678 prohibition of French imports into England, and the subsequent imposition of heavy customs duties on French wine, all led to a crisis for the Bordelais.

The solution was to abandon the production of low quality wines, and to invest in specialist wine production which would bear the high costs imposed by war and customs barriers. Gradually, at the beginning of the eighteenth century, more and more names of individual wine producing properties begin to appear, and despite trade restrictions during the War of the Spanish Succession (1702-13), these wines found their way onto the English market, many arriving from Dutch ports or as prizes captured from French fleets. A remarkable source of information concerning the wine purchasing habits of the English nobility at this time, survives in the Book of Expenses, referring to the years 1688-1742, and written by

John Hervey, first Earl of Bristol (Hervey, 1894). During the late seventeenth century he was purchasing numerous different types of wine, including Hermitage (1690, 1691, 1692), Rhenish (1692), Gallicia (1692), Navarre (1696, 1699), Sack (1696), Palme (1697), Canary (1697), Burgundy (1697), Languedoc (1699) and St Laurent (1699), and he was apparently well used to bottling his wines. Thus, in an entry for 16th December 1692, he records the following: 'Paid my share to Mr. Long for ye hogshead of Gallicia wine at £17.10 & ye charges of corking, bottelling etc., £11.1.10' (Hervey, 1894). In 1700 he first recorded a purchase of champagne, and then from the early years of the eighteenth century he began buying wines from single named properties around Bordeaux. In 1702 he purchased four hogsheads of 'obrian' (Haut-Brion) for £101.10.0., in 1703 he bought a single hogshead of 'Margoose clarett' (Margaux) for £27.10.0., and in 1705 he bought a further two hogsheads of 'Obrian' (Haut-Brion) and one of 'white Langoon' (Langon). Thereafter Haut-Brion is mentioned in various spellings in 1707, 1709 and 1714, and Margaux in 1706, 1716, 1718, 1723, and 1724. Other properties mentioned slightly later include Pontac (1714, 1717, 1724), 'La Tour Claret' (1720) and 'La ffittee Clarett' (1721 and 1724). Prices for such wines were high, with Margaux selling for £43 a hogsead in 1706 and Haut-Brion for £48 a hogshead in 1709, and by the second decade of the century, Hervey seems to have preferred to buy them in smaller quantities by the flask.

Previously, most of the wines exported from Bordeaux had come from the south or east of the city, but these wines, known at the time generally as New French Clarets, were from the Médoc, to the north. Here, on the poor quality sands and gravels, the first half of the eighteenth century saw a veritable explosion of vineyard development, as money was invested in the production of high quality wines (Higounet, 1974; Pijassou, 1980; Enjalbert, 1983). Many of the owners of these new vineyards in the Médoc were lawyers and magistrates who had acquired their positions and wealth from their membership of the Bordeaux Parlement (Penning-Rowsell, 1985), and Forster (1961: 23) has noted that during the eighteenth century 73 per cent of the aggregate income of the sixty-eight families of this *noblesse de robe* came from the sale of wine. Gradually, improved methods of vinification, gained from practical experience rather than any real theoretical understanding, were introduced. By 1710 at Château Margaux black and white grapes were being separated prior to vinification, and by 1740 new oak barrels were being used, with the wine regularly being racked in April and then fined using egg whites (Faith, 1980; Lachiver, 1988). Moreover, sulphur wicks were being used to sterilise the barrels, and among the main properties a distinction was being made between the best and the second quality wines. The best vineyards were generally cultivated using day labour, but 'for vineyards of less value the owners preferred to hire a peasant family for the entire year to work a certain number of acres' (Forster, 1961: 26). Most of the wines were, in the first instance exported to Amsterdam or Rotterdam, but considerable amounts also went to the ports of Brittany, from which some at least may have been smuggled into the west coast ports of Britain (Simon, 1926). The New French Clarets were the preserve only of the very rich, who could afford the high duties that were levied on them, and they thus became one element in the new symbolism of the rich and powerful of early eighteenth century England.

The creation of champagne

Wines from the region around Reims and Epernay had been well known in Paris during the medieval period, and had been widely purchased by the nobility of neighbouring countries, such as the Counts of Hainaut (Sivery, 1969). However, during the seventeenth century, as in Bordeaux, numerous wine producers in Champagne began experimenting with new techniques of viticulture and vinification. The most famous of these was Dom Pierre Pérignon, cellar master of the Abbey of Hautvillers from 1668–1715 (Figure 35) (Gandilhon, 1968). Eager to improve the quality of the abbey's wines, he reorganised its vineyards and began selecting specific grape varieties, most notably Pinot Noir, for its wines. Moreover at the vintage, he chose individual panniers of grapes, which were blended together to produce the particular wine that he wanted. These grapes were pressed rapidly, to minimise contact with the skins, and the fermentation produced an almost white still wine which was known as *vin gris*. The result was a high quality wine, which could again sell at premium prices.

Following the Restoration in Britain, the wines of Champagne achieved increasing popularity in London, partly through the proselatysing activities of the French soldier, philosopher and courtier, St-Evremond, who came to the Court of Charles II in 1660 (Simon, 1962). These still wines were usually imported to England in barrels during the late autumn or winter, and following the growing fashion, they were then normally bottled on arrival. This bottling often took place in the great houses of the nobility, rather than in the merchants' cellars, and numerous accounts, such as those of the Duke of Bedford at Woburn Abbey dating from the mid-1660s, indicate that regular purchases were made of bottles and corks together with wine from Champagne (Simon, 1962: 49). It was soon discovered, however, that the wines tended to develop an effervescence after several months in the bottle, and this new sparkling wine rapidly became extremely popular among those who could afford it. The sparkling nature of such wines had certainly been established by 1676, when Sir George Etheridge mentioned 'sparkling Champagne' in his play *The Man of Mode; or Sir Fopling Flutter* (Act IV,

Figure 35 The village of Hautvillers
Source: author 24th April 1987

scene 1). It seems that the natural fermentation of the wines made around Epernay and Reims was not usually completed before the cold winters in the region had set in, and the fall in temperature caused the fermentation to cease while there was still a substantial amount of sugar left within the wines. Thus, in the spring, when the temperature began to rise again, the fermentation would recommence. If this was allowed to take place in sealed bottles the carbon dioxide given off during the fermentation could not escape, and the resultant wines became sparkling (Enjalbert and Enjalbert, 1987).

The popularity of these wines grew rapidly, and ways were sought, both by the merchants and those who regularly imported their own wines, to guarantee the success of this second fermentation, particularly through the use of stronger bottles and better fitting corks. The development in England of coal-burning glass furnaces during the seventeenth century, following the proclamation of 1615 which forbad the use of wood for glass production, had led by the early 1660s to somewhat stronger glass than had previously been available. Then in 1675 George Ravenscroft perfected the use of lead oxide in its manufacture, and this further stabilised and strengthened the glass (Charleston and Angus-Butterworth, 1957; Charleston, 1984). By the end of the century, therefore, English manufacturers were able to make strong glass bottles well able to withstand the pressures involved in the production of sparkling wine. As the practice of making champagne became more sophisticated, better fitting corks also came to be used, and their removal was facilitated by the development towards the end of the seventeenth century of the bottle-screw, or corkscrew (Younger, 1966: 352). English merchants had traditionally also not been averse to adding a variety of ingredients to the wines that they imported, and when it was realised that the addition of sugar before bottling greatly increased the amount of sparkle in the wine, this led to the quite widespread production of adulterated sparkling wines.

Both strong bottles and close-fitting corks were uncommon in Reims and Epernay until about 1695 (Bondois, 1929), but as early as 1673 Dom Pérignon had ordered a new wine cellar to be excavated into the rocks above the Abbey of Hautvillers (Figure 35), and it seems that he too soon began to experiment with the new bottling techniques in order to create a sparkling wine (Gandilhon, 1968). Lachiver (1988: 278–9) thus notes that, having racked them, he usually bottled his wines around 15th March and then left them in the cellar for about 18 months before selling them. The outbreak of war between England and France, however, severely curtailed the export of champagne to England, and it was not until the eighteenth century, when the wines became fashionable at the French Court, first during the Regency of the Duc d'Orléans (1715-23) and then under Louis XV (1723-74), that their popularity became assured (Enjalbert and Enjalbert, 1987). The extravagance of the court provided a ready market for quality wines from throughout France (Fizelière, 1866), but the proximity of the Champagne region to Paris, and the unique effervescence of its wines, made them among the most popular wines drunk by such figures as Madame de Pompadour, the poet Voltaire and the Duc de Richelieu. This encouraged further investment in the vineyards and cellars of Reims and Epernay, and gradually there emerged a group of companies devoted solely to the production of champagne. The oldest champagne company, Gossart, dates from 1584, but of the new eighteenth century creations the earliest company still surviving dates from some time before 1729, when there occurs the first entry in the ledger of a cloth merchant and vineyard owner from Epernay called Nicolas de Ruinart (Hautefontaine, 1979–80). By 1769 his tenth son, Claude, had divested the cloth interests to other members of the family, and established his own champagne company, Ruinart Père et Fils, at Reims. Other famous companies were established soon afterwards, most notably the House of Moët founded in Epernay in 1743 by Claude Moët, and the company created by François Delamotte at Reims in 1760, which through the association of Jean-Baptiste Lanson early in the eighteenth century subsequently became known as Lanson Père et Fils.

Port

The third new wine created towards the end of the seventeenth century was port, named after the town from which it was exported, *Porto* in Portuguese, and Oporto in English. As with claret and champagne, its development required the use of bottles in which the wine could mature, but unlike these other wines the success of its longevity lay in its fortification with brandy. Moreover, its rise to fame owed much to the political treaties established between England and Portugal at a time when the former was at war with France (Sellers, 1899; Simon, 1934; Croft-Cooke, 1957; Francis, 1966, 1972; Howkins, 1982; Bradford, 1983).

By the middle of the seventeenth century trade between England and Portugal was on the increase, with English merchants or factors seeking to sell English cloth in exchange for the luxury products of Portugal's surviving overseas territories. During the previous centuries some Portuguese wines, particularly the sweet wines produced in the Algarve and around Lisbon, had found their way to England, but during the reign of Charles II relatively little Portuguese wine appears to have been imported into the country, even though he was married to the Portuguese Catherine of Braganza. English Factories had gradually been established at Lisbon (Walford, 1940) and Oporto (Sanceau, 1970; Delaforce, 1979), following the 1654 Treaty between England and Portugal which provided for commercial privileges and religious freedom for the English Protestant merchants (Delaforce, 1982). To the north of Oporto, English merchants had long been based at Viana do Castelo, from whence they exported the local light wines, which were probably similar to the *vinhos verdes* still produced in the Minho. These, however, did not travel well, and during the 1670s and 1680s a number of merchants began to search further afield for fuller bodied, more alcoholic wines. Portuguese protectionist policies made it difficult for English merchants, and in 1677 they petitioned the English Parliament unsuccessfully to reduce the duties on Portuguese wines, so that they would be able to import them more readily in exchange for English cloth exports, which at that date were worth in the order of £400,000 a year (Simon, 1926; Francis, 1972).

With the outbreak of war with France in 1678, the opportunity was provided for the English merchants in Portugal to increase the scale of their activities, and imports of Portuguese wines into England reportedly rose from 1,013 tuns in 1679 to 16,772 tuns in 1683. Nevertheless, as Francis (1972) has pointed out, exports of wine from Oporto and elsewhere in Portugal do not seem to have risen commensurately, and much so-called Portuguese wine imported to England at this time was probably really from France. With the renewal of hostilities in 1689, and the subsequent dramatic increases in duties on French wines imported into England, genuine exports of Portuguese wines to England did gradually begin to increase. Much of this wine came from Madeira, Algarve and the region around Lisbon, which produced heavier wines better able to survive the voyage to England. Some wines were exported from the north of the country, but in many cases it seems that these were adulterated and mixed with other wines to give them more body.

By the end of the 1670s some enterprising British merchants had begun to explore further inland for fuller bodied wines, and legend has it that the wines of the Upper Douro were first discovered by two sons of a Yorkshire shipper, who had been sent out to learn about the wine trade, and had been given some wine from Pinhão by the Abbot of a monastery in Lamego (Francis, 1972; Robertson, 1982; Bradford, 1983). After persuading them to buy some of the wine, the abbot reportedly told them that he had added some brandy to it during fermentation in order to conserve its sweetness and increase its body. Whatever the truth of the legend, it appears that by the 1670s the practice of fortifying wines with brandy had been established in northern Portugal, possibly having been introduced from neighbouring Spain, and from the 1680s onwards the British merchants turned increasingly to these wines of the Upper Douro valley around Régua for their supplies. There is little evidence of merchants buying their own vineyards at this date, and it was the local landowners who expanded the area under vineyards, and then sold their wines to the British merchants based in Oporto.

It was not until the early eighteenth century that the port trade really began to expand, and its success was largely as a result of the political and commercial treaties negotiated between England and Portugal by John and Paul Methuen in 1703 (Francis, 1966). The former of these, the Offensive Quadruple Treaty and the Defensive Triple Treaty signed on 16 May 1703, brought Portugal into the Grand Alliance, and the latter Commercial Treaty signed on 27 December 1703 provided for the export of English cloth to Portugal in return for the importation of Portuguese wines to England at a preferential rate of duty one-third less than the duties on French wines. The effects of this were to encourage more British merchants to settle in Portugal, and for the landlords of the Upper Douro considerably to expand the area of their vineyards and their production of wine. By the middle of the eighteenth century approximately two-thirds of all wine imported into England came from Portugal, and much of this was supplied from the Douro valley (Francis, 1972). Most of the British merchant houses established during the seventeenth century were dealing in other products as well as cloth and wine, but it is from this period that many of the main port shippers date their foundation. Interestingly the oldest company, C.N. Kopke & Co was founded by a German, but a number of British companies such as Warre & Co established by John Clark in 1670, Croft & Co founded in 1678, Quarles Harris founded in 1680, and Taylor, Fladgate and Yeatman founded in 1692, were established before the end of the seventeenth century. Other companies came into being early in the eighteenth century, with Morgan Bros dating from 1715, Offley Forrester from 1729, Butler & Nephew in 1730 and Hunt Roope in 1735 (Robertson, 1982).

Much of the wine produced from the Douro at the turn of the eighteenth century was, however, harsh and despite its relatively cheap price it did not immediately attract a strong following in England. The astringency of the wines was caused partly by the grape varieties used in its production, but also by the high temperatures of the Upper Douro valley which, not only produced grapes with a high sugar content, but also caused rapid rates of fermentation. Moreover, many producers added elderberries to the wine in order to improve its colour. Gradually, during the first third of the eighteenth century, it was realised that the addition of brandy prior to the end of fermentation made sweeter and yet stronger wines, more suitable to the English palate. Nevertheless, the brandy and wine took time to blend, and when drunk young these fortified wines remained relatively unpleasant. Consequently, many shippers began ageing the wines in cask before export, to allow the wine and brandy time to blend together.

The first half of the eighteenth century witnessed continual conflict between the shippers and the landowners of the Upper Douro valley. As well as producing the wines, the landowners controlled the river traffic, and were therefore in a strong bargaining position with the foreign merchants seeking to purchase their wines. In 1727 the British merchants formed an Association and formulated a set of agreements in order to improve their position by acting together to force down the prices that they paid for the wines (Croft, 1787). Despite these regulations and attempts to improve the quality of the wines, they were still generally thought to be inferior to other wines for sale in London. In 1754 the British merchants used this as an excuse to write to their agents in the Douro accusing the landowners and growers of a range of abuses, including the addition of elderberries and poor quality brandy to the young wines, which they described as being 'nothing less than a diabolic procedure' (Francis, 1972: 203). The growers replied that they had been forced to add brandy and elderberries to satisfy the demand of the shippers for wines strong in colour and high in alcohol. There was, therefore, no agreement on the best way to produce port wines from the Douro valley, and the debate continued well into the nineteenth century when Forrester (1845) still maintained that the wines should be left natural and unfortified. In 1755 a Spaniard, Bartholomew Pancorvo, attempted to establish a new company to break the English monopoly and to export port wine to other countries of northern Europe, but he had insufficient capital to compete with the British merchants who combined forces to exclude him. The disorder and low repute of the port trade at this juncture provided the opportunity for the Portuguese government to enter the dispute, and in 1756 Sebastião Joseph de Carvalho e Mello, the Prime Minister of Portugal, established the *Companhia Geral da Agricultura das Vinhas do Alto Douro* (The General Company of Agriculture of the Vineyards of the Upper Douro), which amongst other things was to have a monopoly of the export of Oporto wines to all destinations apart from the United Kingdom (Croft, 1788; Marques, 1972: 388; Bradford, 1983). Its main objectives, as defined in its foundation decree (1756, Section 10), were to support the reputation of the wines and the cultivation of vines, and to benefit the wine trade through the establishment of a stable price structure. The *Companhia Geral* was to control all wine and brandy production in the Upper Douro, the transit of such wine, and its storage in Oporto. However, the establishment of the Company was not only designed to improve the organisation of the port trade, and in subsequent decrees taxes were imposed on wine, brandy and vinegar, as well as on other products, in order to benefit elementary schools in the Alto Douro region. In order to fulfil these requirements, the areas in which the Company was to operate had to be delimited, and this led to the Upper Douro valley becoming the first demarcated vineyard area in the world. Moreover, the activities of British shippers were considerably restricted, and thereafter they 'could only export wines grown in the Upper Douro, passed by the company as Factory wines and approved in Oporto for shipment by one of the company's tasters' (Francis, 1972: 209). While these actions were initially unpopular, and decrees in 1777 annulled the *Companhia Geral*'s monopoly of the sale of wine from parts of the Alto Douro, and also permitted the proprietors of vineyards to bottle and sell their wine without interference from the Company, they did generally lead to an improvement in the quality of the wines, particularly through the banning of additives such as elderberries and through the Company's monopoly of brandy production, which ensured that only good quality brandy was added to the wines.

The young wines, in which the brandy was as yet poorly integrated, remained harsh, and towards the end of the eighteenth century the British merchants increasingly began to mature their wines in barrels at Vila Nova de Gaia, opposite Oporto near the mouth of the river Douro (Figure 36). It was soon realised that wines so matured mellowed considerably, and took on subtle nuances of flavour, which could be enhanced further by ageing them in bottles. At the beginning of the eighteenth century most bottles had been bulbous in shape, and although corks were used as stoppers they soon dried out and let air in to spoil the wine. Gradually, through the century the design of wine bottles was improved, so that eventually by 1775 elongated bottles with short necks had been developed, which could be lain on their sides thus enabling the wine to stay in contact with the cork (Allen, 1961). It is thus from approximately this period that the first true vintage ports date, and thereafter their success was assured. The significance of this has been well captured in Simon's (1926: 143) eulogistic words,

> during the latter part of the same century, the quality, the price, the imports and the popularity of port all increased simultaneously, so much so that at the dawn of the nineteenth century the superiority of port over all other wines had become part and parcel of the creed of every true-born and true-hearted Englishman.

THE SEVENTEENTH CENTURY: NEW WINES AND NEW ATTITUDES

During the course of the seventeenth century, the spread of viticulture to parts of the world where it had previously been absent, and the development of new types of alcoholic beverage, including new wines, represented a considerable change in the organisation of the wine trade. However, signs of these changes were foreshadowed during the previous century, and the reorganisation was perhaps not as abrupt as has sometimes (Enjalbert and Enjalbert, 1987; Lachiver, 1988) been suggested. In the sixteenth century, for example, there had already been a number of attempts to mature wines for periods of longer than a year, and the selection of particular grape varieties in certain parts of Europe was

Figure 36 Port lodges at Vila Nova de Gaia
Source: author 21st March 1980

quite widespread. Moreover, the technical innovations, such as distillation and glass manufacture, were already generally known in much of Europe by 1600. The most significant change during the seventeenth century was thus the conscious investment of capital in the processes of wine production, built upon the assumption that this would produce higher quality, or more fashionable wines, from which higher profits could be reaped.

Such capital investment took place in three ways: through the creation of new types of alcohol, most notably brandy and gin; through the development of viticulture in new places, such as Madeira and the Cape Colony; and through the creation of new wines, such as champagne and port. All of these involved technical changes, but more importantly they also involved a change of attitudes, in which the generation of profit by trade became supplemented by the creation of even greater profits through the introduction of new methods of production. Moreover, these economic changes were closely related to political actions taken by European governments and designed to increase their economic and political power. The establishment of vineyards in Virginia and the Cape Colony thus owed much to the colonial policies of the British and Dutch governments, and the wine producers in Madeira and the Upper Douro valley likewise owed their increasing fortunes largely to the introduction of heavy customs duties by the British government on French wines.

Significantly, though, it was generally not the merchants themselves who undertook this direct capital investment in the production of new wines. With the exception of the investment by Dutch merchants in distilleries in France, it is surprising how few examples there are of either indigenous or foreign merchants buying land to produce their own wines. Thus in the Canaries, Madeira, Champagne and the Upper Douro valley, the expansion of viticulture and the production of new wines was mainly undertaken by the previously existing landowners, and in Bordeaux, while some rich merchants did invest in the purchase of estates, it was largely the new aristocracy, the *noblesse de robe*, who invested capital in the vineyards of the Médoc.

The rise in quality of many of the resultant wines, and their ability to improve with age, led to the reorganisation of the international wine trade. In essence, most of the new wines had greater capital invested within them for a longer period of time than had previously been the case. It was therefore impossible for small and poor producers, who needed to realise their capital in the short term, to profit from the new wines. This was readily acknowledged by Arthur Young (1792) who during his travels in France between 1787 and 1789 noted that the most successful vine cultivators were those with the most capital (Forster, 1961). Only the landlords and the merchants, who could afford to invest over the longer term, were able to benefit from the greater profits that accrued from capital tied up in the production, storage and maturation of wine. This profit, however, could only be realised if it was possible to sell these new wines at prices considerably above those prevailing for wines sold in the year in which they were produced. That they were able to do so is clearly apparent from the differences in price that emerged in the Médoc during the eighteenth century, with the top growths being sold at up to ten times the price of lesser quality wines (Lachiver, 1988). In the London market fashion also played its role in determining the fortunes of individual wines and their producers, and among the elite, seeking a symbol to distinguish themselves from the mass of the population, the new quality wines available in small quantities and at a high price found a ready market.

An alternative strategy for many wine merchants was to seek out, or create, wines which could be sold cheaply to the mass of the population at the lower end of the market. While highly priced French wines were served at the tables of the nobility in early eighteenth century England, it was rough cheap port that was sold in the taverns of London and the other cities and towns of the realm. With low prices being the critical factor in their success, all such cheap wines were liable to systematic adulteration, and indeed the creation of port, through the addition of brandy, can reasonably be interpreted as merely being one other form of such adulteration. For the merchants trading in such wines, it was essential to ensure their acquisition at the lowest possible purchase prices, and, with the attempted creation of a monopoly by the British merchants in 1727, the history of port again reflects the typical method by which this was achieved.

This distinction between expensive, high quality and high status wines on the one hand, and cheap, rough wines for the mass of the urban population on the other, was one which pervaded European viticulture and the wine trade throughout the eighteenth century. In France it was reflected in the growing distinction between the *Grands Crus* and the *vins populaires* (Lachiver, 1988), which by the early nineteenth century had become particularly noticeable. In a manner remarkably similar to the evolution of the Roman wine trade 1800 years previously, wine producers throughout France and Germany increasingly turned their attention to the production of low quality wines for the rapidly growing urban population.

9

CRISES AND EXPANSION: THE RESTRUCTURING OF VITICULTURE IN THE NINETEENTH CENTURY

La vigne seule en France et dans les régions où elle peut mûrir ses fruits, a le pouvoir de créer la richesse dans les terrains pauvres et délaissés; seule elle y rendre 10 pour 100 au capital avancé, seule elle y fondera à perpétuité de grands et de riches domaines. [In France and in the regions where it is able to mature its fruits, only the vine has the ability to create wealth in poor and neglected soils; alone it can yield 10 per cent on the capital advanced, and alone it will be capable of maintaining great and rich estates.]

(Guyot, 1860: 203)

. . . le capitaliste, ce puissant agent du progrès, ne dissipera point ses forces en cultivant sur des espaces sans fin de misérables produits à basse ou même à moyenne main d'oeuvre. Il saura bientôt qu'un hectare de Château-Laffitte ou de Clos-Vougeot produit plus de richesses pour tous que cent hectares de landes, de friches ou de savarts laissés en pâturage, planté en bois ou mis en culture de ferme. [The capitalist, this powerful agent of progress, will not waste his energy in cultivating worthless products over endless expanses employing partially qualified, or unqualified, labour. He will soon find out that one hectare of Chateau-Laffitte or of Clos-Vougeot, produces more wealth for everyone than a hundred hectares of heath, of fallow or of waste left to pasture, put down to wood or turned into an ordinary farm.]

(Guyot, 1860: 5)

Two central processes characterised the fortunes of viticulture and the wine trade during the nineteenth century. These were the spread of various fungal and insect parasites to Europe from North America, and the introduction and expansion of viticulture into new areas of the globe, most notably Australia and California. Both of these processes reflected increasing economic and social connections between different parts of the world, influenced not only by the growing demands of capital, alluded to

270

by Guyot (1860) in the above quotation, but also by the political and ideological aspirations of the European colonial and imperial powers. Moreover, within Europe itself a series of viticultural changes, engendered during the eighteenth century by both environmental and socio-economic factors, found their expression in the 1850s in the first widespread formal classifications of vineyard quality.

QUALITY OR QUANTITY: THE POLARISATION OF DEMAND

The technical changes in viticulture and vinification that had taken place during the seventeenth century laid the foundations for the emergence in the ensuing centuries of two distinct types of wine production, on the one hand the creation of high quality, luxury wines, and on the other the manufacture of cheap, poor quality wines for the mass market. However, this division also depended on the development of two distinct markets. Central to any understanding of these changes was the resurgence in demographic growth that occurred during the eighteenth century, and in particular the increasing percentage of the population living in towns.

The eighteenth century restructuring of wine production

From its nadir in the early 1700s the population of Europe slowly began to expand, with the rate of growth accelerating appreciably in the second half of the eighteenth century (Pounds, 1979: 67). Moreover, this general increase in population was accompanied by an even more rapid increase in urban population. Thus Clout (1977a: 490) comments that 'the proportion of town dwellers in France doubled during the eighteenth century from about one-tenth to one-fifth of the total population, which grew from 19 millions to 26 millions over the same period'. In absolute terms this represented an increase in the urban population of France from about 2 million to around 5.7 million, and provided a massive increase in demand for low quality wine. By the end of the century Paris alone had a population of over half a million, which was more than five times that of either of the next largest cities, Lyon and Marseille. The dominance of Paris, due largely to the centralised political structure of France, con-trasted markedly with the smaller towns and fragmented states of Germany, where the total population in 1800 of about 20 million was only 5 million more than it had been in 1600. Whereas the demand of Paris acted as a major integrating force in the French economy, leading to the expansion of viticulture in many parts of the country, the smaller German towns and cities failed to serve such a role. Moreover, the devastation of the Thirty Years War, which had reduced the German population to perhaps 10 million in 1650, coincided with a widespread recession in viticulture on the Rhine and its tributaries. The lack of a strongly growing internal market for wine, coinciding with a period of climatic deterioration and expansion in beer production, led to a considerable reduction in the extent of vineyards in Germany during the seventeenth century (Enjalbert and Enjalbert, 1987), but the low total demand for wine meant that it was nevertheless possible for many German towns to be supplied with wine from their immediate vicinity without having to resort to imports from further afield. Similarly in Italy, where 'political fragmentation prevented the emergence of large capital cities' (Pounds, 1979: 121), much of the wine trade remained relatively local in nature.

The effects of urban population growth on wine production during the eighteenth century were therefore most marked in the case of Paris (Dion, 1959). In Lachiver's (1988: 337) words, in the Ile-de-France in the immediate vicinity of the capital,

> dés la fin du XVIIᵉ siècle, les vignerons sont à la recherche de plants susceptibles de donner en grande quantité des vins d'un niveau plus médiocre, mais capables de se vendre moins chers et donc de s'imposer encore plus facilement au petit peuple de la capitale voisine. [from the end of the seventeenth century, the wine producers went in search of plants capable of giving a considerable quantity of low quality wines, but which could be sold less expensively and would therefore be more readily attractive to the poor of the neighbouring capital.]

The vine they turned to was the Gamay, which was high yielding but produced a relatively mediocre quality wine, and particularly around Argenteuil the area devoted to its cultivation increased dramatically.

Moreover, the improvement of communication links between Paris and the remainder of France also permitted producers from further afield to turn to mass production of low quality wine for the *guinguettes* of the capital. This was nowhere better exemplified than in the restructuring of viticulture in the region around Orléans. The construction of the Briare canal between 1608 and 1642, linking the Loire with the Loing, a tributary of the Seine, had greatly facilitated the provisioning of Paris with supplies of wine, coal, timber and salt from the Massif Central and the Loire valley during the second half of the seventeenth century (Clout, 1977b; Pounds, 1979), but it was the completion of a second canal from the Loire near Orléans to Montargis, also on the Loing, in 1692 that provided the immediate context for the rapid expansion of viticulture in the Orléanais. As a result, in the words of Enjalbert and Enjalbert (1987: 58) 'Toute la région proche d'Orléans s'était transformée en un vaste vignoble, dont la prospérité se prolongea pendant tout le XVIIIᵉ siècle'

[All of the region near Orléans was transformed into a vast vineyard, whose prosperity was extended throughout the eighteenth century]. Considerable areas of wheat fields were turned over to vineyards, and excess production of wine was converted either into vinegar or into a distilled product known as *eau-de-vie d'Orléans* which had become popular in Paris in the last decades of the seventeenth century. Elsewhere in France, Lyon came to be provided with low quality wines for general consumption from the Beaujolais area in a similar way to the provisioning of Paris with the wines of Orléans.

The growing demand for wine by the mass of the urban population in the eighteenth century provided one option for wine producers with ready access to this particular market. For others, with sufficient capital to invest, the production of high quality wines for export was another alternative. Indeed, as wine became increasingly the drink of the mass of the population, the aristocracy sought to create new types of wine as a symbol of their distinction. In Lachiver's (1988, 330) words,

Comme le vin est devenu un bien de consommation courante, il est impossible aux riches de boire le vin de tout le monde d'où, dès la fin du XVIIe siècle, la volonté de produire de grands vins et la réussite éclatante de quelques châteaux en Médoc et de quelques clos en Bourgogne. [As wine became an item of everyday consumption, it was impossble for the rich to drink the common wine. Thus, from the end of the seventeenth century, there was a desire to produce great wines, which led to the astounding success of several châteaux in the Médoc and several clos in Burgundy.]

Gradually, through the course of the eighteenth century, producers in different parts of France began to specialise in wines for the two contrasting markets. Quality wine production came to dominate in the Médoc, Burgundy and Champagne, whereas much of the rest of the country, particularly in the Loire basin and the south, focused on the supply of the urban market.

Throughout the eighteenth century, though, viticulture also remained an important part of the peasant polyculture of southern and western France, as it did throughout the countries along the northern shores of the Mediterranean. Indeed, in France in the 1780s, when prices for wine were low, contemporary commentators widely blamed the poverty of such areas on the predominance of vineyards (Fel, 1977). This view was well expressed by Thomas Jefferson (Lawrence, 1989), at that time Minister Plenipotentiary to the Court of France and later to become the third President of the United States of America, who in a letter to William Drayton from Paris dated 30 July 1787, argued that 'The culture of the vine is not desirable in lands capable of producing anything else. It is a species of gambling, and of desperate gambling too, wherein, whether you make much or nothing, you are equally ruined' (Randolph, 1829: 195). Jefferson went on in this letter to suggest that vines are 'a resource for a country, the whole of whose good soil is otherwise employed, and which still has some barren spots, and a surplus of population to employ on them. There the vine is good, because it is something in the place of nothing' (Randolph, 1829, 196). This poverty, though, was not necessarily a direct result of vine cultivation. Another contemporary, Arthur Young (1792: 390), concluded from his travels in France between 1787 and 1789,

> The poverty is obvious; it is connected with vines, and for want of proper distinctions, it is considered as necessarily flowing from vineyards; but in fact it is merely the result of small properties amongst the poor: a poor family can no where be better situated than in a vine province, provided he possesses not a plant. Whatever may be the reason, they are sure of ample employment among their richer neighbours, and to an amount, as we have above seen, thrice as great as any other arable land.

Young's judgement here neatly illustrates the social contrasts that existed in France on the eve of the revolution of 1789, and also emphasises his own dislike of peasant agriculture. As an advocate of technological change in the hands of capitalist landlords and tenants, he saw little to be admired in the maintenance of small peasant properties.

Climatic change and vineyard restructuring

The extent of vineyards in different parts of Europe, and the dates of grape harvests have for long been used as indicators of climatic change (Le Roy Ladurie, 1972; Le Roy Ladurie and Baulant, 1980; Lamb, 1982). While there is some indication based on these sources of a slight cooling in temperatures at the end of the sixteenth century, the survival of vineyards as far north as parts of England throughout the sixteenth, seventeenth and eighteenth centuries (Barty-King, 1977) indicates that the climate was never sufficiently harsh during these period to prevent the cultivation of vines in northern France and the Low Countries. Nevertheless, particularly severe winters could be catastrophic. One such disaster, which led to a significant restructuring of viticulture in parts of northern France, occurred in January and February 1709. On the 13th January the temperature in Paris fell to −20°C, and remained there for ten days before thawing, only to freeze heavily again during February (Lachiver, 1988). The effects were dramatic, with rivers freezing, and the sea at Dunkerque remaining ice-bound until April. It was so cold that wine froze in the cellars. In Pijassou's (1980: 104) words 'l'hiver historique de janvier 1709 causa des dommages si importants qu'il laissa de durables souvenirs dans toute la France' [the historic winter of January 1709 caused such important damage that it left lasting memories in the whole of France].

The immediate effect of the cold was to lead to the widespread destruction of vines throughout northern Europe, particularly in low lying valley areas. One of the most seriously affected parts of France was the western Loire valley, where most of the vines in the region surrounding Nantes were killed. Elsewhere, none of the major quality producing areas of Bordeaux, Burgundy or Champagne escaped damage. Yields plummeted, with, for example, the Abbey of Saint-Denis in Paris producing only 14 muids, each of 268 litres, in 1709, compared with 130 muids in 1706 (Lachiver, 1988). The price of wine quadrupled, but the rapid resumption of production levels, with the Abbey of Saint-Denis recording a yield of 167 muids in 1711, suggests that in parts of the country the effects were relatively short term. The high price of wine in 1709 and 1710 nevertheless encouraged extensive replanting in areas where vines had been killed, and also the further expansion of viticulture into new areas. It also provided a clear opportunity for producers to replant their vineyards with vines destined for one of the two markets noted above, either with low yielding high quality cultivars, or with vines designed to produce wine for the mass urban market.

In the Paris region the extensive replanting with Gamay has already been noted, and the winter of 1709 provided the ideal opportunity for producers to replant their vineyards with this high yielding variety destined to produce wine for the rapidly growing population of the capital. Elsewhere, one of the most marked transformations took place at the mouth of the Loire, in the Pays Nantais, where the traditional varieties killed by the frost were replaced mainly with the Melon de Bourgogne, which was reputed to be a much hardier variety and which formed the basis for the subsequent production of Muscadet wines in the region. In the Bordelais, Pijassou (1980) furthermore suggests that the damage inflicted on the vineyards of the Graves may well have favoured the rapid spread of viticulture that was then taking place in the Médoc.

The extension of vineyards throughout France during the decade following the winter of 1709 was also, though, fuelled by the increasing demand for wine, both as a result of the growing urban population and also the increasing demand for quality wines. While in average years production and demand were approximately in balance, abundant vintages, such as that of 1724, caused a collapse in the price of wine. As a result the *Conseil d'Etat* took measures to prevent the further expansion of viticulture. On 16th January 1725 all further planting of vines was prohibited in Touraine and Anjou (Lachiver, 1988), and on 27th

February a similar regulation was passed for the Bordeaux region (Pijassou, 1980). These measures had been preceded by the actions of the Parlement of Metz which in 1722 had forbidden the planting of new vines in its area of jurisdiction and ordered the uprooting of all vines which had been planted outside the traditional vineyards since 1700 (Dion, 1959: 598). In part, these restrictions were introduced at the behest of the owners of quality vineyards, who saw their profits tumbling through the extensification of vineyards throughout France, and their pleas achieved some success on 5th June 1731 when a royal edict extended the ban on new vineyards to the whole of the kingdom (Dion, 1959). This edict, however, reminiscent to that of Domitian in AD 92, proved impossible to enforce, and, following a general enquiry concerning the efficacy of the edict initiated in 1756, it was effectively cancelled in 1759 when the Royal Council lifted the constraints on the use to which land was to be put.

Wine fraud and the emergence of wine classifications

Taken as a whole, the period from 1700 until 1778 was one in which the price of wine remained remarkably constant, reflecting an approximate balance between increasing levels of supply and demand, and as Lachiver (1988: 331) has commented, this 'preuve que la production sup-plémentaire est absorbée sans trop de difficulté' [proves that the sup-plementary production was absorbed without too much difficulty]. However, one outcome of the growth in demand for wine was that the amount of falsification and corruption increased correspondingly. The 'treatment' of wine was nothing new, and the various additives used by the Romans have already been mentioned, but during the seventeenth century it achieved the status of a veritable art. The various remedies used at this period in England were clearly outlined in a paper delivered to the Royal Society in November 1662 by Walter Charleton (1669) and entitled *The Mysterie of Vintners. Or brief discourse concerning the various sicknesses of wines, and their respective remedies, at this day commonly used.* This also included a section on *Some observations concerning the ordering of wines* by Christopher Merret, and by the time of the third edition published in 1692 a further section entitled *The art and mystery of vintners, and wine coopers: containing approved directions for conserv-ing and curing all manner and sorts of wines; whether Spanish, Greek, Italian, Portugall, or French: As it is now practised in the City of London* had been added. Charleton's discourse revealed a clear knowledge of basic methods of vinification, recognising, for example, that cold temperatures slowed down fermentation and also clarified new wine (Charleton, 1669: 166-8). Moreover, he also commented on the use of sulphur to prevent the spoiling of wine or must (Charleton, 1669: 163-4), and he noted the general use of isinglass and egg whites to clear wines (Charleton, 1669:

162). Alongside these practices, however, he also described the following more disengeneous activities undertaken by vintners

> They transform poor *Rochel* and *Cogniak* White wines into *Rhenish*; Rhenish into *Sack*; the Laggs of *Sacks* and *Malmsies* into *Muskadels*. They counterfeit *Raspic*-wine, with *Flower de Luce roots*; *Verdea* with decoctions of *Raisins*; they sell decayed *Xeres*, vulgarly *Sherry*, for *Lusenna* wine: in all these impostures deluding the palate so neatly, that few are able to discern the fraud; and keeping these *Arcana Lucrifera* so close, that fewer can come to the knowledge of them.
>
> (Charleton, 1669: 195-6)

Charleton's (1669: 176-8) description of some of the ways in which claret could be improved is worth quoting at length:

> To amend *Claret* decayed in *Colour*, first they rack it upon a fresh Lee either of *Alicant*, or *Red Bordeaux* wine; then they take 3 *pound* of *Turnsol*, steep it in all night in two or three gallons of the same wine, and having strained the infusion through a bagg, pour the tincture into the Hoggshead (Sometimes they suffer it first to fine of itself in a Rundlet) and then cover the bung-hole with a tile, and so let it stand for 2 or 3 dayes; in which time the wine usually becomes well-coloured and bright.
>
> Some use only the tincture of *Turnsol*.
>
> Others take half a bushel of full-ripe *Elder-berries*, pick them from their stalks, bruise them, and put the strain'd juice into a hoggshead of discoloured Claret and so make it drink brisk, and appear bright.
>
> Others, if the Claret be otherwise sound, and the Lee good, overdraw 3 or 4 gallons; then replenish the vessel with as much good *Red Wine*, and rowl him upon his bed, leaving him reversed all night: next morning turn him again so as the bung-hole may be uppermost, which stopt, they leave the wine to fine. But in all these cases they observe to set such newly recovered wine abroad, the very next day after they are fined, and to draw them for sale speedily.

While such practices were widespread in the importing countries of northern Europe, where wine often needed 'reviving' after its ocean voyages, they also became increasingly common in the producing countries further south. There were two main reasons for this. On the one hand, the rising urban demand for cheap wine provided an incentive for producers and merchants to stretch their supplies of genuine wine with whatever else they had to hand, and, on the other, if poor wine could be passed off as something better, through the use of various additives, it could be sold for greater profit. In particular, with the increasing development of quality wine production, there came a need to guarantee its provenance.

The first formal attempt at demarcating a wine region took place in Portugal, as part of the Marquis de Pombal's reorganisation of the port wine trade in 1756. One of the key measures associated with the foundation of the General Company for the Agriculture of the Vineyards of the Upper Douro was the establishment of a demarcated area in the Upper Douro valley upon which the prices paid to the farmers for their wines was determined. Thus, the maximum price fixed for best quality wines from the Upper Douro was to be 30 milreis a pipe, with the Company being empowered to take any first quality wine at 25 milreis a pipe and second quality wine at 20 milreis a pipe (Francis, 1972). Maximum prices for wines grown outside the delimited area ranged from 12 to 20 milreis a pipe, with the wines from the Minho near Oporto being only 4 milreis a pipe. While this was in part intended to improve the quality of the wines, it is evident that the measure was also designed to enhance the profits of the merchants and the large landowners in the Upper Douro at the expense of poorer farmers elsewhere. Writing early in the nineteenth century, Redding (1833: 211) had no doubts about the effects of the Methuen Treaty and the subsequent establishment of the Company:

> The history of no country in the world furnishes an example of greater political absurdity than our own, in the conclusion with Portugal of what is commonly called the Methuen Treaty, better characterized as the Methuen or wine merchants' job. By this treaty Englishmen were compelled to drink the fiery adulterations of an interested wine company, and from the coarseness of their wines, exposed to imitations of them without end.

In Francis's (1972: 210) words, 'In practice the large owners in the Upper Douro and certain interests in Oporto were those benefited. The small peasant proprietors of the Upper Douro benefited a little and those of the Minho not at all'.

The desire by wealthy landowners to guarantee the provenance of their wines in order to ensure their continued profits also lay behind the emergence of wine classifications in other areas, most notably in Bordeaux and Burgundy. By the end of the eighteenth century a clear hierarchy of estates had emerged in these regions classified largely on the basis of the prices that could be obtained by their wines. During his journey through France and Italy in 1787 Thomas Jefferson thus noted that there were three main classes of quality red wines recognised in Bordeaux, all of which sold for considerably more than the five hundred livres a ton of one thousand bottles obtainable for the best of the common wines (Randolph, 1829: 153). In the first class he included the wines of

'Chateau Margau' (Margaux), 'La Tour de Segur' (Latour), 'Hautbrion' (Haut-Brion), and 'Chateau de la Fite' (Lafite), which were then selling at 2,000 livres a ton for the 1783 vintage, and 2,400 livres for the better quality 1784 vintage. The second class of red wines consisted of those from 'Rozan' (Rauzan), 'Dabbadie or Lionville' (Léoville), 'la Rose' (Gruaud-Larose), 'Quirouen' (Kirwan), and 'Durfort' (Durfort-Vivens), and these sold at 1,000 a ton, and in the third class, selling for between 800 and 900 livres a ton were the wines of 'Calons' (Calon-Ségur), 'Mouton' (Mouton-Rothschild), 'Gassie' (Rauzan-Gassies), 'Arboete' (Château Lagrange), 'Pontette' (Château Langoa), 'de Terme' (Marquis de Termes) and 'Candale' (Château d'Issan) (Randolph, 1829; Pijassou, 1980). Among the white wines of the Graves, the best noted by Jefferson were 'Pontac', 'St Brise', and 'De Carbonius'; in Sauternes he recommended the wines of M Diquem, M de Salus, and then M de Fillotte, while the best wine in Prignac was that of President du Roy, and the best in Barsac was that of President Pichard (Randolph, 1829: 153-4). While this classification has been criticised as being only partial, reflecting the shortness of Jefferson's five-day visit to Bordeaux in May 1787 (Pijassou, 1980), it does nevertheless reflect that by the eve of the French revolution certain landowners in the Médoc had been able to propagate a quality image for their wines, enabling them to sell them at more than four times the price of the best common wines, and twenty times the price of the poorer quality wines which sold at only 120 livres a ton.

Jefferson also noted a similar, but less detailed, classification for the red wines of Burgundy during his visit to the Côte d'Or on 7 and 8 March 1787, with the 'strongest' being those of 'Chambertin' (Gevrey-Chambertin), 'Vougeau' (Vougeot) and 'Beaune', which all sold at forty-eight sous the bottle, followed by 'Voulenay' (Volnay) selling at twelve sous a bottle (Randolph, 1829: 117). The best white wines were those of 'Monrachet' (Montrachet) selling at forty-eight sous a bottle, followed by the top wines of Meursault, those of the Goutte d'Or, which sold at six sous a bottle (Randolph, 1829: 117). As in Bordeaux, the importance of a quality name for particular wines in Burgundy was also noted by Jefferson when he observed that

> It is remarkable, that the best of each kind, that is, of the red and white, is made at the extremities of the line, to wit at Chambertin and Monrachet. It is pretended that the adjoining vineyards produce the same qualities, but that belonging to obscure individuals, they have not obtained a name, and therefore sell as other wines.
>
> (Randolph, 1829: 117-18)

Elsewhere in his diary, Jefferson commented on numerous other French and Italian wines, particularly praising those of Hermitage (Randolph, 1829: 121), the Muscats of Lunel (Randolph, 1829: 143-4), and the wine produced in the vicinity of Turin, which he described in the following terms: 'There is a red wine of Nebuile made in this neighbourhood, which is very singular. It is about as sweet as the silky Madeira, as astringent on the palate as Bourdeaux, and as brisk as Champagne. It is a pleasing wine' (Randolph, 1829: 134–5).

Within thirty years of Jefferson's travels, Abraham Lawton, a Bordeaux *courtier* (wine broker), had greatly refined the art of vineyard classification, and in 1815 he distinguished 323 separate crus in the Médoc, fifty-one of which were identified as grands crus, divided into three first growths, six second growths, nineteen third growths and twenty-three fourth growths (Pijassou, 1980). Other classifications followed, including ones by Jullien (1816) and Franck (1824), until in 1855 the *syndicat des courtiers* was charged by the Chamber of Commerce of Bordeaux with producing an official classification of the wines of the Gironde on the occasion of the 1855 *Exposition Universelle* in Paris. This was submitted on 18th April, and included fifty-eight red wines, divided into four *premiers crus*, twelve *deuxièmes crus*, fourteen *troisièmes crus*, eleven *quatrièmes crus*, and seventeen *cinquièmes crus*, together with twenty-one white wines, subdivided into one *premier cru supérieur*, nine *premiers crus* and eleven *deuxièmes crus* (Table 6). Until then the flexibilty of the unofficial classifications left the pricing of wines open to some degree of manipulation by the producers (Penning-Rowsell, 1985: 433), which was well described by Cyrus Redding (1833: 151), who in 1833 commented that

> The wines are classed by the brokers, who decide to which class the wine of each grower shall belong. The latter use all their efforts to place their wines in a higher class, and thus emulation is kindled, and they are justified in their efforts by the profits. The price of their wines too, is less governed by particular merit, than by the number which they occupy in the scale of classification. It often costs them sacrifices to reach that object. They will keep their wine many years to give it a superior title, instead of selling it the first year according to custom. By this means an individual will get his wine changed from the fourth to the third class, which he had perhaps occupied before for many successive years.

However, since 1855, and despite much debate, the classification has remained substantially the same, with the only alteration being the elevation of Château Mouton-Rothschild to a first growth in 1973. While the 1855 classification was based largely on the wine prices then pertaining, it nevertheless reflected the continued economic power of the *courtiers* as well as the political strength of the wine producers in different parts of the Gironde in the middle of the nineteenth century, with all of the red wines in the classification apart from Haut-Brion in

Table 6 The 1855 classification of the wines of the Gironde

RED WINES

Premiers Crus
Château Lafite
Château Margaux
Château Latour
Haut-Brion

Deuxièmes Crus
Mouton
Rauzan-Ségla
Rauzan-Gassies
Léoville
Vivens Durfort
Gruau-Laroze
Lascombe
Brane
Pichon Longueville
Ducru Beau Caillou
Cos Destournel
Montrose

Troisièmes Crus
Kirwan
Château d'Issan
Lagrange
Langoa
Giscours
St.-Exupéray
Boyd
Palmer
Lalagune
Desmirail
Dubignon
Calon
Ferrière
Becker

Quatrièmes Crus
St.-Pierre
Talbot
Du-Luc
Duhart
Pouget-Lassale
Pouget
Carnet
Rochet
Château de Beychevele
Le Prieuré
Marquis de Thermes

Cinquièmes Crus
Canet
Batailley
Grand Puy
Artigues Arnaud
Lynch
Lynch Moussas
Dauzac
Darmailhac
Le Tertre
Haut Bages
Pedesclaux
Coutenceau
Camensac
Cos Labory
Clerc Milon
Croizet-Bages
Cantemerle

WHITE WINES

Premier Cru Supérieur
Yquem

Premiers Crus
Latour Blanche
Peyraguey
Vigneau
Suduiraut
Coutet
Climens
Bayle
Rieusec
Rabeaud

Deuxièmes Crus
Mirat
Doisy
Pexoto
D'arche
Filhot
Broustet Nérac
Caillou
Suau
Malle
Romer
Lamothe

Source: Penning-Rowsell (1985), Stevenson (1988)

the Graves being from the Médoc. With the subsequent price of the wines being determined largely by the status accorded to a property by the *courtiers* in 1855, the classification in effect provided a barrier preventing the emergence of new wine estates, and guaranteeing the high prices of wines from those at the head of the classification.

The effects of the revolution of 1789 differed substantially in Bordeaux and Burgundy. In Bordeaux, while a number of estates were confiscated and sold, 'this was not so much owing to the extermination or expropriation of the owners as to their emigration, as happened at Lafite' (Penning-Rowsell, 1985). Consequently, the majority of properties remained substantially intact. In contrast, in Burgundy, where many of the vineyards had been owned by the Church and monastic houses, the subsequent sale of lands taken into the possession of the State led to their fragmentation and subdivision. The purchase of small areas of vineyard by members of the Dijon Parlement during the seventeenth century had itself led to some reorganisation in the structure of landholding in the region, but after 1789 there was a comprehensive redistribution of the land. Despite these trends, the eighteenth and nineteenth centuries gradually saw the emergence of a number of high quality vineyards in Burgundy (Gadille, 1967). One of the first indications of this hierarchy can be found in Arnoux's description of the region in 1728, where, for example, he picks out the vineyards of les Fèves, les Grèves, Clos du Roi, and aux Cras in Beaune, Champans in Volnay, and Clos de la Commeraine in Pommard as being of particular repute (Lachiver, 1988). In 1833 Redding (1833) also produced a broad classification of the wines of Burgundy based upon the prices then obtainable. However, the most famous classification, coinciding in date with that of the Gironde undertaken by the *courtiers* of Bordeaux, was Lavalle's account of the vineyards and wines of the Côte d'Or in 1855. Lavalle's classification consisted of three main categories: *hors ligne* (outstanding), subdivided into *tête de cuvée n° 1* and *tête de cuvée n° 2*; *première cuvée*; and *deuxième cuvée*. He began his classification of the *hors ligne*, *tête de cuvée n° 1* red wines with those of Romanée-Conti at Vosne, Clos de Vougeot, and Chambertin and Clos de Bèze at Gevrey, and these were then followed by Clos-de-Tart, part of Bonnes-Mares and Lambrays at Morey, part of Corton at Aloxe, Musigny at Chambolle, Richebourg, Tâche and part of Romanée-Saint-Vivant at Vosne, and Saint-Georges at Nuits. Among the white wines only Montrachet at Puligny was classified as outstanding, although Bâtard-Montrachet at Puligny, Perrières at Meursault, and Corton at Aloxe were at the top of the *premières cuvées*.

Through the nineteenth century, as improvements in transport made a greater range of wines ever more accessible to the consumer, these classifications, providing information on the reputed quality of wines and thus the price at which they could be sold, became increasingly important. The expansion of the rail network in the second half of the nineteenth century played a fundamental role in the commercialisation of the wine industry. In France it enabled the burgeoning population of the capital, Paris, to be supplied with wine from the furthest corners of the country, and in particular it led to the rapid expansion of low quality, but high quantity, wine production in areas such as the Midi. Once the Parisian market had been made accessible through the construction of the railways, land prices rose in such peripheral regions as a new capitalist monoculture system of wine production began to replace the traditional peasant polyculture. These changes, though, coincided with the appearance in European vineyards of the first of a number of diseases and pest infestations which were to bring viticulture virtually to its knees by the end of the century.

THE NINETEENH CENTURY CRISES IN EUROPEAN VITICULTURE

The advent of oïdium

A decade before the 1855 classifications of Bordeaux and Burgundy, a certain Mr Tucker, gardener to John Slater of Margate, a small town in south-east England, noticed a powdery substance on the leaves of some of his vines, and sent a sample to the Rev. M.J. Berkeley of King's Cliffe, Northamptonshire, for identification (Ordish, 1987). Berkeley decided that this was a new species of fungus and named it *Oïdium Tuckerii* in honour of the gardener. Soon afterwards the effects of this powdery mildew were noticed throughout Europe, drastically reducing the yields of grapes. In 1846 it was identified on vines at the Palace of Versailles and in 1851 the vineyards of southern France were badly attacked, by which date it had also been noted in Algeria, Greece, Hungary, Italy, Spain, Switzerland and Turkey (Ordish, 1987). The precise origins of oïdium, and the reasons why it swept through Europe in the 1840s and 1850s remain uncertain, but the great susceptibility of *vinifera* varieties of vine to powdery mildew and the resistance of certain American varieties suggest that it originated in north America. Its first identification on greenhouse vines in Europe, in turn, suggests that it was probably introduced on some ornamental vine brought across the Atlantic. Moreover, the 1840s coincided in Britain with the expansion of interest in foreign plants and artefacts that characterised the early years of the reign of Queen Victoria, and that was exemplified in the dramatic expansion of the Royal Botanic Gardens at Kew following the appointment of Sir William Hooker as the first Director in 1841 (Turrill, 1959). This could well explain why it was at this period and not earlier that oïdium was apparently introduced. The rapidity of its spread was also enhanced by the expansion of the railway network throughout Europe in the 1840s, and particularly in France where 3,600 kilometres of new line linking Paris with six provincial destinations had been authorised in 1842 alone (Clout, 1977b: 469).

The dramatic reduction in yields caused by oïdium precipitated a flurry of activity in the search for a cure (Lachiver, 1988), and by 1852 Grison, the head of the greenhouses at the Palace of Versailles, had achieved some success in preventing its spread by spraying the vines with a boiled lime and sulphur mixture, known as *Eau Grison* (Heuzé, 1852; Ordish, 1951). Such careful treatment, however, could not be duplicated over large areas of vineyard, and, recognising that certain American vines were resistant to attack by oïdium, a number of producers began importing and cultivating American vines. One of the most notable proponents of this method was Laliman, who had used American vines on his own estate since 1840 and who published a book in 1860, arguing that their wider use could counter the threat of oïdium throughout France. By the early 1860s, however, another solution, the dusting of vines with fine sulphur was found to be successful by Henry Mares, and this became the standard form of treatment throughout Europe.

The effect of oïdium in reducing the yield of the vines led to a considerable reduction in wine production throughout Europe, with French production, for example, falling from 45 million hectolitres in the 1840s to 29 million hectolitres in 1852 and 11 million hectolitres in 1854 (Lachiver, 1988). The price of wine rose correspondingly, and, although some wine producers, particularly in the north, abandoned their vineyards, the relatively rapid discovery of a solution to the problem meant that by the late 1850s yields had recovered to their earlier levels. Despite the additional costs of the sulphur treatments, and the extra labour involved in its application, the continued high demand for wine meant that price increases were readily absorbed. Indeed consumption per head of wine in France in the ensuing years rose from fifty-one litres per inhabitant in 1848 to eighty litres by 1880 (Lachiver, 1988: 410). Unwittingly, though, the mass introduction of American vines to Europe, initiated largely by the attempt to control oïdium, led to a crisis of far greater proportions, which was to have grave social and economic implications for the wine producers of the continent.

The arrival of phylloxera in Europe

In the early 1860s vineyard owners in the south of France began to notice that some of their vines were dying, with there being no apparent cause for the drying up of the leaves and the subsequent death of the vines. In 1863, for example, at Pujault in the Côtes du Rhône it was noticed that several vines suddenly died, with the cause of their death being put down to poor protection against oïdium (Penanrun, 1868). By 1867, however, there was no doubt that a serious new problem was affecting the vineyards of Provence and the Rhône. In that year five hectares of dead or dying vines were recorded by Delorme at a vineyard at Saint-Martin-de-Crau, between Arles and Salon, and further problems were reported in Narbonne, the Gard and Vaucluse (Penanrun, 1868; Ordish, 1987; Lachiver, 1988). Meanwhile, in 1863 a sample of insects on a vine leaf from a greenhouse in Hammersmith, to the west of London, had been sent to Oxford University, where they were later identified by Westwood (1869) as the aphid phylloxera, which had unwittingly been brought to Europe on American vines, and which, unknown to American vine growers, had been the cause of the lack of success they experienced in attempting to grow European varieties of *Vitis vinifera* in their own country. It was not until 1868, though, that the connection between this aphid and the destruction of the southern French vineyards was made as a result of the work of a Commission convened by the Société Centrale d'Agriculture de l'Hérault (Bazille *et al.*, 1868). The cure took longer, and within two decades phylloxera had spread to virtually all of France, and to much of the rest of Europe, destroying the vines as it went (Gervais, 1904). It was confirmed in Portugal and Turkey in 1871 (Ordish, 1987), in Austria-Hungary in 1872, in Switzerland in 1873 (Fatio and Demole-Ador, 1875; Fatio, 1879; Comision Organizadora del Congreso Internacional Filoxérico de Zaragoza, 1880) or 1874 (Ordish, 1987), in Spain by 1875 (Graells, 1881), in Italy by 1879 (Dalmasso, 1937), and in Germany by 1881 (Ordish, 1987). In most of these instances, though, it is probable that phylloxera had become established several years before it was officially recognised, and despite the imposition of strict control measures it seems likely to have been transmitted mainly through the widespread distribution of American vines from France.

The most generally accepted argument for the explanation of the timing of the arrival of phylloxera in Europe in the late 1850s and 1860s is that until then it could not survive the long crossing of the Atlantic by sailing ship (Ordish, 1987; Stevenson, W.I., 1980; de Blij, 1983). This has been succinctly summarised by Ordish (1987: 5) as follows:

> The time taken for the Atlantic voyage before the coming of the steamship explains why the phylloxera did not reach Europe, or at least was not established there, sooner than 1863. The insect is so destructive that a long interval between its presence and its discovery is unlikely, so that its successful establishment cannot have been much before 1863.

He goes on to argue that if earlier American vine introductions in the form of cuttings, rooted cuttings and seeds

had been infected with phylloxera aphids they would have died by
the time the long sea voyage had been completed. But the steamship
carried the plants far more quickly and the railway reduced the time
of the inland voyage from port to garden, so that the pest survived
its long journey and, once established, spread rapidly.

(Ordish, 1987: 5-6)

This argument, however, is not entirely convincing for two main reasons.
First, the gradual replacement of sail packets by steam ships from the
middle of the nineteenth century only reduced the Atlantic crossing time
from about three to two weeks. Secondly, it seems perfectly possible that
phylloxera aphids were able to survive the Atlantic voyage even before the
arrival of the steam ships. Ordish's arguments are based essentially on the
supposition that if American vines carried by sailing ships had been
infested with phylloxera, the aphids would have killed the plants during
the crossing and, having nothing upon which to feed, would consequently have died themselves. However, if only a small number of
infected vines had been included among many other samples, there
would probably have been sufficient vine material upon which the
aphids could feed even during a voyage of three or more weeks. Moreover,
overwintering eggs or hibernating larvae could easily have been
transported across the Atlantic, without there being any need for a
sufficient supply of food, and in any case the time between oviposition
and adult emergence of phylloxera on certain American vines can be well
over 35 days (Granett et al., 1983). Consequently it is necessary to search
elsewhere for an explanation of the devastating spread of phylloxera in
Europe during the 1860s.

One possibility is that the source of American vines imported into
Europe changed during the 1850s from one that was free of phylloxera to
one that was infected. Indeed small differences in the biology of phylloxera in different locations in eastern northern America might have meant
that not all American populations of phylloxera could readily have
established themselves in Europe (Williams and Shambough, 1988). The
outbreak of phylloxera in the 1860s could therefore have resulted from the
introduction of a very specific strain of phylloxera to which European
cultivars of *Vitis vinifera* were particularly susceptible. Granett et al.,
(1983) suggest that the population density of phylloxera is a factor
influencing the rate of survival of vines, and early small scale introductions of phylloxera may therefore simply have killed the vines in their
immediate vicinity, and spread no further. This seems unlikely, though,
as a general explanation for the early distribution of the disease, given the
high fecundity of phylloxera. Planchon (1877), who first identified
phylloxera in southern France, was of the opinion that the importation
of a considerable number of rooted American vines had taken place into

the region between 1858 and 1862, and that it was possibly this that had led to the infestation of Europe's vineyards. Some combination of a new source of vine supply and the introduction of a particular strain of phylloxera may therefore have been the explanation. Given that the arrival of phylloxera came close on the heals of the outbreak of oïdium, it is also possible that the European *Vitis vinifera* vines had been in some way weakened, either by oïdium or by treatments to control it, making them more susceptible to destruction by the particular strain of phylloxera that had been introduced. Furthermore, the coincidence of the spread of phylloxera with a series of warmer than average years between 1857 and 1875 (Le Roy Ladurie and Baulant, 1980) may also be of some significance.

The spread of phylloxera within France and the search for a cure

The pattern of the spread of phylloxera within France is well established, and is illustrated in Figure 37 (Stevenson, 1980; Lachiver, 1988). Until 1869 it was confined to six *départements* in the lower Rhône valley as well as the *département* of the Gironde, and this spatial separation suggests that there were therefore at least two independent introductions, one in the *département* of Gard prior to 1863 and the other by 1869 in the Gironde. From these two foci, phylloxera spread throughout most of southern and central France by 1880, and to most other wine producing areas of the country by 1890. The effects in infested *départements* were devastating: in Gard which had 88,000 hectares of vines in 1871, there were only 15,000 hectares in 1879; in Hérault, which had 220,000 hectares of vineyard in 1872, there remained only 90,000 in 1881 (Lachiver, 1988: 416). In France as a whole, though, the area devoted to vineyards continued to grow until the 1870s. Based on the series of statistics collated by Lachiver (1988: 582), it can be seen that the total area of vines in France rose from 2,190,000 hectares in 1852 to 2,465,000 in 1874 and only thereafter fell to 2,197,000 hectares in 1882 and 1,800,000 hectares in 1892. Of these areas, though, only 1,777,644 hectares in 1882 and 1,386,303 hectares in 1892 were in full production. Figures for total wine production paralleled those of area, rising from 38,100,000 hectolitres in 1852 to a peak of 84,500,000 hectolitres in 1875 and then falling to a low of 23,400,000 hectolitres in 1889. At a national scale the continued success of viticulture well into the 1870s, therefore provides one insight into the reasons why efforts made to combat the troubles were relatively slow in getting off the ground. Indeed, while the low point in production in 1889 was only 28 per cent that of the peak in 1852, it was actually 61 per cent of the yield in 1852 and more than double the yield of the disastrous year of 1854 which had been affected by oïdium.

At a more detailed scale, Stevenson, (1980, 1981) has traced the spread of

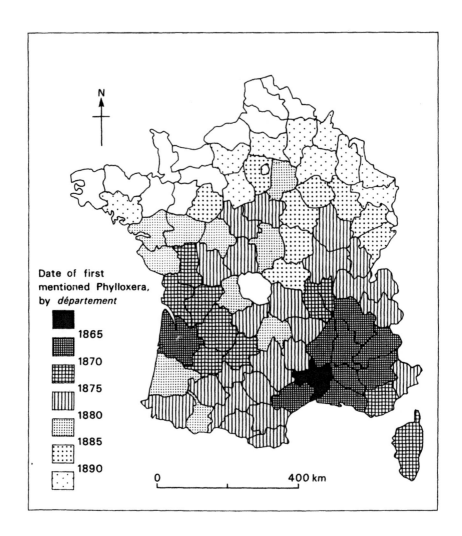

Figure 37 The spread of phylloxera in France
Source: derived from data in Lachiver (1988)

phylloxera in the *département* of Hérault, which had the greatest extent of vineyards of any *département* in France prior to the outbreak of phylloxera. While phylloxera generally spread from east to west across the *département*, Stevenson (1980: 53) notes that 'periods of relative dormancy alternated with those of rapid expansion'. Indeed,

Some areas of long-standing infestation, such as the eastern *garrigue,* seem to have proved relatively resistant to the pest while others of much more recent infection, particularly in the middle Hérault valley near Clermont, appear to have succumbed much more quickly and completely.

(Stevenson, 1980: 54)

He concludes that soil texture was a major factor determining the spread of phylloxera, noting that while winged phylloxera could spread easily from vineyard to vineyard across the plain, the closely-packed deep soils to be found there limited the movement of the wingless root-based phylloxera from vine to vine. Interestingly, Stevenson (1980: 60) also points out that many ordinary vinegrowers remained highly complacent in the face of the threat, arguing that this 'almost hysterical under-estimation of phylloxera's destructiveness, was a recurrent feature of the viticultural state of mind in the Hérault and this apathetic and incredulous attitude seems to have contributed in part to the propagation of the pest in the *département*'. In particular, great faith seemed to be placed in the belief that the pest would be constrained by physical barriers such as hills.

The search for a cure for the damage inflicted by phylloxera was slow. The outbreak of the Franco-Prussian War in 1870, and the proclamation of the Third Republic in September of the same year took considerable attention away from a problem which at that time appeared to be confined to a few parts of France, and had yet seriously to affect the supplies of wine to the capital. Indeed, the expansion of the railway system during the 1850s and 1860s had enabled Paris to be supplied with wine from many areas from which it had previously not been profitable to send it. In particular this had led to a considerable spread of viticulture in the two southern *départements* of Hérault and Gard, where the total area devoted to vines had risen from 184,279 hectares in 1852 to 256,372 in 1862.

In 1869 the Société des Agriculteurs de France had established a Commission, which included Planchon as a member, to investigate the problem of phylloxera, and in 1870 the Minister of Agriculture had offered a small prize of 20,000 francs for a remedy, but it was only four years later in July 1874 that the government as a whole became sufficiently concerned to offer a prize of 300,000 francs for the inventor of a cure. In order to evaluate the proposed remedies the School of Agriculture at Montpellier set aside an infested vineyard, known as Las Sorres, where the Commission départementale de l'Hérault pour l'étude de la maladie de la vigne (1877) tested 317 of the 696 remedies submitted to them in the period before October 1876 (Ordish, 1987; Lachiver, 1988). In only two experiments did the treated plots show any marked advantage over the control plots, and these involved the treatment of the vines with potassium sulphide in human urine and the application of sulphide with colza cake. Other series of experiments included the flooding of vineyards, and in 1876 the Commission also planted thirty-five different American vines in order to investigate their response to phylloxera. Flooding proved to offer a solution in areas where it was practicable, and in 1874 Faucon argued that phylloxera could be successfullly eradicated if the vineyards were completely flooded to a depth of 10 cm for at least forty days in autumn or winter. Elsewhere in the country, other learned and agricultural societies also established committees to investigate the problem of phylloxera, with, for example, the Société d'Agriculture de la Gironde forming a Commission Générale du Phylloxéra in March 1874 and the Comice Agricole de Saint-Emilion creating a special association charged with the fight against phylloxera in 1875 (Teisseyre, 1978).

The most successful chemical treatment for infected vines proved to be the application of carbon disulphide (CS_2). This highly toxic and inflammable chemical had been found to be successful in the eradication of other pests during the 1850s, and despite early failures in the late 1860s subsequent experiments in the 1870s involving its injection into the soil around the roots of vines were effective in eradicating the phylloxera aphid (Ordish, 1987). However, the method was expensive, there was considerable concern about its influence on the taste of wine, its effectiveness varied with soil type, and it only offered a temporary solution since it did nothing to prevent the subsequent reinfestation of the vines. Numerous other chemical treatments were tried (Mouillefert, 1876), but to little avail, and gradually during the 1870s those fighting phylloxera began to fall into two conflicting schools of thought. On the one hand were those still advocating the use of chemicals, and on the other those, following Laliman (1879, 1889), who supported the use of American vines. Early experiments had shown that American vines withstood the onslaught of phylloxera, but the wines produced from them had a distinctly foxy flavour. There was, therefore, considerable resistance within France to the idea of grafting indigenous French vines onto American rootstocks, for fear that the wines produced from the grafted vines would carry traces of the taste of the rootstocks. Another problem for those advocating the use of American vines was that it was widely recognised that it was these very vines that had initially introduced phylloxera to Europe, and several previously uninfested *départements* sought to ban their introduction. Thus a decree passed in Burgundy in September 1874 prohibited the import of all types of vine in an attempt to prevent the spread of phylloxera into the region (Ordish, 1987).

Eventually, two laws, promulgated on 15th July 1878 and 2nd August 1879, provided a framework within which action against phylloxera was to be taken (Lachiver, 1988). These laws authorised the creation of *syndicats antiphylloxériques* (anti-phylloxera syndicates), which proved to be spectacularly unsuccessful (Stevenson, 1981), and they also divided the country into three zones in an attempt to restrict the distribution of American vines. In the southern infested zone there was to be free use of French and American vines; in a second zone, where there were as yet few signs of phylloxera, these were to be treated with chemical insecticides and American vines were to be prohibited; and in the third zone, believed to be free from phylloxera, traditional methods of viticulture were to be continued, with the entry of American vines being strictly prevented. Despite these measures being reinforced by a decree of 1882, they failed to prevent the spread of phylloxera. This was largely due to the prolific fecundity of the phylloxera aphid, and the continued use of somewhat ineffective chemical measures against it. In an analysis of the development of anti-phylloxera syndicates in a single *département*, that of Loir-et-Cher, Baker (1983) has illustrated that, although phylloxera had been identified in 1876, there was mounting resistance by some wine producers during 1880 and 1881 to the compulsory chemical treatment of their vines. Baker (1983: 216) also argues that the main reason for the subsequent failure of the anti-phylloxera syndicates in Loir-et-Cher was the limited effectiveness of the chemical treatment that they were promoting, and that this in turn led 'to a reinforcement of the traditional peasant *mentalités* which considered struggle against the whims of government as inevitable as that against the vagaries of nature'. However, the costs of the treatment as well as the scepticism and pride of the farmers also played their part in preventing the wider spread of the syndicates.

The turning point in the fight against phylloxera came in 1881 at the International Phylloxera Congress held in Bordeaux (Fitz-James, 1889), when it was eventually accepted that the best solution was the grafting of French vine scions onto American rootstocks. This engendered numerous experiments to select the best rootstocks for different soil types and the best method of grafting. However, there was a marked social differentiation in the ability and willingness of farmers to adopt this solution. In the Hérault, for example, Stevenson (1978) has shown how it was the rich owners of large vineyards, such as Henri Bouschet and members of the Bazille and Lichtenstein families, together with Planchon who had recently returned from his visit to the United States of America in 1873, who were at the forefront of the adoption of American vines in the late 1870s. The vast majority of vineyard owners held but small parcels of land, and were unable to afford the expenses involved in tearing up their old vines and replanting them with grafts on American rootstocks. It was only through the strenuous efforts of Gustave Foëx, Director of the Ecole d'Agriculture de Montpellier, who in 1882 produced a small booklet strongly recommending the use of American vines, and written in a clear style specifically designed for the small vine grower, that adoption became more widespread (Stevenson, 1978).

The pace of replanting of French vineyards with grafted vines accelerated in the 1890s, and reaching its conclusion during the 1920s, by which time virtually all of the vines cultivated in France were grafted onto American rootstocks. The phylloxera crisis nevertheless initiated a decline in the area devoted to vineyards in France, which has continued to this day. From 1,800,000 hectares in 1892 the area of vineyards fell to 1,409,000 hectares in 1922, 1,299,000 hectares in 1962 and 982,000 hectares in 1987 (Lachiver, 1988: 583–4). The new grafted vines also proved to be much less tolerant of chalk, and tended to develop chlorosis on soils with a high lime content. In some instances this led to a shift in vineyard location from areas of chalk hillslope to the deeper and more acidic soils of the plains (Degrully, 1896; Stevenson, 1981). Moreover, the widespread introduction of American vines brought with it yet another fungal parasite, downy mildew, which was first noted in France in 1878. By 1882 it had affected most of the major wine producing areas of the country, further contributing to the decline in yields initiated by phylloxera. Fortunately, the discovery of a remedy was much more rapid, and following successful experiments by Millardet using copper sulphate sprays in the Gironde in 1883 and 1884, the use of this 'Bordeaux mixture' became universal by the end of the decade.

The spread of phylloxera beyond France

During the 1870s phylloxera spread rapidly beyond Europe. In California there is some debate as to the date of its introduction, with Davidson and Nougaret (1921) suggesting that it was introduced as early as 1858, although other authorities suggest that it was not clearly identified there until 1873 (Carosso, 1951; Ordish, 1987). It is also uncertain whether it was the result of the importation of French vines or Concord vines from the eastern States (Winkler, 1962). As in France, farmers were slow to recognise its effects, but, in contrast to the European situation, the absence of the winged varieties of phylloxera in California meant that its spread was somewhat slower there than it had been in Europe (Hutchison, 1984). Phylloxera was first recorded in Australia as early as the mid-1870s, when large areas of Victoria began to be infested, and by the mid-1880s it was also detected in New South Wales (Gregory, 1988; Hardie and Cirami, 1988). However, rigid quarantine restrictions imposed by the Parliament of South Australia proved effective in preventing phylloxera from reaching this state's vineyards.

Although by the 1880s grafting was becoming generally accepted as the solution to the problem, the expense involved meant that farmers in many of the poorer countries of Mediterranean Europe simply abandoned the cultivation of vines. Moreover, the spread of American vines from France, while providing the eventual cure, also appears to have carried the phylloxera aphid to previously uninfected vineyards (Laliman, 1889). Different countries adopted a variety of strategies in the face of phylloxera. Thus, in Switzerland the immediate destruction of infected vineyards and compulsory anti-phylloxera insurance were introduced, but as elsewhere this failed to halt its spread (Fatio and Demole-Ador, 1875). In southern Spain, the diversity of traditional vine varieties necessitated considerable experimentation to identify the most suitable rootstocks (Rubio, 1978), and in Jerez it led to a concentration on higher quality grape varieties with the abandonment of more peripheral vineyards (Naranjo, 1978). Further east in Hungary, phylloxera caused a reduction in the vineyards of Tokaj-Hegyalja from 6,000 hectares in 1870 to 1,725 hectares by 1890, although differences in soil type meant that some regions were more seriously affected than others, with vines grown on the high quartz content loess soils of Mount Kopaszhegy being relatively unaffected (Berenyi, 1978).

By the early years of the twentieth century nearly all European vineyards had been uprooted and replanted with grafted vines, and most other wine producing areas of the world which had previously escaped infestation had begun to report the presence of phylloxera. There were, however, a number of notable exceptions where phylloxera failed to take hold, and which today remain free of the pest, including parts of Australia, New Zealand, Russia, California, Crete, Cyprus, and Rhodes (Ordish, 1987). Moreover, the whole of Chile is widely reported to be free of phylloxera. Stevenson (1988: 410) thus argues that all of Chile's vines owe their origin to imports of European varieties made in 1851 by Silvestre Ochagáva, and that 'With no subsequent need to import vines, and with Chile's unique situation, bounded by the Pacific Ocean to the west, the Andes to the east, vast deserts to the north and the Antarctic to the south, the dreaded phylloxera has not found its way to any of this country's vineyards'. Elsewhere, the use of careful quarantine measures, as in Australia, and the existence of adverse environmental conditions have given rise to the existence of the other pockets where phylloxera has not yet penetrated. In particular, phylloxera is unable to survive in very sandy soils, and this has enabled producers in areas such as Colares in Portugal to continue to use ungrafted vines.

Much still remains unknown about the spread of phylloxera, but it seems that both environmental and biological factors contribute to determine the susceptibility of a vineyard to attack. The most important environmental factor would appear to be soil type, but climatic conditions at critical times in the insect's life cycle may also be significant. Among the biological factors, it is apparent that different varieties of vine, and indeed different clones, have different levels of resistance to phylloxera, and this may well be the explanation for the survival of some scattered pockets of ungrafted *Vitis vinifera* in Europe. Moreover, the existence of different strains of phylloxera each adapted to a particular type of vine now seems likely, and recent reports suggest that roots of American species of *Vitis*, such as St George and Ganzin I, which are generally considered to be resistant to phylloxera, are by no means completely immune to its effects (Granett *et al.*, 1983).

The implications of the spread of phylloxera

The spread of phylloxera throughout the vineyards of the world had four main social and economic influences. First, as has been seen above in the case of France, it caused a significant reduction in the area devoted to vineyards and also to the overall production of wine. It was not phylloxera alone, though, but rather its combination with oïdium, downy mildew and other later pests and diseases that initiated this decline. Moreover, if subsequent economic conditions had been conducive to the continued expansion of wine production it seems likely that many abandoned vineyards would have been replanted. Instead, the increased costs of the new methods of viticulture required, involving grafting and regular spraying, coinciding with an increase in cheaper beer and cider production, meant that the terms of trade moved against wine in the aftermath of the phylloxera crisis. While the total area devoted to vines decreased the average productivity per vine increased, and this to some extent compensated for the declining area of vineyards, although it had adverse effects on the quality of the wine. In France as a whole, for example, yields per hectare increased from 13.5 hectolitres in 1880 to 38.6 hectolitres in 1920 (Lachiver, 1988: 582-83).

Secondly, the spread of phylloxera led to increased differentiation in the structure of wine production (Loubère, 1978b). While large, capital intensive enterprises, such as the great châteaux of Bordeaux and the major port producers, could withstand the increased costs associated with grafting and spraying, small peasant producers found it increasingly difficult to make a satisfactory return from viticulture. Thus, as in the Jerez region of Spain (Naranjo, 1978) and in many parts of France and Portugal, numerous small wine producers left their farms and either became day labourers (Stevenson, 1981), or emigrated to neighbouring towns or overseas in search of a new life. This was particularly apparent in the Upper Douro valley where numerous vineyard terraces were abandoned towards the end of the nineteenth century, despite government experiments to use them for tobacco, coffee, tea and citrus fruit production (Robertson, 1982). In France many thousands of small wine producers likewise left their farms in the Midi (Galtier, 1960), and ports in wine producing areas such as Bordeaux (Roudié, 1985, 1988) also experienced a considerable rise in emigration. The effects that this had on peasant society in the Hérault have been well captured in Baisette's (1956) novel *Ces Grappes de ma Vigne*, which traces the ever increasing misfortunes to befall the poorer members of that society as a result of the arrival of phylloxera and the subsequent restructuring of viticulture in the region.

Thirdly, the reduction in supply of wine, particularly in France in the 1880s, but also elsewhere at a later date, led to a considerable expansion in the amount of fraud and chemical adulteration of wines. The most typical frauds involved the addition of sugar before fermentation to increase the alcohol content of the wines, known as chaptalisation after the French chemist Chaptal (1756-1832), the production of wine from the addition of water to dried raisins, which received a considerable impetus following the publication of a book in 1880 by Audibert entitled *l'Art de Faire le Vin avec les Raisins Secs*, and the addition of dyes and flavourings (Ordish, 1987; Lachiver, 1988). While none of these practices was in itself new, the reductions in the supply of genuine wine reinforced and expanded their use. The importation of sugar and raisins through the southern port of Marseille, made this a focus for the fraudulent manufacture of wine. Likewise, the proximity of the southern French ports to Algeria provided the ready opportunity for unscrupulous merchants to blend the stronger Algerian wines with those of Languedoc, and Sète became an important centre for this activity, as well as for the creation of new blended alcoholic beverages such as Vermouth. It was not only Algerian wine, though, that was used in blending, and the deliberate falsification of the origins of wines enabled considerably greater profits to be made at the expense of the unsuspecting customer. During the first three decades of the twentieth century the growing problems of fraud eventually led to the emergence of legislation designed to guarantee the origins of French wines, and similar laws were later promulgated in most of the countries of western Europe.

Fourthly, the crisis experienced in traditional viticultural regions led to a dramatic increase in wine production both in areas which were only influenced relatively late by phylloxera, and also in places from which the vine had previously been entirely absent. Loubère (1978a, b) has thus noted how the collapse in French wine production and exports between 1860 and 1870 led to a considerable growth of viticulture in south-eastern Italy, and to an increase in both Spanish and Italian wine exports before these countries were in their turns devastated by the spread of phylloxera. The country initially to benefit most from the crisis in Europe was the French colony of Algeria (Galtier, 1960), where the area devoted to vines increased from 10,500 hectares in 1865 to 60,400 hectares in 1885 and 168,000 hectares in 1905 (Lachiver, 1988: 586). This vast increase was made possible largely by the relatively late arrival of phylloxera in

Algeria, which was first reported only as late as 1885 (Isnard, 1975; Stevenson, 1981).

These effects of phylloxera presented a substantial challenge for European wine makers, but by the end of the nineteenth century viticulture had also become firmly established in the Antipodes and California. This in turn provided a new framework of competition in the global wine economy, and set in motion the forces that were to shape much of the wine industry in subsequent years.

THE SPREAD OF VITICULTURE TO THE NEW LANDS OF AUSTRALIA, NEW ZEALAND AND CALIFORNIA

The examples of the introduction of viticulture into Latin America and southern Africa described in the previous two chapters illustrated that a range of economic, social, ideological and political processes were involved. The Dutch development of vineyards in the Cape reflected a greater concern with the investment of capital in the expectation of profit, than was the case in the Spanish introduction of wine production in Mexico, where social and cultural factors appear to have been of most significance. The subsequent spread of viticulture to California and Australia during the nineteenth century (Figure 33) can likewise be interpreted in the context of the rising dominance of industrial capital, and the emergence of the imperial age in the latter quarter of the century (Hobsbawm, 1975, 1987; Baumgart, 1982). However, in both cases it is impossible to see the spread of viticulture and the consumption of wine purely as a response to economic necessities. While the increasing needs of industrial capitalists to buy labour and raw materials as cheaply as possible, and to realise their profits through the sale of ever increasing quantities of commodities to an expanding market, forced the European powers to expand their political control to the furthest corners of the world during the nineteenth century (Avineri, 1969; Luxemburg, 1972; Corbridge, 1986), the early developments of the wine industry in these areas owed little directly to the economic requirements of capitalism. Instead, they were often the outcome of the pioneering activities of small numbers of individuals who were determined to introduce the vine into their new homelands.

Viticulture in Australia and New Zealand

As with the introduction of viticulture to southern Africa by the Dutch, the first vines planted in Australia were brought there, not by settlers from the Mediterranean homeland of the vine, but rather by denizens of a traditionally wine importing country of northern Europe. The British decision to create a penal colony in Australia implied the introduction of

296

an agricultural enterprise to support it, and carried with it the expectation that the new colony might one day furnish produce for Britain which had previously been imported from elsewhere. It is nevertheless somewhat surprising that vines were to be found among the plants carried in the holds of the eleven ships under the command of Captain Arthur Phillip that arrived in Australia in January 1788 to establish the first colony. These vine cuttings had been obtained from Rio de Janeiro and the Cape, and were planted at Farm Cove soon after the establishment of the convict settlement at Port Jackson (Bishop, 1980; Mayo, 1986; Gregory, 1988). In the high humidity the vines grew rapidly, but were subject to disease and only produced small amounts of fruit of poor quality. As a result, Captain Phillip ordered that a new vineyard be planted further inland on a three-acre site by the Parramatta River in 1791, by which date another one-acre (0.4 ha) vineyard had also been planted by a settler named Schaffer (Laffer, 1949).

Thereafter, the expansion of viticulture was slow, but following his crossing of the Blue Mountains in 1813, Gregory Blaxland established a new vineyard with vines from the Cape at Brush Farm, Parramatta, in 1816 (Laffer, 1949). Although the wines produced were not of a particularly high quality, Blaxland was awarded a silver medal by the Society for the Encouragement of the Arts, Manufactures and Commerce in London for a red that he had submitted in 1822, and later in 1828 he was also awarded a gold medal for a further shipment. Meanwhile, another early pioneer, Captain John Macarthur, who had obtained a land grant of 8,500 acres (3,440 ha) in 1805, but had subsequently been forced to leave the colony because of his role in the Rum Rebellion (Rickard, 1988), used some of his time abroad to visit the vineyards of France and Switzerland in 1815, where he gathered practical information about the production of wine. Macarthur is best known for his influence on the Australian sheep industry, but on his return to Australia in 1816 he brought with him a selection of French vine cuttings, and used these to establish a vineyard at Camden Park (Halliday, 1985). While some doubt has been expressed as to how many of these cuttings actually arrived in Australia (Ramsden, 1940), by 1839 Macarthur's son William is recorded as having introduced German vine growers as colonists, and the Macarthurs can be seen as the first people to have developed a commercial winery in the continent (Mayo, 1986). At the beginning of the nineteenth century, though, New South Wales society still largely reflected its convict origins, and within this community beer and spirits were by far the most popular alcoholic beverages, leaving little demand for the wine that was beginning to be produced. Moreover, the rapid success of wool production and the dramatic expansion of sheep numbers in the 1820s (Rickard, 1988) provided a far more profitable source of income for the new settlers of the continent than did viticulture.

These early experiments with wine production in Australia neverthe-less coincided almost exactly with the rise to power of Napoleon Bonaparte in Europe between 1797 and 1815, and to some extent can be seen as reflecting the concerns of the British government to increase production of all kinds of basic commodities in the colonies, particularly, as with wine, where they could replace those previously imported from France. Indeed the disruption of trade between France and Britain consequent on the outbreak of war between the two countries in 1793, acted as a major incentive for the small number of wine producers in Australia. During this period Napoleon sent a commission to Australia to report on the colony, and it is interesting to note that one of the Commissioners, a certain Monsieur Peron, reported in 1803 that 'In spite of the fact that Britain's consumption of wine, both at home and on her Fleets, is immense, she grows none of it herself. Australia must therefore become the "Vineyard of Great Britain" ' (Laffer, 1949: 8). Despite this forecast, it was not until the 1830s that Australian viticulture became firmly established, and this was largely as the result of the enterprising activities of James Busby, who is widely considered to have been the 'father' of Australian viticulture (Hyams, 1987; Mayo, 1986; Gregory, 1988).

Before setting out for Australia Busby had spent some time in France, and he also visited the vineyards that had been established at the Cape. Based on these experiences, and derived largely from the writings of Chaptal, he spent much of the voyage to the colony compiling *A Treatise on the Culture of the Vine, and the Art of Making Wine*. . . (Busby, 1825). In the introduction to this Treatise, Busby argued that vineyards could supply 'the great *desideratum* of a *staple article of export*, to which the colonists of New South Wales might be indebted for the future prosper-ity' (Busby, 1825: xviii). This conviction was based on what he saw as the profits which could be derived from the cultivation of vineyards, the favourable climate and soils of New South Wales, and the extensive market which could be found. His arguments nevertheless apparently fell on deaf ears, and for two years he was employed as a teacher at the Male Orphan School at Cabramatta in New South Wales, where he had the task of teaching the orphans viticulture (Ramsden, 1940; Mayo, 1986). When the Church Corporation took over the orphanage in 1826 his services were dispensed with, and he sought a variety of other jobs in the public service. Meanwhile, in 1824 he had acquired a small property between Branxton and Singleton in the Hunter Valley, which he called Kirkton, and by 1830 it is reported that vines were being cultivated there (Halliday, 1985). His second book on wine, *A Manual of Plain Directions for Planting and Cultivating Vineyards and for Making Wine, in New South Wales*, was also published in 1830, and unlike his *Treatise*, which was aimed at the higher classes, the *Manual* was designed for the class of smaller settlers in order to 'convince them that they, and each member of their families may, with little trouble and scarcely any expense, enjoy their daily bottle of wine, the produce of their own farms . . . and not strong drink' (Busby, 1830: 7). Busby was evidently appalled by the high consumption of spirits and beer that he had encountered in his five years in New South Wales, and he hoped that his book would lead to a more sober and happy population. Thus, in his dedication to the Governor of New South Wales he argued that

> Those who have witnessed the temperance and contentment of the lowest classes of the people in the Southern Countries of Europe, where wine is the common drink of the inhabitants, and have contrasted them with the unhappy effects produced by the consumption of spirits, or of malt liquors, among the same ranks, in less favoured climates, will easily perceive, how much it would add to the happiness of the Colonists of New South Wales, if their habits were assimilated, in this respect, to those of the inhabitants of wine countries; and will appreciate the importance of introducing the one beverage, and diminishing the use of the others, in a Community constituted like this, – in which the high price of labour is calculated to allow the almost unlimited use of ardent spirits, and where the excitement they produce, is more likely than in most other countries, to terminate in mischievous results.
>
> (Busby, 1830: v-vi)

Copies of the *Manual* were distributed to the District Constables from whom it could be purchased at the same price as it could be obtained in Sydney, and it does appear to have had some influence on the planting of new vineyards. Thus, by 1832 the number of vineyards in the Hunter Valley had increased to ten, and thereafter they expanded considerably, with fifty vineyards having been established there by the end of the 1840s (Halliday, 1985).

Busby, however, was not satisifed with the employment offered to him in the colony, and he returned to London in 1830 to plead his case. From there he went on a four-month tour of France and Spain in the winter of 1831 in order to collect vine varieties growing in different soil and climatic conditions, which might be suitable for cultivation in Australia. In his *Journal*, published in 1833, he recorded details of the trip, including conflicting reports on the number of vine cuttings that he collected. It seems that he obtained cuttings of 437 varieties from the Botanic Gardens at Montpellier (Busby, 1833: 72-3), 110 different varieties from the Luxembourg Gardens in Paris (Busby, 1833: 107), seventy-four varieties from vineyards in Rousillon and Languedoc (Busby, 1833: 112-20), seventeen from 'Xeres and Malaga' in Spain (Busby, 1833: 120-6), and a further forty-four varieties from Syon House near London.

These were then sent to Sydney where they were established at the Botanic Gardens, with 362 varieties reported still to be alive in January 1833 (Busby, 1833: 126). However, later that year Busby took up his new appointment as British Resident in New Zealand, and his connections with the Australian wine industry ceased. Although some of his cuttings were made available to the aspiring vine growers of the colony, the collection was poorly maintained and by the 1840s it had been allowed to fall into ruin.

The expansion of viticulture in Australia thereafter followed the increasing pace of colonisation and settlement, with vines being planted almost immediately on the foundation of each new colony. Vineyards were established in Tasmania in 1823, in Western Australia in 1829, in Victoria possibly in 1834 but certainly by 1838, and in South Australia in 1837 or 1838 (Halliday, 1985; Mayo, 1986). This early wine industry was given a considerable boost during the gold rush of the 1850s when numerous immigrants from France, Switzerland, Italy and Germany settled in the colonies. Many, having failed to make a success of prospecting turned to farming and, in particular, to viticulture. Wine nevertheless remained a minority drink in Australia, and as in Britain beer and spirits prevailed as by far the most popular alcoholic beverages until well after the end of the Second World War in 1945. The first official exports of Australian wine to Britain were in 1854, but only once between then and 1879 did they exceed 50,000 gallons (Laffer, 1949). Despite the support of British Members of Parliament during the 1870s (Briggs, 1985), Australian wines failed to make a significant impact on the British market until the inter-war period between 1918 and 1939. Eventually, though, by 1940 Ramsden (1940, 367) was at last able to say that 'Australia is now the vineyard of the British Empire', 137 years after Peron's initial prediction.

The first introduction of vines to New Zealand preceded Busby's appointment as Resident and was a direct result of the missionary activities of Samuel Marsden, the chief Anglican Chaplain to the Government of New South Wales. In 1814 Marsden had established the first mission in New Zealand at Rangihoua, and in 1817 he appointed Charles Gordon as superintendent of agriculture there. Numerous varieties of crop were introduced by Gordon, and it seems likely that these included vines. The first specific reference to the planting of vines occurs in 1819 when Marsden returned to establish a further mission at Kerikeri and recorded in his Journal that he planted about a hundred grape vines of different kinds (Thorpy, 1983: 8). It is uncertain whether wine was produced from these early vineyards, but by 1840 the French Commander of the ship Astrolabe visited Waitangi, where Busby had planted a vineyard on his arrival in 1833 (Ramsden, 1940), and praised the light white wine made from the local grapes (Wright, 1955). Meanwhile, in

1838 Bishop Pompallier, the first Catholic Bishop of the South Pacific, had arrived on the North Island to establish a series of Catholic missions, and he brought with him a number of vines which were planted and began to produce sufficient fruit for the production of wine by 1843 or 1844 (Thorpy, 1983: 12). Two years after Pompallier's arrival the French government, who did not recognise the British claim to New Zealand, established a small settlement at Akaroa, where the settlers immediately began cultivating vines and producing their own wine (Thorpy, 1983).

However, throughout the nineteenth century viticulture and wine production remained relatively unimportant in New Zealand, and it was not until 1894 when Romeo Bragato, the viticultural adviser to the Government of Victoria, was invited by the Prime Minister of New Zealand to write a report on the state and future of the country's wine industry, that any real interest was taken in its advancement. Bragato's (1895) report was enthusiastic, and led to the establishment of an experimental vineyard at Te Kauwhata in 1898. By then, Dalmatian imigrants from what is now Yugoslavia had also begun to grow vines, and in 1906 Bragato reported that there were some 550 acres (223 ha) of vines in the country as a whole. By 1909 the area under vines had increased to 668 acres (270 ha), but subsequently, with lack of Government support and the rising political power of the prohibitionists, this collapsed to only some 179 acres in 1923 (Thorpy, 1983).

The origins of Californian viticulture

While the early development of Australian viticulture owed much to the desire by the British government to produce wine in its colonies, the first cultivation of vines in California was the direct result of the expansion of Franciscan missions in the late eighteenth century. Although viticulture was firmly established by the Spaniards in Mexico in the sixteenth century, it was not until the increasing threat of French, Russian and English expansion in the eighteenth century led to the Spaniards establishing a series of presidios, pueblos and missions in Alta California (Billington, 1974; Brogan, 1985), that viticulture spread northwards (McKee, 1947; Carosso, 1951; Pinney, 1989). Jesuit missionaries had settled in Baja California as early as 1669, but it is widely reputed to have been the Franciscan Junípero Serra, who brought the Mission or Criolla vine north into Alta California (Hyams, 1987; Adams, 1984). Although the San Diego mission was first established in 1769 (Krell, 1979), there is no evidence that vines were immediately planted, and instead, it seems that together with the other early missions established in the 1770s, the wine that was required for the celebration of the Mass was supplied from Mexico (Pinney, 1984a). This system of supply was, however, haphazard and unreliable, and in 1777 Serra wrote to the Viceroy in Mexico City suggesting that grapevines could usefully be introduced into California. His request appears to have been granted, since in 1779 a friar at the mission of San Juan Capestrano mentioned the recent arrival of some vine cuttings in a letter to Serra, and it seems likely that these were used to produce wine for the first time in 1782 (Brady, 1984). Subsequently, most of the missions began to cultivate vines and produce their own wines, with the Mission of San Gabriel becoming particularly famous for its fine wines (Krell, 1979). Most mission vineyards were, however, quite small and although there were some 53,000 non-mission vines being cultivated in the vicinity of Los Angeles in 1818, it was not until the 1830s that commercial viticulture really began in California.

In 1833 the Mexican government secularised the missions (Krell, 1979), depriving the Franciscans of all of their temporalities, and opening up their lands to settlement by those with enough influence to acquire them (Billington, 1974). This paved the way for a number of immigrants to develop new vineyards, and one such aspiring Frenchman was Jean Louis Vignes, a native of Cadillac near Bordeaux, who bought land to the east of Los Angeles and began to produce both wine and brandy at his newly established El Aliso ranch (Teiser and Haroun, 1983; Adams, 1984). During the 1830s Vignes experimented with a number of imported European vine cuttings, but apparently with little success, since the dominant grape variety remained the Mission (Pinney, 1984a). Elsewhere, other ranch owners, such as William Wolfskill, the Englishman William Workman and the Mexican Tiburcio Tapia, also planted vineyards, and in 1847 when the United States annexed California, vineyards were to be found throughout the State from Sonoma and Napa counties in the north to Los Angeles in the south.

The discovery of gold in 1848, and the subsequent Gold Rush led to a massive increase in demand for all kinds of alcohol throughout the gold counties of Amador, Calaveras, El Dorado, Nevada, Placer and Tuolumne. This encouraged the expansion of Californian vineyards already in existence, and once the first fever of the Gold Rush was over in the mid-1850s, a number of new immigrants also turned to wine production as a more reliable source of income (Carosso, 1951). The subsequent expansion of viticulture was dramatic: by 1857 there were reported to be about 1,500,000 vines in California, and only three years later, following legislation in 1859 which exempted vineyards from taxation, this had risen to some 6,000,000 vines (Pinney, 1984b; Adams, 1984). Much of the wine produced, however, was of poor quality, and the massive increase in supply led to a crisis with both wine prices and the value of vineyards falling appreciably. This crisis was temporarily resolved during the 1860s by the introduction by immigrants from Europe of varieties of vine more suited to the Californian environment and of improved methods of vinification. Among the most enigmatic of these immigrants was the Hungarian born Agoston Haraszthy de Mokesa, who apparently came to the United States as a refugee in 1840 and then purchased the Vineyard Farm at Sonoma in 1857, renaming it Buena Vista (Bartholomew, 1947; Hutchison, 1984). In 1861 he persuaded the Governor of California, John G. Downey, to send him to France, Germany, Switzerland, Italy and Spain to study European methods of viticulture and to 'report to the next Legislature upon the ways and means best adapted to promote the improvement and culture of the grape-vine in California' (Haraszthy, 1862: xv). He recorded the details of his journey in his book *Grape Culture, Wines and Wine-making, with Notes upon Agriculture and Horticulture*, which was published in 1862, including appendices quoting from German authorities, such as J.C. Leuch, on viticulture, and it is upon this together with his introduction of vines that his reputation as the father of Californian viticulture largely rests. During his travels, Haraszthy (1862: xx) claimed to have purchased 'in different parts of Europe 100,000 vines, embracing about 1400 varieties', and these were then shipped to California, where they appear to have been planted somewhat haphazardly. However, Haraszthy was not alone in importing vines. A number of other immigrants such as Charles Krug and Mariano Vallejo, also imported European grape varieties, and in the long run they turned out to be much more successful wine producers than Haraszthy (Teiser and Haroun, 1983; Hutchison, 1984). Indeed, California's most widely planted red grape variety, the Zinfandel, which was for long considered to have been introduced by Haraszthy, has now been shown to have been first grown in America as early as 1830 on Long Island, New York by a certain William Robert Prince (Adams, 1984: 547). Haraszthy's description of his travels in Europe is nevertheless of particular interest for the light it sheds on the development there of a capitalist wine industry, and on his hopes for the development of a similar industry in California. In his report to the Governor, he thus noted that he 'endeavoured to induce capitalists to come among us and establish business places, to purchase the grapes from the samll producers as in Europe, and to erect manufactories for making wine and extracting sugar from Sorgho, beet-root, and Imphee' (Haraszthy, 1862: xix). Typical of the factories in Germany that had impressed him was the 'Champagne manufactory of the Joint-Stock Association' at Hocheim, whose Director was Herman Dressel, and which he described as follows

> This is one of the largest establishments in Germany. It employs eighty men, and makes daily three thousand bottles of Champagne. The capital invested is 1,000,000 güldens (about $400,000). It makes very good sparkling wine, and imitates excellently the French Champagne. Some of the immitations are really much better than

the brands they pretend to immitate. The establishment makes money.

(Haraszthy, 1862: 61)

The experiments of wine producers such as Haraszthy, Krug and Vallejo led to an improvement in the quality of Californian wine, and by 1870 there were 139 wineries with over 26,500,000 vines producing more than 75,000 hectolitres of wine (De Blij, 1983). Californian wine was also being exported as far afield as Australia, China, Hawaii, Peru, Denmark and England (Hutchison, 1984), and with the opening of the first transcontinental railway in 1869 the vast east coast market, which had previously been the preserve of producers in Ohio, Missouri and New York, was opened to Californian producers. A typical Californian vintage scene of around this time is illustrated in Figure 38, which shows Chinese labourers treading the grapes on top of large wooden butts. As in France, though, the inability of producers to satisfy demand with reasonable quality wine led to considerable fraud, and a fall in the reputation of Californian wines. The economic depression of the early 1870s also hit Californian viticulture severely, and in 1876 grapes sold for less than the cost of picking them, with wine prices falling to 10 cents a gallon (Hutchison, 1984). Only forty-five wine producers survived, but with phylloxera taking its toll of European wine production the late 1870s and early 1880s saw another boom in Californian viticulture, with production reaching a record 10.2 million gallons in 1880. Nevertheless, phylloxera had also been introduced into California some time before 1873 (Carosso, 1951), and although its spread was slow, by 1890 most of California's vineyard districts had been affected. The remedy of grafting European cultivars onto American rootstocks, already discovered in France, was adopted more rapidly than in Europe, and despite a depression from 1886-92 a new period of prosperity for California's wine producers emerged in the 1890s and the first decade of the twentieth century. The recurring problem of overproduction had not yet been overcome, though, and in 1911 'an oversupply of grapes broke the wine market for the fourth time' (Adams, 1984: 229). Moreover, throughout the previous three decades the anti-alcohol lobby had been growing in strength, and in 1914 at the outbreak of the First World War thirty-three American States were dry. In 1919 the Wartime Prohibition Act was passed and the Eighteenth Amendment to the Constitution was ratified, leading in 1920 to the passing of the National Prohibition Act, named after its sponsoring congressman Andrew J. Volstead (McCarthy and Douglas, 1949). This stated that the manufacture, sale or transportation of intoxicating liquors within the United States was prohibited, and provided the context for yet another crisis for the country's wine makers.

Prohibition immediately led to the destruction of some vineyards, but

Figure 38 The vintage in California: wood engraving after a drawing by Paul Frenzeny, illustrated in *Harper's Weekly*, 5th October 1878

Source: Baird (1979)

one section of the Volstead Act, Section 29, which had initially been put in as a concession to the traditional cider producers of Virginia, provided a loophole which was widely exploited by grape growers throughout the United States. Section 29 stated that the penalties would not be applied 'to a person for manufacturing nonintoxicating cider and fruit juices exclusively for use in his home', and this rapidly led to vineyard owners selling grape juice and packages of pressed grapes known as wine bricks to home winemakers and bootleggers (Adams, 1984; Teiser and Harroun, 1984). Moreover, some wine was allowed to be made under permit for sacramental and medicinal use as well as for conversion into vinegar. The great demand for grape juice in the early 1920s led to renewed planting of vineyards, and by 1923 vineyard prices had risen from $200 to $2,500 an acre. Once again, though, overproduction in 1925, when grapes 'rotted at the eastern terminals waiting for buyers who had already had enough' (Adams, 1984, 24), caused yet another crash, and vineyard prices tumbled to just $250 an acre in 1926. One effect of the crash was the establishment of the California Vineyardists Association late in 1926 in an attempt to stabilise prices and lobby the government. Despite the attempts of the anti-alcohol lobby, many vineyards survived and of the 700 wineries reported in California on the eve of prohibition, 140 were still in existence at the end of 1932 (Teiser and Harroun, 1984). By the time of the Twenty-First Amendment proclaimed on 5th December 1933, which repealed the Eighteenth Amendment and made the manufacture of intoxicating liquors legal once more, the number of bonded wineries had again risen to 380, but the industry was in a very poor condition. Furthermore, the public's taste had changed from a preference for dry wines to one for wines with a high sugar and alcohol content, which the country's vine growers were in a poor position to supply.

THE CRISES AND RESTRUCTURING OF GLOBAL VITICULTURE

By the end of the nineteenth century viticulture and wine production had become truly global in scale, reflecting the momentum of the capitalist mode of production which had come to embrace the whole world within its set of productive relations (Avineri, 1969). Seven main implications of this transformation can be seen in the crises that faced viticulture during the nineteenth century and in the consequent transformations of the structure of wine production.

The most obvious difference between viticulture in 1900 and in 1800 was the far higher levels of capital invested in both the production and storage of wine at the end of the century than at its beginning. This was reflected not only in the need for greater investment in the actual cultivation of vines, as a result of the necessity for comprehensive spraying programmes and the use grafted vines, but it was also illustrated by the expansion in some regions, such as the Douro valley, Bordeaux and Burgundy, of the production of quality vintage wines which required the investment of capital in their storage. In the latter case, much of the capital was provided by *négociants* or merchants, who invested in the cellar space and were able to take a longer term view of the circulation of their capital, unlike the producers who regularly needed an annual return on their labour.

A second consequence of the changes that were taking place, and particularly of the crisis induced by phylloxera, was the inability of many small wine producers to survive in the new world of higher capital inputs. The immediate result was the replanting of many areas of vineyards with other more profitable crops, and the emigration of the peasantry from certain wine producing areas such as the Midi and the Upper Douro valley.

Moreover, while in southern Mediterranean Europe viticulture had always formed part of the economic trilogy of wheat, olives and vines, the population increase and extensive rural to urban migration that occurred during the nineteenth century led to a considerable rise in demand for the low quality wines produced in these areas. This increasing urban wine market was generated by the demand of the many new town dwellers who came from families which had traditionally produced their own wines, but it was also a result of a general increase in the level of alcohol consumption through the century, in itself partly a result of the poor living conditions to be encountered in the nineteenth century city. The rising demand for low quality wines, though, could only readily be met by the urban wine merchants, who, unlike the small scale peasant producers in the rural areas, were able to purchase sufficient quantities of wine to match and supply the demand. This in turn reinforced the economic strength of the merchants, and shifted the balance of power from the producers to the retailers of wine. All of these changes were facilitated by the construction of the railways which enabled Paris and other major towns to be supplied with wine from much further afield than had previously been possible.

Furthermore, the high prices obtainable for wine following the devastations of the vineyards caused by oïdium, phylloxera and mildew, led to considerable speculation by those with sufficient capital to invest in the planting and cultivation of new vineyards. Typical of such enterprises was the Compagnie des Salins du Midi in southern France which between 1880 and 1890 built up a series of properties between Sète and Aigues-Mortes covering some 800 hectares and capable of producing more than 100,000 hectolitres of wine a year (Galtier, 1960; Lachiver, 1988: 465). Similar capital investment, but on a smaller scale, was also characteristic of California where the rapid expansion of viticulture in the fifty years following the Gold Rush was fuelled by numerous immigrants seeking to profit from the rising price of wine.

Fifthly, however, this very process of vineyard expansion in response to high demand and prices, was in itself largely responsible for the periodic crises of overproduction and price collapse that were encountered in places as far afield as France and California. Such crises were enhanced by two important characeristics of viticulture, namely the variable yields of vines from year to year, and the delay between vine planting and wine production. These meant that wine producers were unable to respond rapidly to high prices by immediately increasing production, and also that while in average years supply and demand could be approximately balanced, there was always the danger of overproduction and a collapse of prices in years of high yields. The most serious of these periodic crises of overproduction occurred in France following the vintage of 1900, which was 40 per cent higher than that of 1899, and which precipitated the subsequent revolt by vignerons in the Languedoc in 1907 (Lachiver, 1988). However, the recurring overproduction crises experienced in California at the end of the nineteenth century and the beginning of the twentieth century were a similar expression of the same underlying cause. It was not until a much more stable balance between supply and demand, and a more uniform price structure was established later in the twentieth century that these crises were averted. However, this was achieved partly by the further restructuring of the wine industry, through a process of filtering out which occurred during each of the overproduction crises, and benefiting the larger, more heavily capitalised producers and wineries. This was particularly visible in California, where many of the smaller wine producers failed to survive, enabling organizations such as the California Wine Association, which was established in 1892, to emerge as the controlling forces in the new industry.

A sixth important feature of the changes that took place during the nineteenth century was the deliberate expansion of viticulture by the major European powers in their colonies. This was not only true in the case of the encouragement given by the British government to viticulture in Australia, but it was also apparent in the emergence and development of viticulture in Algeria by the French. While viticulture cannot be said to have played an important role in the imperial scramble for Africa at the end of the nineteenth century, it is nevertheless evident that, as in other sectors of the economy (Blaut, 1975), the one imperial power which was not a major wine producer in its own right, namely Britain, sought strenuously to import wines from its colonies in southern Africa and Australia to the detriment of its imperial competitors France and Germany. Similar motives can also be seen in President Jefferson's encouragement of viticulture and the wine industry in the eastern States of north America early in the nineteenth century.

By the end of the nineteenth century, though, there were as yet few signs of monopolist interests in the production and sale of wine. In general, the industry was dislocated, with the cultivation of the vineyards and the production of wine remaining in separate hands from the commercial enterprises responsible for the subsequent sale of the wine. There were, however, important exceptions, notably the acquisition of some vineyards in Burgundy by the wealthiest *négociants* of Beaune, and also the establishment by two German musicians, Charles Kohler and John Frohling, of an integrated wine production and retailing enterprise in California in 1854. This last example presaged the whole future direction that was to be taken by the wine industry in the twentieth century. Needing only a small amount of capital, Kohler and Frohling purchased a vineyard in Los Angeles, and also rented a cellar in San Francisco from which they sold their wines. In Pinney's (1984a: 6) words

> Their success allowed them to organize a large part of the Los Angeles industry; through contracts with local growers to buy their harvests, and with local wineries to make their wine, their firm was able both to manage the supply in an orderly way and to secure uniformity in the standards of production.

By the mid-1860s the company was shipping wine in 100,000 gallon containers to New York, and by the mid-1870s Kohler was able to boast that he supplied wine to all of the largest towns in the United States. The wider intergration of the production and sale of wine, and the emergence of the large trans-national wine and spirit corporations in the twentieth century, nevertheless awaited the closer integration of interests between different fractions of capital, and in particular the rise to dominance of finance capital.

10

THE DOMINANCE OF CAPITAL IN THE TWENTIETH CENTURY: DEMARCATION, SECTORAL INTEGRATION AND THE CREATION OF FASHION

There is a notion gone abroad that there is something fixed and unchanging in an Englishman's taste with respect to wine. You find a great number of people who believe, like an article of Christian faith, that an Englishman is not born to drink French wines. Do what you will, they say; argue with him as you will; reduce your duties as you will; endeavour even to pour the French wine down his throat, but still he will regret it . . . What they maintain is absolutely the reverse of the truth, for nothing is more certain than the taste of English people at one time for French wine. In earlier periods of our history French wine was the great article of consumption here. Taste is not an immutable, but a mutable thing.
(The Chancellor of the Exchequer, W.E.Gladstone,
Hansard, 10th February 1860: 847)

During the late nineteenth century the wine industry was plagued by recurrent crises of overproduction, which created short-term collapses in wine prices. However, the continuing global increase in demand for wine meant that prices recovered relatively quickly, particularly when an abundant grape harvest was followed by a period of relatively low yields. The critical factor influencing the price of the lower quality wines produced for the mass market was the variable quantity of the annual grape harvest, whereas, in contrast, the price of higher quality wines was determined largely by the quality of the vintage. Indeed, in its susceptibility to price variations caused by annual fluctuations in the weather, wine is notably different from other alcoholic beverages, such as beer and spirits. Many of the changes that have taken place in the wine industry during the twentieth century have been designed to reduce this level of variability in order to guarantee the successful accumulation of profit by wine producers. This has been achieved in three main ways: through the delimitation of demarcated wine regions designed to guarantee the source of origin and quality of a wine, through the vertical sectoral integration of production and retailing, and through technical advances in viticulture and vinification which have permitted the production of wines of greater consistency. Above all, though, the twentieth century has seen the balance of power shift away from individual producers and merchants, and into the hands of a few large trans-national corporations with sufficient capital to invest in new technology and marketing.

SCANDAL AND DEMARCATION: THE DELIMITATION OF PRIVILEGE

The source of origin of wines

By the beginning of the twentieth century, despite the attempts that had already been made to classify and demarcate wines, two fundamental problems remained. On the one hand many wines contained chemical additives designed to mask their flavours, and on the other hand, wines purporting to come from one source of origin frequently contained wine from elsewhere. For the producer, both strategies were designed to reduce the costs of production in order to maximise profits, and for the consumer the only relatively safe option was to purchase wines from the most reputable merchants. Moreover, there was a fundamental structural problem with the classification and demarcation of wines, since no sooner had a wine been identified as being of a particularly high quality than unscrupulous producers began to stretch the limited supplies of such wines through the addition of other wines or chemicals, while at the same time merchants also sought to pass off lesser wines as being those of more prestigious repute.

While such strategies could quite easily be legislated against at a national scale, a further problem encountered by those seeking to guarantee the origins of their wines was the use by foreign producers of European regional names to denote their own particular wines. Haraszthy's (1862) comments on German 'champagne' have, for example, already been noted, but although there were several examples of deliberate fraud, the main reason for the use of such nomenclature was the desire by producers in the new wine lands to describe their wines in a way that would be familiar to their consumers. Thus 'sherry' and 'port' style wines were made in southern Africa and Australia, most sparkling wines were known as 'Champagne', and 'Burgundies' could be found produced as far afield as Australia and California. In many cases the wines were indeed made from the same grape varieties as their names would suggest, but increasingly during the 1950s and 1960s quality names came to be used primarily for marketing purposes. Thus 'Spanish Chablis' and 'Spanish Sauternes' became popular in England, although they tasted little like the French wines after which they were named. Moreover, in the 1970s the expansion of wine production in Australia and New Zealand was combined with a proliferation of German sounding wine names. Hallgarten (1987) thus comments that while this was relatively harmless before the wines began to be exported, and reflected the nostalgia of German wine producers living in Australia for their homeland, the appearance of such wines in Europe and in the traditional international markets for German wines could cause confusion. More deliberate cases of obfuscation can be found in the example of Japanese wine labelling. Hallgarten (1987: 101-2), for example, again comments that

> From a European point of view the scandal with Japanese wines is that they are made up to appear more of European than of Japanese origin. Bottles are mostly of identical shape with those used for French Burgundy or Bordeaux wine and the labels display French names. To mislead the Japanese consumer still more, the labels sometimes show a château, or have a château name printed on them followed by the customary words 'mis en bouteille au château'. Names like Château Semillon, Château Lafife or Château Lafutte are used, also 'Estate Wines' or 'Estate Bottled but originating in Japan'.

As international agreements on the labelling of wines have been signed, many of these practices have disappeared, and during the 1970s and 1980s there has been a widespread change in non-European wine producing regions from the use of regional appellatives to varietal names for their wines. Despite this, Californian 'ports', 'Chablis', 'sherries', 'Burgundies' and 'champagnes' are still widely encountered in the 1990s, and as the Austrian diethylene glycol and Italian methyl alcohol wine scandals in 1985 and 1986 indicate, considerable fraud still exists, even in the European heartland of viticulture.

The essential characteristic of all wine demarcations based on the territorial origin of wines is that they attempt to guarantee the quality of a wine through reference to the land upon which the vines are grown. By purporting to guarantee quality, however, such classifications also enable owners of such land to reap greater profits than would otherwise be realised from their vineyards in the form of a monopoly rent (Bruegel, 1975: 41). Each demarcated area becomes its own monopoly, with the owner, or owners, able to extract a greater profit than was the case before demarcation. This process was clearly summarised by Marx in the third volume of *Capital* as follows:

> A vineyard bears a monopoly price if it produces wine which is of quite exceptional quality but can be produced only in a relatively

small quantity. By virtue of this monopoly price, the wine-grower whose excess over the value of his product is determined purely and simply by the wealth and the preference of fashionable wine-drinkers can realize a substantial surplus profit. This surplus profit, which in this case flows from a monopoly price, is transformed into rent and accrues in this form to the landowner by virtue of his title to the portion of the earth endowed with these special properties. Here, therefore, the monopoly price creates the rent.

(Marx, 1981: 910)

Although Marx was specifically concerned with high value wines, and he failed to consider the artificial creation of monopoly prices through the legal demarcation of vineyards, his observations have two important implications. First, the delimitation of a boundary between vineyards can have a significant influence on the levels of profit obtained by farmers on either side of the boundary of demarcation. Precisely the same levels of inputs can be applied to each type of vineyard, but the wine from a demarcated vineyard can be worth many times that of the wine from a vineyard outside the limits of the defined area. The precise delimitation of each demarcated area therefore offers considerable scope for corruption and social disruption, as different groups of wine producers vie with each other for the opportunity to gain access to monopoly rents. Secondly, however, by controlling the level of production of a particular demarcated wine, for which demand outstrips supply, the producers can extract further profit. The classic example of this occurs with the declaration of port vintages, where the producers only declare vintages in certain years, thus restricting the supply of vintage port and maintaining its high value. The conscious restriction of production levels throughout most demarcated wine regions, aimed in part at maintaining the high quality of the wines, does have as a concomitant effect the restriction of supply and thus the maintenance of profits over and above those obtained from non-demarcated wines.

The emergence of the French *Appellations d'Origine Contrôlées*

Although the Portuguese delimitation of the Upper Douro valley under Pombal was the first modern example of vineyard demarcation, it is the French who have developed the most comprehensive system of demarcation for wines based on their places of origin. De Blij (1983: 109) notes that in the wake of the phylloxera disaster four main deleterious practices had become widespread in France by the beginning of the twentieth century: the use of hybrid vines, an emphasis on quantity rather than quality of grapes, 'the addition of water to "stretch" the wine', and the use of sugar and water to produce more wine from the leftover *marc*. To

WINE AND THE VINE

these Lachiver (1988) also adds the widespread practice of adding lead oxide to wines in order to halt acetification. Prior to the introduction of legislation in France, a number of producers had formed organisations to promote the qualities of their wines. Thus 79 producers in Chablis formed an association in 1900 to guarantee the authenticity of their wines (George, 1984), and in 1901 in the Bordelais the Union Syndicale des Propriétaires de Crus Classés du Médoc was established (Penning-Rowsell, 1985). In response to this activity the French government passed a general law on 1st August 1905 designed to counter the fraudulent production of agricultural products, and to give the consumer some redress if wines were not what they claimed to be (Capus, 1947; Warner, 1960). While this did not in itself specify demarcated wine regions it is widely considered (de Blij, 1983; Lachiver, 1988) to represent the first step towards the emergence of the *Appellation Contrôlée* system. On 29th June and 15th July 1907 further laws were passed, requiring producers among other things to declare the volume of their annual vintage and their stocks, and although this was primarily designed to counter the problem of overproduction, the laws also included measures to limit fraud, particularly that involving the addition of sugar to musts (Warner, 1960; Lachiver, 1988). Later in the year, on 3rd September, in order to tighten up on this legislation, wine was formally defined as coming 'exclusivement de la fermentation alcoolique du raisin frais ou du jus de raisin frais' [exclusively from the alcoholic fermentation of fresh grapes or of the juice of fresh grapes] (Lachiver, 1988: 476–7).

Between 1908 and 1912 decrees were then passed in order to regulate the areas from which Champagne, Cognac, Armagnac, Banyuls and Bordeaux wines could be produced. The most controversial of these decrees was that for Champagne, dated 17th December 1908, which accorded the *Appellation* of Champagne to wines harvested and produced in all of the *communes* and *arrondissements* of Châlons-sur-Marne, Reims and Epernay, and some of Vitry-le-François in the *département* of the Marne, as well as some of the *communes* and *arrondissements* of Château-Thierry and Soissons in the *département* of the Aisne (Capus, 1947). However, none of the vineyards of the Aube *département* were included, even though their owners claimed that Troyes had traditionally been the capital of Champagne, and that their wines were regularly sold to *négociants* in Reims and Epernay. Less scrupulous champagne *négociants* had regularly been buying raw wines in the cheapest market possible, which was usually the Midi, and had then transformed them into champagne (Simon, 1962). Two poor vintages in Champagne in 1909 and 1910, during which the small wine makers had difficulty in selling the limited amounts of wine that they had produced to the *négociants*, led to unrest as Midi wines were still being brought to Epernay. During December 1910 and January 1911 there were protests by

314

the *vignerons* of the Marne, who broke into casks of wine imported from the Midi and poured their contents into the river. However, the passing of a further law on 10th February 1911, specifying that only grapes produced in the delimited area could be used for champagne, and that separate cellars had to be used for other wines, temporarily pacified their protest. Until then the producers of the Aube had hoped that their wines could still be sold as champagne, but thereafter this was no longer possible. Consequently they formed the Ligue de Défense des Vignerons de l'Aube, and in a movement reminiscent of the 1907 protests in the Languedoc (Bechtel, 1976), took to the streets, with demonstrations at Bar-sur-Seine on 19th March, at Troyes on 9th April, when 20,000 people were present, and at Bar-sur-Aube on 30th April (Lachiver, 1988). Meanwhile, in the Marne, the *vignerons* were also outraged at the thought of further competition being allowed from wines from the Aube. Their protests were directed at the *négociants*, whom they accused of fraudulently bringing in wines from elsewhere, and on 11th April they destroyed cellars at Damery, Dizy and Ay, before troops arrived to disperse them. The next day some 5,000 protestors marched on Ay and then Epernay, leading to the dispatch of 15,000 troops who remained in the region until the autumn in order to put down any further protest. In an attempt to calm the situation a further decree was promulgated on 7th June, creating a *Champagne 2ᵉ Zone*, including many vineyards in the Aube, but this failed to satisfy the protestors. It was only the plentiful vintage of 1911 that restored some peace in the region, and although new legislation was promised, the outbreak of war in 1914 prevented any further political unrest in the region. It was not until the law of 22nd July 1927, which allowed 71 communes in the Aube to bear the *Appellation* champagne, and which also specified the three grape varieties to be used for the production of champagne, that the issue was eventually settled.

Following the resumption of peace at the end of the Second World War, a new law was passed on 6th May 1919, which required anyone who claimed that an *Appellation d'Origine* was applied to a product to their prejudice, contrary to the origin of the product, or of local, loyal and constant use, could initiate legal proceedings against the use of the *Appellation* (Capus, 1947; Lachiver, 1988). Concentration simply on the area of production, however, was no guarantee of quality, and subsequent laws were therefore passed to specify the various methods of viticulture and vinification which accorded to the best traditional practice in each *Appellation*. Before any national legislation was introduced, however, the wine producers of Châteauneuf-du-Pape, led by Baron Le Roy Boiseaumarié, devised their own charter to guarantee the quality of their wines, specifying a delimited area, the grape varieties to be used, the training and pruning methods, the alcohol content of the wine, and the quality of the grapes (Livingstone-Learmonth and Master,

1983). Similar legislation subsequently came to form the basis of all French wine laws. Thus the law of 22nd July 1927, forbad the use of hybrid vines, and included regulations concerning the cultivars which were required for a wine to be included within an *Appellation*, and eventually, on 30th July 1935 the law creating the *Appellations d'Origine Contrôlées* (A.O.C.) was promulgated, specifying not only the area of production, but also the choice of vine cultivars, the minimum level of alcohol in the wine, the methods of viticulture and the vinification processes to be used (Kuhnholtz-Lordat, 1963). A Comité National des Appellations d'Origine was created, becoming in 1947 the Institut National des Appellations d'Origine, and this began work on the enormous task of demarcating the French vineyards. However, by 1950 only about one tenth of total French wine production was classified as A.O.C., and it was not until the 1960s that the great expansion of such wines began.

The German wine laws

In contrast to the situation in France the central issue in German wine legislation has been the problem of the addition of sugar to the musts in order to raise the alcohol content of the wine. In its northern vineyards, where the grapes ripen well much less frequently than in France, the addition of sugar is often necessary, but the key task of the German legislators was to devise a system which would indicate whether or not a wine was naturally sweet or whether it had been artificially sweetened. Consequently, unlike France, 'Germany does not base the classification of its wines on the geographic origin but on the finished product, the quality in the glass' (Hallgarten, 1987: 39). In theory, the monopoly created by the demarcation of specific areas, does not exist in Germany, and every vintner could produce the highest possible quality of German wine. Nevertheless, in practice, German wines are also described by place of origin, and the reputation of certain vineyards places a premium on their wines.

Since the first modern German wine law in 1879 a wide range of legislation was introduced during the early part of the twentieth century, designed to control the addition of sugar, grape juice, and wines from other countries. This culminated in a new law in 1930, laying down the regulations which controlled the industry until 1969 (Sichel, 1983; de Blij, 1983; Hallgarten, 1987). This specified that vineyard names were to be registered in each community, and that only European varieties of vine grafted onto American rootstocks could be cultivated. As a result some 30,000 vineyards were registered, and a great diversity of labelling practices adopted. Fraud and corruption remained rife, however, and Hallgarten (1987: 46) quotes a Public Prosecutor Oberstaatsanwalt Bohr

316

who in 1964 wrote a report on the state of the German wine trade, in which he noted that 'Rhine wine is not always pure; many German wines contain more water than permitted and wine growers and merchants adulterate and mix wines to such an extent, that the criminality is quite significant'. In referring to the criminals, Bohr went on to comment that

> Many of them are the 'Great Ones' able to pull many strings to obtain a decision favourable to them. These 'honourable merchants' have intimate connections with all the highly placed people in the economic and political life of the community. When necessary such people are persuaded to put pressure on the Public Prosecutor, who is but an official who must look after his career so that he can progress to a higher income. Such a criminal plays his many cards in turn, his money, his position in society, his membership of Associations and Official organisations, such as the Chamber of Commerce and Chamber of Agriculture, and his many aspects of public life . . . The reason why he commits crimes is because he has substantial means and needs even more to extend and strengthen his social position and his influence in commercial, cultural and political affairs.
>
> (Hallgarten, 1987: 46-47)

In an effort to cope with the chaos, a new wine law was promulgated in 1969, but its introduction coincided with wider discussions within the European Community concerning the introduction of a new European wine law. Eventually a compromise agreement was reached, and on 14th July 1971 a further German wine law was passed, specifying both a hierarchy of wine regions and a hierarchy of wine qualities, which, satisfied the requirements of Germany's European Community neighbours, and in particular France.

The 1971 law divided the country into eleven *Anbaugebiete*, or specified regions, within which there were two or more *Bereiche*, or districts. These were further subdivided into *Grosslagen*, or collections of vineyards sharing similar geological and climatic conditions, and *Einzellagen*, or individual vineyards, each of which had to be more than five hectares in size (Baumann and Michel, 1976). The 30,000 vineyard names that had existed before 1969 were thus reduced to approximately 130 *Grosslagen* and 2,600 *Einzellagen*. The overall effect of this was often to increase the size of prestigiously named vineyards so that after 1971 they included a greater amount of less good land, with a consequent overall diminution in the quality of the wines produced, although those from the old core areas still remained as good. These site specifications, unlike those of France, bear no direct relation to the quality of the wines, which are divided into two broad categories, *Tafelwein* and

Qualitätswein. The basic *Deutscher Tafelwein* must be made from prescribed grape varieties in four *Tafelwein* regions (Rhine-Mosel, Main, Neckar, and Oberrhein), and sugar can be added before fermentation to raise the alcohol level. In 1982 a higher quality *Tafelwein*, called *Deutscher Landwein*, was designated, and this can come from any of fifteen Landwein regions and must be either *trocken*, dry with less than 9 grams per litre of residual sugar, or *halbtrocken*, semi-dry with less than 18 grams per litre residual sugar.

The German quality wines are further divided into two types, *Qualitätswein bestimmter Anbaugebiete* or QbA and *Qualitätswein mit Prädikat* or QmP, with the addition of sugar being legal only in the former. It is these wines which are labelled according to the hierarchical regional structure of *Bereiche, Grosslagen* and *Einzellagen* outlined above. The *Qualitätswein mit Prädikat* category is further subdivided depending on the time of the vintage and thus the natural sugar content, or Oechsle must rating, into the following categories, noted in ascending order of quality: *Kabinett,* made from normally ripe grapes; *Spätlese,* made from grapes picked at least seven days after the normal harvest; *Auslese,* a sweeter wine made from selected bunches of grapes; *Beerenauslese,* a rich dessert wine made from selected grapes affected by *Botrytis cinerea,* known in Germany as *Edelfäule; Eiswein,* made from frozen Beerenauslese quality grapes; and *Trockenbeerenauslese* wines made from dried up botrytised grapes. The precise Oechsle rating of each level varies depending in part on the grape variety and its area of origin, but in general *Kabinett* wines have a minimum Oechsle rating of 67–85° whereas *Beerenauslese* wines have an Oechsle of 110–128° (Stevenson, 1988)

Despite this apparently rigorous classification, a wine with a year of vintage on the label may legally contain up to 15 per cent of wines from other years, a wine specified as being of a single grape variety may contain up to 15 per cent of wines from other grapes, and a wine said to be from a specific village may contain up to 15 per cent of blended wine from another part of the same region (Hallgarten, 1987). These regulations are designed to enable producers to even out annual variations, and, as a guarantee of their quality, the best wines must be analysed, tasted and granted an *Amtliche Prüfungsnummer,* official approval number, before being allowed to bear a name in the *Qualitätswein* categories. As in France, though, the German regulations introduced in 1971 failed to stop the fraudulent production of wine, with well publicised cases including that of the dilution of foreign wines with water before their conversion to sparkling wine which came to Court in Mainz in 1976, the false use of approval numbers by a firm of wine merchants who were expelled from the Mosel-Saar-Ruwer Wine Trade Association and subsequently prosecuted in Trier also in 1976, and the prosecution in 1980 of 200 sugar merchants and 1,800 wine growers in the Rhineland Palatinate for the illegal use of liquid sugar for sweetening wines between 1977 and 1979 (Hallgarten, 1987).

International agreements on the naming of wines: champagne and sherry

The demarcation of vineyard areas in a single country is of little consequence in the increasingly important export market unless other countries abide by the same naming principals. Consequently, once the French had firmly established their system of *Appellations d'Origine Contrôlées*, it became necessary to guarantee the names of French wines on an international basis. In 1956 a sparkling wine made at Perelada in Spain began to be imported to England by the Costa Brava Wine Co. Ltd. and sold under the name of Perelada Spanish Champagne (Hallgarten, 1987; Faith, 1988). British wine merchants had for a long time been importing 'Spanish Sauternes' and 'Spanish Graves', but the introduction of 'Spanish Champagne' was greeted with some disquiet, and following a debate in the wine trade press the French Comité Interprofessionnel du Vin de Champagne together with the Institut National des Appellations d'Origine des Vins et Eaux-de-vie decided to take action in the British Criminal Courts under the Merchandise Marks Act of 1887 (Simon, 1962).

The case eventually opened on 17th December 1958 in Court No 1 at the Old Bailey and after six days, with fifteen witnesses having been called on each side, the jury returned and found the defendants not guilty (Simon, 1962). At that date the British government did not recognise the French *Appellation d'Origine Contrôlée* regulations, and at a time when increasing amounts of cheap low quality wine were being sold in England, the verdict was welcomed by all those involved in the downmarket trade (Hallgarten, 1987). However, in 1959 twelve major champagne companies, led by Bollinger, issued a writ against the Costa Brava Wine Company in the Chancery Division of the High Court. In November of that year, after listening to counsel from both sides, Mr Justice Dankwerts decided that the plaintiffs did indeed have a right in Law to protect the name 'champagne', and a year later on 29th November 1960 the case came to Court again. In the Chancery Division of the High Court the case hinged on the question of whether or not someone who bought a bottle of 'Spanish champagne' might be deceived into thinking that they were buying champagne. Eventually on 16th December 1960 the Judge declared his verdict, and granted the champagne houses an injunction against the Costa Brava Wine Company from selling their Perelada, or any other wine, under any name which included the word champagne. Central to the plaintiffs' success was a brochure published by Perelada called *Giving a Champagne Party*, and the absence of the

319

adjective 'Spanish' both here and on a menu, where Perelada was referred to simply as 'Perelada (Champagne)', seems to have been critical in determining the outcome of the case. This major victory established a precedent, confirming the use of the name champagne solely for wines produced in the Champagne district of France, and it also had widespread repercussions elsewhere in the international wine trade, although it is significant to note that many producers of sparkling wine in California continue to use the word champagne on their labels.

Within seven years another major wine nomenclature case came to trial in Britain, this time concerning the name 'sherry'. In 1924 the Spanish government had granted the proprietorship of the mark *Jerez* to the Municipality of Jerez, 'in favour of all those growers, traders and shippers resident in Jerez, which concession was also registered at Berne according to a certificate issued on 5th September 1924' (Gordon, 1972: 44). Subsequently, in both 1925 and 1934 merchants in London had been fined under the Merchandise Marks Act of 1887 for using the single word 'sherry' to describe wines not produced in the Jerez district of Spain (Gordon, 1972; Hallgarten, 1987). Nevertheless, from well before the nineteenth century a product known as 'British sherry' had been made in England from a range of ingredients including raisins and rhubarb, and when concentrated grape juice began to be imported into the country at the beginning of the twentieth century its production expanded considerably. The Sherry Shippers' Association accepted that the usage of this term to indicate wines of a sherry character made in Britain began prior to 1887, and they therefore did not bring any cases to Court under the Merchandise Marks Act. During the 1920s increasing amounts of other types of 'sherry', including Australian sherry, Cyprus sherry and South African sherry then appeared on the market, but until the victory of the champagne houses in 1960 the sherry shippers did not seek resort to legal procedures against such wine unless the single word 'sherry' was used to describe it. However, during the 1960s numerous advertising campaigns suggested that British and South African sherries were indeed real sherry, and this prompted the sherry shippers into action. Eventually in 1967, in a civil action between Vine Products Ltd, Whiteways Cider Co Ltd and Jules Duval and Beaufoys Ltd, all members of the Showerings Group, on the one hand, and the four largest sherry shippers, Mackenzie and Co, Williams and Humbert Ltd, Gonzalez Byass & Co and Pedro Domecq SA on the other, the judge, Mr Justice Cross, declared that the word 'sherry' by itself should only be used to refer to wines of a particular characteristic made in Jerez. He also held that other terms such as 'British sherry' and 'Cyprus sherry' 'had been in use for so long without objection that it was too late to stop them now' (Keeling in Gordon, 1972: 48), but he recommended that they should not be advertised or described as simply 'sherry'.

European wine legislation and the constraints of demarcation

With the accession of Spain and Portugal to the European Community in 1986, the problem of nomenclature over the name of sherry came to the fore again, and as a temporary measure, British, Irish and Cyprus sherries were authorised provisionally until the end of 1995 (Niederbacher, 1988). This was not the only wine issue facing the Community, though, and with more than 60 per cent of the world's wine production and employing more than 10 per cent of the Community's total agricultural labour force, the wine industry has presented a serious problem for the Community's legislators. As Niederbacher (1988: 28) has stressed, viticulture

is important in Community agriculture both because of its sheer volume in terms of farming statistics, and because in certain areas there are no valid agricultural or industrial alternatives and it is indispensable if the landscape is to be preserved, people kept in employment, infrastructures used to their full potential and the very fabric of society held together.

At the time of the signing of the Treaty of Rome in 1958, and the establishment of the Common Agricultural Policy, the central problem concerning wine was how to bring together the two largest wine producing countries in the world, France and Italy, whose industries were organised in entirely different ways. It was also recognised that 'wine was subject to frequent imbalances between supply and demand and thus to recurrent crises' (Niederbacher, 1988: 41), and that it was necessary to introduce policies to reduce these tendencies. The contrast between France and Italy could hardly have been greater: in France vineyards were registered and wines were classified into carefully controlled categories, whereas in Italy even the ordinary land register was out of date, and there were no rules designating the origin of wines; in France new vineyards were prohibited, but in Italy a State subsidy existed for people planting new vines; and in France chaptalisation was permitted, whereas in Italy it was severely penalised. In short, the danger facing France was that large quantities of cheap Italian wine would flood the market, leading to a fall in French wine prices which would in turn cause a collapse in the value of the vineyards.

These problems have still not been fully resolved, and they were compounded by the accession of Spain to the Community, but four main provisions were specified in European Community Regulation 24 of 4th April 1962, which laid the foundation for a common market in wine. This required that each country establish a vineyard register (Regulation 143/62, Regulation 26/64), that annual production levels be notified to a central authority (Regulation 134/62), that strict rules be established concerning quality wines produced in specified regions, and that future estimates of resources and requirements be compiled annually. This led to the Italians establishing laws in 1963 by which quality wines were granted *Denominazione di Origine Controllata* status, based on their region of origin, but until 1970 little further progress was made at the Community level. In that year two new Regulations were eventually adopted, providing for the common organisation of the wine market (Regulation 816/70) and creating special provisions for quality wines produced in specified regions (Regulation 817/70). In practice the new rules of 1970 meant that planting and replanting would only be subject to quality constraints, and not to quantitative restrictions as had been the case in France, and that existing practices in member states concerning the marketing of wine could be retained, although there would still be free movement of wine within the Community. However, with a harvest of 154 million hectolitres in 1970, instead of the expected 135–40 million hectolitres, the first signs of a new crisis of overproduction became apparent (Niederbacher, 1988). The intervention mechanisms established in 1970 enabled some of this wine to be purchased for distillation, but by 1973, with a European harvest of 171 million hectolitres, the fragile balance between French and Italian domestic prices crumbled, and increasing amounts of cheap Italian wine poured onto the French market. With another large harvest in 1974, French wine producers blockaded the port of Sète, the source of entry for much of the Italian wine, and the situation was only partially relieved by the imposition by France of an import tax on Italian wine, breaking the general agreement on the free circulation of wine.

The central problem of European legislation until the early 1970s was that it failed to restrict overproduction. Therefore in 1976 a further Regulation (1163/76) imposed a ban on new vine planting, and instituted a programme of subsidies for the conversion of vineyards to the cultivation of other crops. This, however, failed to prevent another record harvest in 1979 of 182.4 million hectolitres, followed in 1980 by one of 163.9 million hectolitres, and despite widespread distillation French wine producers again attacked boats and lorries containing Italian wine in and around Sète. Further legislation providing a premium for the permanent abandonment of vineyards was promulgated (Regulations 457/80 and 777/85), and in 1982 a major shift in policy made distillation, which had previously only been an exceptional measure, the basic instrument for regulating the market and eliminating surpluses (Voss, 1985; Niederbacher, 1988). Eventually in 1987 new laws designed to introduce compulsory distillation with the aim of deterring surplus wine production were introduced (Regulation 822/87). Further subsidies for the conversion of vineyards to other types of land use were also provided, and in another regulation of the same year (823/87) new rules were specified for the production of quality wines.

The enforced distillation of wine, though, is an expensive alternative, with, for example, the cost to the intervention agency of buying Italian wine being approximately 135,000 lire per hectolitre in 1988 and the cost of storage being between 6,000 and 8,000 lire per hectolitre a year, at a time when the proceeds of the sale of such alcohol to non-Community countries for fuel were only in the order of 8,000 to 10,000 lire per hectolitre (Niederbacher, 1988). The situation is exacerbated further by an overall decline in wine consumption in the Community of the order of 1 per cent a year, while production is still increasing at 0.58 per cent a year, and the prospects for a successful resolution of the problem by 1992, the target date for full integration of the Community's wine sector excluding Portugal, seem remote (Voss, 1985). In Niederbacher's (1988: 91) words, 'The area under vines in Europe needs to be reduced by at least 20 to 30 per cent, but that would entail traumatic, dramatic surgery in the social and economic fabric of wine-growing, and indeed the very landscape itself'.

Demarcation and privilege

The attempts at demarcation and control of wine production in the European context noted in the above examples have generally been designed to reduce the wide annual fluctuations of production, and to ensure the livelihood of wine producers by restricting the quantity of wine produced. Moreover, they have also been associated with efforts to raise the quality of the wines produced, thus legitimising an increase in the price of such wines paid by the consumer. In essence the demarcation of wine regions, and the control of prices, has created a series of monopolies enabling producers to extract a greater profit from their vineyards than would otherwise be the case. In this they have been broadly successful. However, the European Community is in practice by no means a closed market, and for north European retailers concentrating on the lower end of the market and eager to undercut the prices of French, German and Italian wines, it is still possible to find cheaper wines elsewhere on the international market, particularly from the countries of eastern Europe where there are considerable state subsidies for the production and export of wine. Indeed between the first half of the 1970s and the first half of the 1980s average annual wine imports into the European Community from Bulgaria, Hungary and Romania more than doubled from 255,192 to 604,675 hectolitres per year (Niederbacher, 1988), and much of this wine was priced well below the level of European Community wines of similar quality.

A further problem with the demarcation of wines in the European context, is that it need not always lead to an improvement in their quality, and this has particularly been true of the much criticised Italian *Denominazione di Origine Controllata* (DOC) system (Anderson, 1982; Belfrage, 1985). Introduced in 1963 the Italian wine law specified for each type of wine the grape varieties allowed, the maximum production level permitted per hectare, the conversion ratio of grapes into wine, the methods of vinification and alcohol levels, and the ageing requirements of a wine. However, there were no tasting panels, and a guarantee of authenticity failed to guarantee quality. Maximum permitted yields were generally much higher than those of France, and several famous denominations, such as Chianti and Valpolicella, were permitted to add up to 15 per cent of wines from other regions. Moreover, within a DOC region there can be considerable variations in relief, soil and microclimate leading to corresponding differences in the wines. By the late 1970s a new system of higher quality wines, under the name of *Denominazione di Origine Controllata e Garantita* (DOCG), was introduced, and in order for wines to achieve this status they must among other regulations be subjected to an analytical test and passed by a tasting panel (Sabellico, 1986). While three of the first four wines to achieve this status, those of Barolo, Barbaresco and Brunello di Montalcino, were undoubtedly worthy of their elevation, the addition of Vino Nobile di Montepulciano, and more recently Chianti and Albana di Romagna is widely considered to have diluted the concept. A central problem remains, in that wines produced anywhere in a DOCG zone subject to the basic requirements of the denomination, can achieve the most elevated official status, regardless of their particular microclimate and relief. There is thus no attempt, as in France, to classify the vineyards within a zone by quality, or to have different quality levels of DOC or DOCG wines. As a result, some producers have dispensed with the official wine law altogether, and are making high quality wines under the theoretically inferior *Vino da Tavola* classification, relying more on their own quality and reputation than any official classification, with typical such wines being Antinori's Tignanello, and Lungarotti's San Giorgio.

The central feature of wine demarcation is therefore not just that it usually generates better quality wines which can be sold for higher prices, but that it is a legislative procedure which creates a privileged position for producers within a demarcated area, thus providing the context within which greater profit can be achieved. Some exceptional producers have, however, worked outside the demarcation system, and concentrated on the production of high quality wines which can achieve exceptional prices based on their individual reputation alone. Demarcation also has significant social implications, with it having been introduced in part to enable wine producers who would otherwise have lost their source of livelihood to remain within the sector. Moreover, political protest can be seen as having played a critical role in determining the shape of the legislation associated with demarcation, whether at the regional level, as in the case of champagne, or at the international level, exemplified by the recent protests by French wine producers against Italian and Spanish wine imports.

FROM WINE MERCHANT TO GLOBAL CORPORATION: THE RESTRUCTURING OF THE RETAIL WINE TRADE IN BRITAIN

Before the twentieth century one of the key characteristics of the wine industry was its lack of sectoral integration. Although there were a few exceptions, it was generally rare for vine growers to be involved in the retail sale of their wines. This situation can largely be explained by the necessary division of labour and skills involved, as well as by the different capital circulation requirements of wine producers and merchants. Particularly as far as the international trade was concerned wine producers had neither sufficient time nor the contacts to travel overseas in order to sell their wines. Moreover, there were few large scale producers, for whom such effort might be warranted, and the majority of growers required an immediate annual return from their vineyards to pay off debts accumulated between vintages. Indeed, many growers regularly owed money to merchants who provided loans guaranteed by the likely yield of their vineyards. In contrast, the merchants had sufficient capital to invest in the storage and ageing of wines, and made it their business to know the requirements of their clients. This lack of sectoral integration does not, though, imply that the wine trade was simply local or national in character. Indeed, another central feature was its international connectivity, with the medieval and early modern periods being characterised by the establishment of wine merchants from northern Europe in most of the major wine exporting regions of Europe.

Typical of the sectoral complexity of the wine trade in the eighteenth and nineteenth centuries was its organisation in Bordeaux, where growers, *courtiers* (brokers) and *négociants* (merchants) all profited from their share of the trade (Pijassou, 1980; Loftus, 1985; Penning-Rowsell, 1985). The owners of the great châteaux had little access to the international market for their wines, and thus relied on the *négociants*, most of whom were English, Irish, or Dutch, for the sale of their wines. Most growers, however, did not deal directly with the *négociants*, but rather sold their wines through an intermediary *courtier*, who usually specialised in the wines of a particular region and knew the specific requirements of each *négociant*. This system grew up in part as a result of the social divisions of French society during the eighteenth century, with the aristocratic proprietors of the great châteaux not wishing to deal directly with the rich bourgeoisie in the personification of the *négociants*, who in turn considered it beneath them to associate with the owners of the smaller properties (Berger *et al.*, 1981). However, it has remained largely intact until the present day, and as Loftus (1985: 80) has remarked, even though 'The top growers could perfectly well bypass both brokers and *négociants*' the system survives because it 'offers them an assured market, for minimum effort, and makes a lot of middlemen very rich'.

During the latter part of the nineteenth century, though, a number of merchants began buying properties of their own in the major wine producing regions of the world, thus initiating the vertical sectoral integration that has become such a characteristic of the twentieth century wine industry. Although the drive to oligopoly in the wine industry lagged behind that of the beer and spirits sector, where 'by the mid-1960s a handful of giant corporations had achieved market dominance in most countries' (Cavanagh and Clairmonte, 1985: 2), it rapidly thereafter became incorporated into the corporate strategies of a small number of highly successful global corporations, such as Seagram and Grand Metropolitan. Cavanagh and Clairmonte (1985), have seen this evolution of corporate power as being illustrative of four central movements within global capitalism: the formation of oligopolies at national and international levels, through capital concentration in specific sectors; the introduction of new corporate priorities, favouring control over processing and marketing rather than primary commodity production; the rapid adoption of technical innovations in order to fragment production processes; and the internationalisation of finance. In comparison with many other major industrial sectors, though, the wine industry is still very fragmented, in part because of the difficulties of producing consistent and uniform wines, and in part as a result of the fragmentation of the market. Moreover, in the United States of America legal restrictions prevent wine producing companies from owning retail outlets other than at their own wineries, and this political intervention has limited the tendency towards vertical integration that is apparent elsewhere.

The underlying reasons for this integration of the wine industry have been the desire to reduce costs, to retain greater control of all stages of the industry, and to expand sales, all of which are essential to the survival of capitalist enterprise. As Taylor and Thrift (1982) have emphasised, though, most models of corporate development fail to situate these changes within their historical context. Indeed, it is impossible satisfactorily to understand the evolution of the modern wine industry without some knowledge of the emergence of the retail wine trade during the eighteenth and nineteenth centuries, of which the English example is particularly salient, reflecting as it does the intricate interconnection between social, political and economic decision making.

Coffee houses, grocers and wine merchants: the retail wine trade in England during the nineteenth and early twentieth centuries

During the medieval period there was a basic division in the retail wine trade in England between wine sold in closely regulated and licensed inns and taverns, and the much freer sale of wine direct to the households of the nobility. While this distinction remained important until early in the twentieth century, the granting of 'Letters Patent of licence "to all citizens brought upon the trade of retailing wines . . . as amply as if they were free of the Vintners' Company" ' by Queen Elizabeth I in 1576 (Crawford, 1977: 65) effectively broke the monopoly of the Company of Vintners and enabled wine to be sold much more extensively. This led to a considerable expansion in the number of taverns in the area between the City of London and the palace of Westminster during the late sixteenth and seventeenth centuries, and it coincided with a general reorganisation of the retail trade which continued into the reigns of the early Stuarts in response to increases in population, wealth and the amount of luxury goods in circulation (Davis, 1966: 55). In this period, taverners were in a fortunate position with respect to the wholesalers of wine, because although anyone could import wine it was only the licensed taverner who could sell it retail, and Davis (1966: 161) argues that they were therefore in a strong position to make very large profits. Indeed, during the seventeenth century taverns not only sold wine for drinking on the premises, but increasingly began to sell it by bottle to customers who wanted to drink it in their own homes.

Following the Restoration, taverns 'experienced serious competition from the fashionable flurry of coffee-houses', where customers could purchase 'not only coffee but cocoa, tea, wine, brandy and punch' (Clark, 1983: 13; Lillywhite, 1963). At the same time, there also emerged a number of grocery establishments, such as that at Number Three St James's Street, known as The Coffee Mill, which supplied the coffee houses around the royal palace of St James in London, with coffee berries, tea and chocolate (Allen, 1950). Through the course of the eighteenth century many similar establishments flourished, and gradually, with the demise of the coffee houses and their replacement by gentlemen's clubs, it was frequently these grocers who began to sell wine retail to the general public. Meanwhile, the gin-shops and alehouses continued to serve those seeking the cheaper spirits and beer. It is with the nineteenth century, though, and what Davis (1966) has called the second retailing revolution, that the structure of the wine retail trade as it is today began to emerge. Typical of the more successful of the new wine companies that emerged were Berry's (Allen, 1950), Harveys of Bristol (Henry, 1986), and Victoria Wine (Briggs, 1985).

In 1803 George Berry, the son of an Exeter wine merchant, moved to

London to work at the grocers establishment at Number Three St James's Street. This property had been acquired by William Pickering early in the eighteenth century, and one of his sons had later formed a partnership with John Clarke, the grandfather of George Berry. By 1810 it seems that George Berry was running the grocery business himself, and with the declining demand for high class tea and coffee in the 1830s for a short while 'he became the accredited agent for the shipment of Bass and Co.'s East India Pale Ale to the Indies' (Allen, 1950: 175). With the demise of this agency, he turned to other goods, and by the mid-1840s wine had begun to challenge tea and coffee as the principal commodity handled by the Berry firm.

The emergence of companies such as Berry's coincided with a period of increasing social concern over the effects of alcohol in Victorian England (Harrison, 1971). This was particularly focused on the so-called evils of gin, with Girouard (1984: 20), for example, commenting that 'In the years around 1830 educated people in England became violently and angrily gin-conscious.' Wine at that time was still largely the preserve of the aristrocracy and to some extent the bourgeoisie, with the preferred drinks of the mass of the population remaining beer and spirits. However, by 1852 with the appointment of a Select Committee to examine the effects of wine duties, a general debate commenced, which was eventually to lead to a reduction in duties on wine (James, 1855; Tennent, 1855; Whitmore, 1853; Briggs, 1985). Those, such as W. Bosville James (1855) who, like Busby in the Australian context, believed that wine drinking was preferable to an excessive consumption of beer and gin, argued that wine was only a luxury because of heavy taxation, and that if the principle of free trade, which had been applied to corn in 1846, was extended to wine it would be possible to develop a new market for cheap wine, which could also lead to an effective increase in national revenue. James was not, though, an uninterested party, and as a wine merchant with 24 years' experience he was keen to create opportunities which would enable an expansion of the wine trade to take place. Other wine merchants, however, believed that a reduction in duty would lower profits and prices, and would break the control that traditional merchants had on the trade (Shaw, 1854).

Earlier in the century in 1813, the preference given to Portuguese wines by the Methuen Treaty was abandoned, and duties on all European wines were thereafter charged at a uniform rate of 5s. 6d. a gallon, being raised to 5s. 9d. a gallon in 1840. Following the general election of 1859, the installation of Gladstone as Chancellor of the Exchequer led to a dramatic reversal in government policy, and in March 1860 the duty on wine imported in cask was reduced to 3s. per gallon (Francis, 1972; Briggs, 1985). From 1st January 1861 a new scale of duties based on the alcoholic strength of the wine was implemented, leading to a further reduction of duty on wines of less than 16° Sykes to 1s. a gallon. Moreover, a new system of off-licences was also introduced enabling general retailers to sell wine in bottles for consumption off the premises on payment of a fee of only 50s. compared with the licence of ten guineas required by merchants dealing exclusively in wines and spirits. A year later the lower duty was then extended to all wines of less than 26° on the Sykes scale. The effects of this legislation on wine imports were considerable, and the clearance of French wines from bond virtually doubled from 1,125,599 gallons in 1860 to 2,227,662 gallons in 1862 (Briggs, 1985: 36).

For Berry's grocers the legislation led to a further concentration on wine. In 1895 the remaining stocks of tea, coffee and cocoa were sold off, and from 1896 Berry Brothers and Co dealt exclusively in wine at Number Three St James's Street (Allen, 1950). It was not only well-established companies such as Berry's that benefited from Gladstone's Acts, though, and newcomers, such as William Winch Hughes, the founder of Victoria Wine, were to seize the opportunity with alacrity. Hughes began his careeer as a clerk in a firm of wine merchants, and by 1851, when he was only 17 he was himself described as a wine merchant of 15 John Street, Minories. He then became a partner in the firm of Weller and Hughes, and in 1861 'he appeared at last on his own as a wine and spirit merchant and broker at 52 Great Tower Street' (Briggs, 1985: 43). In 1866 his business was styled the Victoria Wine Company, and by 1869 there were nine branches of the company in London. A year later a new branch was opened in Bristol, and soon afterwards further branches appeared in Brighton and Birmingham as well as in London. Even then, though, he did not restrict his merchandise solely to wine, and, like Berry's, he regularly sold tea, coffee, beer and mineral waters. During the 1870s Hughes established his business upon four main principles: that cash payment was required, that his wines were unadulterated, that they were bottled carefully, and that deliveries of a dozen bottles or more were free in London. He advertised widely in publications as varied as the *Illustrated London News*, *Punch* and local papers such as the *Hackney Express and Shoreditch Observer* (7th April 1877), in which he claimed to be offering 'wines at prices hitherto unheard of in England, thereby placing an expensive luxury within the reach of all classes'. By 1880 his headquarters at Osborn Street had sixteen floors, with a frontage of 160 feet, and from here his eighty-three shops were supplied with 438,226 bottles of wine and spirits, and 432,464 bottles of beer in the week leading up to 18th December (Briggs, 1985: 50-1). The 1870s had been a boom period in wine consumption, and although spirit and beer consumption rose again above the levels of the 1870s in the 1890s and 1900s, following a slight trough in the 1880s, the same was not true of wine consumption. Wine had become relatively common among the middle classes, but, despite the efforts of Hughes, it was still beyond the reach of the poor.

When Hughes died in 1886, his widow Emma Hughes succeeded him as manager of the business until her death in 1911, at which time there were ninety-six Victoria Wine branches. Although the range of wines sold was quite wide, and much as it had been in the 1870s, Briggs (1985: 84) notes that 'Within the range there was a heavy concentration on a limited number of lines.' On Emma's death Frank Wood, who had worked for the company since the mid-1870s bought out the other beneficiaries of her will, and acquired complete control of the firm. This led to a change of strategy, whereby the former policy of leasing properties was replaced by the outright purchase of premises in particularly profitable locations, and in 1920 he also turned Victoria Wine into a private limited company. Although there was a brief post-war boom in the consumption of all alcoholic beverages, the 1920s and 1930s saw a general fall in wine demand, and more importantly it also saw a restructuring of the nature of that demand. In particular the amount of wine being purchased by the rich for their private cellars plummetted, with many going even further and selling off their wine by auction. The future of the wine trade thereafter depended upon the opening up of the wider middle and working class market. Following Wood's death, Victoria Wine was purchased in 1924 by Sir Charles Edward Cottier, Chairman of Booth's Distillery and John Watney and Company, who turned it into a public company (Briggs, 1985). The company's basic policy of selling cheap wine to the mass market, however, remained the same, and gradually a range of brand names, including Golden Galleon and Victoria Tarragona, were established. A year after Cottier's death, the brewers Taylor and Walker then acquired control of Victoria Wine in 1929, setting in motion a trend of purchases of wine merchants by brewers which was to gather pace after the Second World War. In 1937 there were about 1,000 independent brewers in the United Kingdom, but subsequent amalgamations and takeovers led to a reduction in this number to some eighty companies operating around 160 breweries by 1982 (Vaizey, 1960; Cavanagh and Clairmonte, 1985). Indeed, by 1989 the Monopolies and Mergers Commission reported that the top six brewers had 75 per cent of the beer market in the United Kingdom. As a result of this restructuring, Taylor and Walker was taken over by Ind Coope in 1959, and eventually became part of the Allied Group, created in 1961 from the brewers Ind Coope, Tetley Walker and Ansell. Victoria Wine thereafter became the retail outlet for most of the wines of the group which were imported and bottled by Grants of St. James's, another subsidiary of Ind Coope.

It was not only London companies such as Berry's and Victoria Wine which grew to prominence in the nineteenth and early twentieth century, for elsewhere in the provinces firms such as Harveys of Bristol also found success. In the eighteenth century Bristol was a flourishing port, and home to a ship's master by the name of Captain Thomas Harvey. His son, also known as Thomas (II), took as his second wife, Ann Urch, sister of Thomas Urch, the junior partner of a wine merchant, William Perry (Henry, 1986). Thomas Harvey (II)'s elder son John, apparently disliked the sea, and thus at the age of 16 entered his uncle's wine business based at Denmark Street near the docks in Bristol. To gain further experience he then worked for one of his uncle's partners, Edward Prichard, in Kidderminster, before returning to Bristol in 1840. Following Thomas Urch's retirement, John Harvey took over the Denmark Street wine importing business, to be followed into it by his eldest son John (II) and his brother Edward, and in 1871 the firm became John Harvey & Sons. Little is known about the emergence of the company during the nineteenth century, but as early as 1882 Bristol Cream sherry, upon which it was later to make its name, was first mentioned (Henry, 1986). The changing structure of demand following the First World War meant that Harveys could no longer continue to supply only private customers and institutions. Rather than developing its own retail trade along the lines of Victoria Wine, though, Harveys began an advertising campaign in order to encourage people to ask for its wines by name, regardless of the wine merchant from which they bought it. This strategy proved highly successful, and a further advertising campaign in the 1950s, particularly for the American market, generated sales which by 1959 totalled £4 million (Henry, 1986). A share issue in 1958 provided sufficient capital for the construction of new premises at the edge of Bristol, and then in 1960 a formal partnership was established with the port firm of Cockburn Smithes, which had supplied Harveys with port since the early nineteenth century. However Harveys' success as a leading port and sherry distributor in both the home and the export market made it an attractive bid prospect, and the cider and wine group Showerings, which had been seeking to expand for some time, gained control of Harveys in January 1966. This acquisition failed to make Showerings invulnerable to the growing power of the brewers, though, and in 1968 it became part of the Allied Breweries Group. An injection of capital then enabled Harveys for the first time to take over a sherry producer in Jerez, Mackenzie & Co.. This 'meant that Harveys could maintain tighter control of the quality and quantity of the wines that were blended in its sherries, and that it could export direct from Spain to the EEC and to countries which could not buy from the UK' (Henry, 1986: 21). The Allied connection also provided a large network of public houses and off-licences, and with an increased advertising budget, Harveys strengthened its hold on the domestic market. This strategy was further developed in 1985 when Harveys bought the Spanish sherry and brandy houses of Fernando A. de Terry and Palomino y Verggara to become the biggest sherry company in the world.

These brief case studies all indicate the complex web of commercial interactions that led to the restructuring of the wine industry during the nineteenth and early twentieth century. Despite their diversity, they were all underlain, however, by two broad structural trends. First, there was an overall tendency for wine companies having become successful in one sector of the drinks industry to be taken over by larger, more heavily capitalised businesses, which then incorporated them into an overall integrated enterprise combining production, wholesale and retailing sections in different parts of the world. Secondly, though, this was associated with changes in the social significance of different types of alcoholic beverage, and with attempts by the government to encourage wine rather than spirits consumption among the middle and working classes. While such efforts in the nineteenth century, reflected in the political sphere by Gladstone's lowering of import duties on wines, were largely unsuccessful, the social turmoil associated with the 1914–18 war created a situation which could successfully be exploited through the use of powerful advertising in the following decades.

From Gilbey's to Grand Metropolitan: the rise to dominance of the global corporation

A central feature of successful capitalist enterprises in the twentieth century has been their continued structural reorganisation as their managers seek to create new markets, minimise production costs, and reduce the effects of competition. By 1980 in the alcoholic beverages sector these processes had led to the existence of 27 global corporations with sales exceeding one billion dollars. The complex web of corporate ownership involved in these enterprises is well exemplified by the largest of these in terms of total sales, Philip Morris of the USA,

> which owns 22 per cent of Rothman's International of Britain, which in turn owns 71 per cent of Rothman's of Pall Mall of Canada, which in turn owns 50.1 per cent of Carling O'Keefe. Thus the number two US brewer is linked up to the number one South African wine producer, which is hooked up to the number three Canadian brewer.
>
> (Cavanagh and Clairmonte, 1985: 44)

By the 1980s the successful brewers had themselves been taken over by larger multi-sector global corporations, most of whose drinks divisions combined spirits, wines, beers and also non-alcoholic beverages. With the major exception of the Canadian company Seagram, which itself is part of the Bronfman family's complex holding companies operation (Newman, 1978), very few of the top 27 corporations, though, were concerned exclusively with alcohol, and indeed only 25.9 per cent of

Philip Morris's total sales in 1989 came from alcoholic beverages (Cavanagh and Clairmonte, 1985).

One of the most successful examples of the transformation of a wine merchant into a global corporation is that provided by the development of the Drinks Division of Grand Metropolitan from its origins in the London wine merchants W. & A. Gilbey (Maxwell, 1907; Waugh, 1957; Faith, 1983). Walter and Arthur Gilbey, the sons of a coach proprietor, established a small business as retail wine merchants in February 1857, with cellars at the corner of Berwick Street and Oxford Street in London. Here they were soon joined by their brothers-in-law Henry Gold and Charles Gold, as well as by their nephews Henry Arthur Blyth and James Blyth (Waugh, 1957). On the advice of their elder brother Henry Parry Gilbey, a wholesale wine merchant based in the City of London, the new firm concentrated on the wines of the Cape of Good Hope and achieved immediate success. Within a few months Waugh (1957: 8) thus notes that the new firm had some 20,000 customers, and in their first full year of business they imported over 55,000 gallons of Cape wine (Faith, 1983: 17). Indeed, by 1860 they were the third largest importers of wine to Britain from all sources, supplying it to individual customers, institutions and retailers. In contrast to other, older established wine merchants, they then seized the opportunity provided by Gladstone's budget in 1860 to begin to import cheap wines from the Médoc, and in so doing they were later described by Gladstone as 'the openers of the wine trade' (Faith, 1983: 21). Gladstone's introduction of off-licences in 1861, enabling grocers to sell wine by the bottle for consumption off the premises, also opened another opportunity for the Gilbeys, who subsequently began to deal primarily through a network of such agencies (Maxwell, 1907). In each major town they selected a few grocers who would act exclusively as their agents. Prices, and thus margins, were closely controlled by the Gilbeys, but in return they offered their agents advice and an efficient system of supply which meant that the agents only had to stock small amounts of the most popular wines. Within a few years they had more than 2,000 such agents, and were supplying over 200 different varieties of drink. In 1863, as a guarantee of the quality and value of their wines, they adopted a new trade mark consisting of a griffin confined within a castle tower, and by 1867 after two previous moves of premises, they acquired the Pantheon in Oxford Street as the centre of their operations. Its cellars, however, proved to be inadequate, and in 1869 they leased warehouse premises in Camden Town from which they began to distribute their wines. Soon afterwards, in 1872 the Gilbeys also opened a distillery in Camden Town, representing a considerable step towards vertical integration in their handling of spirits.

During the 1860s the Gilbeys regularly visited wine producing regions to identify those wines that they wished to purchase, and they established close relationships with many of their suppliers, but in 1875, in a remarkable departure from usual practice, they purchased an estate of their own, Château Loudenne, in the Médoc (Maxwell, 1907). This was to serve as a depot from which they could then send the wines that they had purchased to England, but it also represented a further important step towards a vertically integrated production and wholesale enterprise. Immediately after the purchase they initiated the construction of new cellars and storage facilities, as well as a small harbour on the Gironde. However, the late 1870s and early 1880s coincided with the arrival of phylloxera and mildew in the Médoc, and the costs of the building work and treatment of the vines amounted to some £70,800 during the first ten years of their ownership of Loudenne (Faith, 1983: 103). As a purely commercial venture the purchase of Château Loudenne was not entirely a success, with the wine producing side of the business making considerable losses, and Faith (1983: 119) has argued that in retrospect it 'can be seen as marking the peak of the Gilbeys' fortunes. Although the firm continued to prosper after 1875, Loudenne's purchase was the last of the bold strokes that had astounded the world of wines in the two previous decades'.

The end of the nineteenth and the beginning of the twentieth century also saw increasing competition from other companies, and particularly from Victoria Wine, which was offering wine at prices below those of Gilbey's. Furthermore, during the 1880s and 1890s the rise of proprietary brands of champagne, brandy, sherry and port, forced the company to abandon its castle label, and to use instead the brand names of the producers whose goods they imported. Having regularly imported large quantities of port over the previous decades, Gilbey's reacted to this growth in proprietary brands in 1910 by buying the long established port shippers Croft, which had first appeared in the shipping lists as Phayre and Bradley in 1678 (Bradford, 1983). This move was apparently welcomed by the Croft family, since not only had they maintained successful trading links with Gilbey's for a number of years previously, but the purchase also considerably benefited the sales of Croft port. Indeed in 1912, Gilbey's began selling its own Invalid Port, produced exclusively by Croft. This was launched with an extensive advertising campaign as a moderately priced port designed for convalescent invalids, and its success eventually led to the demise of their earlier system of agents. In Waugh's (1957: 72) words, 'There was clearly no point in offering a branded commodity to the general public if individual members of that public could not obtain the advertised commodity at the particular store they patronized'. Meanwhile, Gilbey's also purchased three Scotch whisky distilleries, Glen Spey in 1887, Strathmill in 1895 and Knockando in 1904 (Waugh, 1957), and were thus in a strong position to profit from the rising demand for whisky at the end of the nineteenth century. In the period following the First World War, particular attention was paid to the Empire as a market, and gin and whisky distilleries were established in Australia, with a further distillery also being built in Canada. In part the increased demand for gin reflected the emergence of the cocktail party in the 1920s, and following the ending of prohibition Gilbey's formed a new company, W. & A. Gilbey Ltd. of Delaware, in partnership with National Distillers Products Corporation through which Gilbeys Gin was marketed in the United States of America (Waugh, 1957). In the wine sector in the inter-war period the preferential duties given to wines from the Empire in Churchill's budget of 1927 led to a resurgence of interest in the wines of Australia and South Africa, on which Gilbey's were quick to capitalise. In 1928 they also acquired a chain of 110 off-licences when they purchased T. Foster & Co Ltd, and these then provided a new outlet for many of their products.

Following the Second World War, attention focused mainly on the development of overseas markets, and the creation of new brands of wine and spirits for the home market. Various agreements were made with foreign drinks producers such as Bacardi and Cinzano, and in 1953 Gilbey's launched Smirnoff vodka on the London market. With the reduction of duties on table wine in the 1949 budget and increasing advertisements released by wine producing countries, wine drinking rose considerably in the post-1945 period. The consumption of imported wine thus grew from 367,400 hectolitres in 1939–40 to 754,500 hectolitres in 1959–60 and 1,245,800 hectolitres in 1964–5 (Table 7), with much of the demand being at the lower end of the market (Briggs, 1985). In comparison with these figures, beer and spirits consumption grew very much less over the same period, and in some years actually declined (Table 7). Gilbey's were well positioned to benefit from the expanding wine demand, importing wine from their own vineyards, and bottling it under their own labels in England. However, although Gilbey's represented a high degree of vertical integration for the wine and spirits industry at the period, the difficulties of raising capital and the growing power of the brewers with their own wine and spirits departments supplying tied public houses, meant that further expansion was constrained.

By 1960 Gilbey's recognised that if the company was going to expand and survive a merger with a similar wine and spirits organisation would be necessary. Some time before this, in 1952, a group of three London wine merchants, Justerini & Brooks, Twiss & Brownings & Hallowes, and Corney & Barrow, concerned about the possibility of being taken over by the ever more powerful brewers, had joined together to form United Wine Traders, and the complementarity between them and Gilbey's made a link between the two groups a sensible proposition. Thus, in the Spring of 1962 Gilbey's and United Wine Traders merged to form International Distillers and Vintners Ltd. (IDV). For Gilbey's this provided an

Table 7 Alcohol consumption in the United Kingdom 1930–88

Year	Wine (thousand hectolitres)		Bear (duty paid bulk barrels)	Spirits (Thousand hectolitres of 100% alcohol)
	Imported	British		
1929–30	701.2	127.1		317.4
1939–40	650.7	314.4	25,914,768	276.1
1949–50	367.4	166.5	26,780,892	238.3
1959–60	754.5	310.4	26,646,775	356.0
1964–65	1245.8	435.0	29,907,803	484.8
1969–70	1479.3	459.1	33,463,444	456.0
1974–75	2846.2	762.0	39,108,762	841.3
	(Wine of fresh grapes)	(Made-wine)		
1979–80	4200.7	589.4	42,114,000	1110.9
1984–85	5735.8	589.7	37,764,238	907.0
1987–88	6226.2	569.2	38,997,000	1001.0

Note: On 1 January 1976 the classifications in statistics produced by HM Customs and Excise for wine changed, and thereafter wine of fresh grapes has been used to refer to wines produced from the naturally fermented juice of the grape, whereas made-wine has been used to refer to fermented beverages made from such things as concentrated grape juice.
Source: dervived from the Brewers' Society (1989), *Statistical Handbook 1988*, London: Brewing Publications.

increased network of outlets for their range of products, and also brought in Justerini & Brooks' highly successful J. & B. Rare Scotch Whisky, which was particularly popular in the American market, where the agent was the Paddington Corporation. However, as the first *Annual Report and Accounts* of IDV (1962: 4) recognised 'Although the merger was a bringing together of two groups basically in the same trade, their structures, methods of operation and administration were in many ways different'. Considerable reorganisation was therefore necessary, and in the home market Gilbeys' Agency Division was integrated with the Agency firm of Twiss & Brownings & Hallowes, with overseas agencies being transferred where possible to a local Gilbey subsidiary.

During 1963, IDV acquired a number of other retail companies, namely Peter Dominic Ltd, Hunter and Oliver Ltd, J.H. & J. Brooke Ltd, P.W. Feather Ltd and J.A. Feather Ltd, bringing its total number of retail outlets throughout the country to approximately 300 (*IDV Report and Accounts, 1963*). At that time the major brands handled by IDV included Gilbeys Gin, Crofts Port, Spey Royal Whisky, Château Loudenne Claret and White Bordeaux, Triple Crown Port, Golden Velvet (Canadian) Whisky, and J. & B. Rare Scotch Whisky, and they were also agents for Hennessy Cognac, Heidsieck Dry Monopole Champagne and Smirnoff Vodka. By 1964, though, fear of being taken over forced the new group into further negotiations, and all of the retail subsidies of Gilbey's, Peter

Dominic and Hunter & Oliver were consolidated into a new company United Vintners Ltd., in which Seagram, the Canadian distillers, acquired a 35 per cent interest for £1,500,000 (*IDV Report and Accounts, 1964*). These proceeds enabled further retail outlets to be purchased in different parts of the country, and provided sufficient capital to see off an unwelcome takeover attempt by the Showerings Group. In the mid-1960s increases in customs duty and taxation on alcohol created a somewhat depressed home market, but the group's consolidated profits after taxation were helped considerably by the export market and rose comfortably from £1,039,290 in 1964 to £1,904,280 by 1967. Critical to the success of IDV was the policy of building up traditional brand names, as well as the creation and development of new brands to fill gaps in the market. Thus, in 1965 after a detailed market research exercise, which 'revealed that most people preferred a pale sherry but at the same time many found the pale sherries already on the market too dry' (*IDV Supplement to Report and Accounts, 1967*), Croft Original Cream Sherry was launched with the colour of a fino and yet the sweetness of a true cream sherry.

In 1967, having successfully fought off the Showerings bid, IDV was reorganised on a divisional basis into a Trading Company rather than a Holding Company, and Seagram's share in United Vintners was bought back for £1,500,000. The retail efforts became concentrated in one company, that of Peter Dominic Ltd, where the style of marketing was directed towards the increasing demand for lower- and middle-priced wines, such as the Carafino brand. With over 300 shops, 'The aim was to create a multiple company of a quality which could win public recognition as *the* wine company – leaders in their field in the way that Marks & Spencer, Boots or W.H. Smith are in theirs' (*I.D.V. Supplement to Report and Account, 1968*). At this time, the quality image was reinforced by the use of advanced wine handling technology, laboratory analysis and quality control measures, the use of which was then in its infancy. During 1968 Watney Mann, the brewers, took a large shareholding in IDV and at the same time IDV acquired the former Watney Mann retail companies Brown and Park Ltd, and Brown, Gore and Welch Ltd. This initiated links which early in 1972 were to lead to the acquisition of IDV by Watney Mann. This was followed immediately by the takeover of Watney Mann itself by Grand Metropolitan Hotels Ltd, which was reported at the time to be the largest industrial acquisition yet seen in Britain (*Grand Metropolitan Hotels Ltd., Annual Report and Accounts, 1972*). In the meanwhile, IDV had acquired the port company Delaforce Sons & Ca Lda in 1968, had launched a new French wine Le Piat de Beaujolais and a Portuguese range of cheap wines under the name Justina, and in conjunction with the House of Diez Hermanos SA had created a new company in Spain called Croft-Jerez SA.

The events of 1972 meant that IDV became part of a much larger multi-sector international corporation, with interests ranging from hotels, such as Grand Metropolitan Hotels Ltd, to gambling in the form of Mecca Bookmakers Ltd, restaurants like Berni Inns Ltd, and dairies such as the Express Dairy Co. Ltd. In the drinks sector, Grand Metropolitan then owned ten breweries, 7,200 public houses and 775 retail off-licences within the United Kingdom alone. During the early 1970s, IDV as a fully owned subsidiary of Grand Metropolitan was therefore in an exceptionally good position to benefit from the considerable expansion of wine and spirits consumption that occurred between 1970 and 1975 when wine imports consumed in Britain rose from 1,479,300 to 2,846,200 hectolitres, and spirits consumption increased from 456,000 to 841,300 hectolitres (Table 7). Typical of the new options made available by the varied nature of the group was the possibility of creating entirely new types of alcoholic beverage. Thus in 1975 Bailey's Irish Cream was launched, using IDV's Irish subsidiary, Gilbeys of Ireland Ltd, as a production base, and the combined research and development expertise of Express Dairy and IDV.

By 1980 Grand Metropolitan was reorganised into six operating divisions, Hotels and Catering, Milk and Food, Brewing and Retailing, Wines and Spirits, Leisure, and Liggett. The acquisition of the American Liggett Corporation brought with it two marketing companies operating in the United States, Carillon Importers and The Paddington Corporation, which had for a long period handled J. & B. Rare. The justification for this purchase was clearly stated in Grand Metropolitan's *Annual Report* (1980: 15) as follows:

> I.D.V.'s main objective is to build international brands, and the United States is by far the most important market place in the world. It was, therefore, vital for us to secure a direct marketing presence there. Many of our major international competitors are increasingly concentrating their main efforts on their own brands. Without a U.S. presence we would have been particularly vulnerable, particularly in developing new brands.

The development of new brands continued apace, and in 1979/80 a new French table wine called Piat d'Or was launched with an extensive advertising campaign, and by 1984/85 it was to become the largest single table wine exported from France. Other new brands put on the market in the same year included Croft Particular Sherry, an orange based aperitif called Primavera, and a rum-based product with a coconut flavour called Malibu.

The 1980s saw an inexorable rise in IDV's profits, built on the successful production, marketing and sale of alcoholic beverages. The growing strength of the anti-alcohol lobby, however, was beginning to cause anxiety, and it is interesting to note that in IDV's *Annual Report* for 1984, the company stated that 'We continue to take an active role in encouraging the beverage alcohol industry to recognise its social responsiblity, especially in combating alcohol abuse'. By the late 1980s, despite almost continual reorganisations of the Grand Metropolitan group, including the $5.8 billion purchase of the Pillsbury Group with its Burger King outlets in 1988, the £180 million acquisition of the Wimpy, Pizzaland and Perfect Pizza chains from United Biscuits in 1989, the sale of Inter-Continental hotels in 1988, and the sale of the William Hill and Mecca betting shops in 1989, its drinks group IDV has continued to expand successfully.

In 1987 it benefited from its parent company's purchase of Heublein Inc, which brought in Smirnoff Vodka and Jose Cuerva Tequila, as well as a significant presence in the United States domestic wine market. Heublein's wine interests in California had been built up following the wine boom of the 1960s. In 1969 Heublein bought two of the most prominent wineries in the Napa Valley: Inglenook, which had been founded by Gustave Niebaum and which had been owned since 1964 by Allied-United Vintners (Parker, 1979), and Beaulieu, founded by Georges de Latour. Beaulieu, with André Tchelistcheff as winemaker, was in particular renowned for the quality of its wines, and Blue (1988: 9) has suggested that in the 1940s, 1950s and 1960s it represented 'the standard by which all other American wines were judged'. Inglenook, however, under the management of the Allied-United Vintners Co-operative had suffered somewhat, and although continuing to produce Napa Valley wines, its name was also used to promote lower quality wine produced in the San Joaquin Valley and known as Inglenook-Navalle. Heublein itself was acquired by R.J. Reynolds Nabisco in 1982, and in 1987 it purchased Almaden Vineyards, the third largest wine producer in California, from National Distillers. This preceded Grand Metropolitan's acquisition of Heublein from R.J. Reynolds Nabisco by only ten days (Blue, 1988: 339), and subsequently there has been a substantial reorganisation of the company's structure, with production of wine for the mass market becoming concentrated at the Inglenook-Navalle facility at Madera, near Fresno. In 1989 Heublein also acquired the Christian Brothers winery and vineyards in the Napa Valley. Although this acquisition was primarily for the company's brandy production, it brought with it 1,200 acres of vineyards, a new winery just south of St Helena on which $30 million had been invested since 1982, and also another small winery, called Quail Ridge. Subsequently, Heublein has established a Fine Wine Group, combining all of its Napa Valley vineyards and wineries, with the intention of preserving the integrity of each of the brands and developing their full potential as prestige wines in the high quality market. These acquisitions have been among the largest in the Californian wine industry, and represent a significant degree of horizontal integration, in a

339

situation where vertical sectoral integration, combining production, wholesale and retail outlets is legally prohibited.

In the United Kingdom in 1987, the addition of Saccone & Speed's 340 outlets strengthened IDV's position in the industry, and the signing of a distribution agreement with Martell enhanced its place in the developing Far East market, which was further strengthened by commercial and trading agreements in the region established with Seagram in 1988 (*IDV Annual Reports, 1987* and *1988*). By the end of 1988, IDV owned fifteen wine and spirit brands, and operated subsidiary companies in thirty-two countries of the world.

At the end of the 1980s a further reorganisation took place in Grand Metropolitan's corporate structure, which had particular significance for its off-licence retail network. Following the introduction of a consumer research programme in 1986, the company's off-licence retailing was reorganised into a single group, the Dominic Group, which in 1989 became part of the retail sector of the corporation rather than continuing to remain in the drinks sector as part of IDV. The Dominic Group now consists of three separate identities, the traditional Peter Dominic stores, the Bottoms Up superstores, and the Hunter and Oliver specialist wine merchants, each targeted at a particular sector of the market. Thus, the Bottoms Up superstores, which increased in number from thirty-one to sixty-nine in 1989, are intended for the mass market buying alcoholic beverages in bulk, and the Peter Dominic stores continue to serve the traditional off-licence trade aimed at the top-up shopper. In contrast, the Hunter and Oliver name was relaunched within the group in 1989 as a specialist wine merchants targeted at the recreational consumer interested in purchasing a good choice of quality wines. This can be seen as a direct response to the increasing market share of small, independent wine merchants, as well as specialist wine chains, such as Oddbins, supplying the growing number of knowledgeable wine consumers in the south-east of the country. The first four Hunter and Oliver premises opened in 1989 were thus in Amersham, Maidenhead, Barnet and Reading, with the next two opening at Henley and Virginia Water, all of which lie in the affluent commuter belt to the west and north of London.

The development of IDV described above illustrates five main features common to the emergence of global corporations as the dominant force in the wine industry since 1945. First, the successful companies all embarked on a process of vertical and horizontal integration, combining production with international trade and the development or acquisition of clearly identified retail outlets. Thus, in a similar way to that in which IDV built up the Peter Dominic chain as its main retail outlet in Britain, Seagram acquired and developed the highly successful Oddbins chain of retail wine shops; as the main outlet for Croft's ports and sherries is Peter Dominic, so Oddbins is the outlet for Sandeman's ports and sherries.

Secondly, this involved a general trend in which individual wine merchants were purchased by brewers, who then developed into drinks conglomerates which were often then taken over by multi-sector global corporations. Faced with stagnating beer sales during the 1960s (Table 7), the brewers turned to wine and spirits in order to increase their turnovers and profits, and the success of these ventures then encouraged further investment in these sectors. There are, however, two important exceptions to this general trend, namely the growth in importance of the supermarket chains in the retailing of wine and the survival of the specialist wine merchant. The increase in the amount of wine sold by supermarkets has represented a major challenge to the global corporations, and by the mid-1980s over half of the retail wine business in Britain was in the hands of supermarkets, such as Sainsbury's, Marks and Spencer, Waitrose and Tesco (Loftus, 1985; Key Note Reports, 1988). This emergence of the supermarkets, and more recently hypermarkets, as major purveyors of alcoholic beverages in Britain dates from the early 1960s when changes in the licencing laws allowed off-licences to keep shop opening hours rather than those of public houses, thus enabling supermarkets to sell wine and spirits throughout the day. It also reflects a change in shopping behaviour from frequent low quantity shopping visits for food which were characteristic of the early 1960s, to weekly or monthly high value and large quantity visits, as well as the increasing amounts of alcohol being purchased by women. Central to the success of supermarkets in the wine trade has been their adherence to own label brands which are seen as guarantees of the quality image established upon their other ranges of products. In contrast, specialist wine merchants have been able to maintain and even expand their trade largely by concentrating on the upper end of the market and by providing specialist services which the larger retail chains and supermarkets are unable to satisfy (Loftus, 1985). Indeed, as wine consumers have become more educated, partly as a result of the expansion of wine journalism during the 1980s, there has been growing dissatisfaction with the service and quality of wines provided by the multiple chains enabling these smaller merchants to establish a niche in the market.

This introduces the third main characteristic of the retail wine trade since 1945 which has been the concentration on brand names and image. If it is possible to identify a single policy which has been at the forefront of management decision-making in the global drinks industry during the 1970s and 1980s, it has been this development of international brand names, through the use of extensive advertising campaigns. However, unlike spirits, it is difficult to produce large amounts of uniform tasting high quality wine, and it has therefore been much less easy to establish brand images in the wine sector. To some extent, this has been achieved by German companies, such as Langenbach and Kindermann, with their

Crown of Crowns and Black Tower 'liebfraumilch' white wines, and it has also been accomplished by IDV with Le Piat d'Or, but branded wines are generally of poor quality and remain relatively few in number. Such wines are designed for mass consumption at the lower end of the market, and by their very nature quality wines can never be produced or marketed in this way. Unlike still table wines, though, higher quality fortified and sparkling wines can indeed be marketed successfully as brand names, and this is well reflected in the amounts currently being expended on their advertising. Furthermore, most successful companies endeavour to include a range of brands within their portfolios in order to cover different sectors of the market, from prestige ports and single malt whiskies, to cheap wines at the lower end of the market.

This attempt to cover a broad range of brands is associated with a fourth feature of the global corporations which has been their attempts to identify niches in the market which can then be expanded. Currently, in the late 1980s and early 1990s, faced with increasing pressure from the anti-alcohol lobby and increasingly health-conscious consumers, there has therefore been a marked trend towards the production of low-alcohol beers and wine coolers, as well as other non-alcoholic drinks. If marketed successfully, such drinks can be the source of considerable profit margins for their producers. However, there have been many other earlier examples of such activity. Moët & Chandon, for example, moved into sparkling wine production in Argentina in 1954 in order to overcome the high import taxes then prevailing which effectively prohibited the sale there of French champagne. Similarly, in 1973, wishing to develop the North American champagne market, but being unable readily to expand production in France, Moët & Chandon purchased 900 acres of land in the Napa Valley, from which the production of sparkling wines under the name Domaine Chandon began in 1977.

A fifth important characteristic has been the growth in joint international ventures between different companies and groups. The wine trade, built upon a product which can only be produced successfully in a limited number of countries, has always involved international trade, and the vertical integration of wine companies therefore implied that there would automatically also be international integration. In contrast, the internationalisation of spirits, which can be produced almost anywhere, resulted in part from the existence of restrictive tariff agreements in various parts of the world. More recently the changing pattern of alcohol consumption in different countries, has acted as a further incentive for global corporations to expand their activities to cover both new markets and new sectors of the drinks industry. As Table 8 indicates, wine consumption in the traditional producing countries of Europe has declined dramatically during the last decade, at a time when beer drinking has tended to increase. Moreover, throughout most countries of

342

Table 8 Changes in the consumption of wine, spirits and beer, 1976–86

Country	Percentage change in consumption 1976–86			Wine consumption per head of total population (litres)	
	Wines	Spirits	Beer	1976	1986
Australia	+52.6	−4.0	−17.0	13.5	20.6
Belgium	+32.3	−1.5	−12.9	16.4	21.7
Denmark	+58.2	−16.0	+5.8	12.5	19.8
France	−22.6	0.0	−17.0	101.3	78.4
Italy	−25.2	−36.8	+38.3	98.0	73.3
Portugal	−27.6	−14.0	+32.4	97.8	70.8
Spain	−36.6	+3.2	+29.4	71.0	45.0
United Kingdom	+78.7	+3.0	−9.1	5.8	10.4
United States of America	+39.6	−21.5	+11.5	6.6	9.2
West Germany	−1.3	−30.9	−2.8	23.6	23.3

Source: derived from Key Note Reports, *An Industry Sector Overview: Wine*, London: ICC Information Group, 1988: 7.

the world the consumption of spirits has also decreased. It has therefore been essential for the survival of French and Italian wine companies, for example, to expand their sales in northern Europe, and for spirits companies in the United States of America faced with a 21.5 per cent decline in spirits consumption to diversify into wines and beer. The most successful global corporations have therefore been those best able to exploit and manipulate these changing patterns of consumption. However, with the very large amounts of capital now necessary for complete takeovers, several of the largest corporations have recently formed joint ventures or agreements in order to dominate the markets. The agreements between Grand Metropolitan and Seagram for the Far East market were mentioned above, but other examples include the creation of European Cellars by Allied-Lyons and Whitbread in 1986 which brought together Grants of St. James's, Stowells of Chelsea, Langenbach, Calvet and Hermann Kendermann GmbH Weinkellerei, and also the establishment of cross-shareholdings between Guinness and the French LVMH Moët Hennessy Louis Vuitton SA group in 1988. This last move, in particular, enabled both Moët & Chandon and Hennessy to benefit from Guinness's global distribution network, and despite the legal wranglings over the deal it has been a considerable financial success for all parties involved.

CAPITAL, TECHNOLOGY AND FASHION

Central to the above processes have been the capitalist imperatives of cost minimalisation and market maximisation. These have led to the investment of capital not only in improvements in technology but also in the creation of specific drinking fashions. Thus, the role of particular drinks as social symbols has come to play an increasingly important part in the advertising campaigns of the global corporations. Moreover, wine itself has entered the realms of finance capital, as bottles of great vintages are stored away as financial investments, some destined never to be drunk.

Technological change in the service of capital

Technological change in the wine industry is nothing new, but following the Second World War, and particularly since the 1960s, a wide range of technological innovations in viticulture and vinification have taken place, designed not only to reduce the costs of production, but also to produce wines of invariant quality best suited to the emergent mass market for such products. These changes have also been associated with an increase in the numbers of people undertaking courses in viticulture and oenology at institutions such as the University of California, Davis, and Roseworthy College in Australia (Winkler, 1973; Bishop, 1980). This has meant that during the 1970s and 1980s the industry has received a large influx of highly qualified wine makers, who have played a significant role in raising the general quality of wine production (Stuller and Martin, 1989).

During the late nineteenth century numerous experiments were undertaken by enterprising wine makers in order to improve the qualities of their wines. Many of these were first attempted in the new environments encountered in north Africa, California and Australia, and were designed to enable successful viticulture and wine making to take place within the different climatic regimes to be encountered in these regions. Two problems in particular needed to be overcome: first, it was necessary to identify the best varieties of vine to be cultivated in each climatic zone, and secondly, it was essential to overcome the problems associated with the high temperatures of the fermentations that took place in the much warmer climates encountered nearer the equator. The first experiments in cool fermentation for wine seem to have been undertaken in Algeria as early as the 1880s, and although these were followed in Australia in 1898, it was not until the 1950s and 1960s that the pace of technological change began to quicken. As Johnson (1989: 452) has noted,

> Those first years of the '60s are the turning point in modern wine history. A radical new idea was born in many places at once: that wine was not an esoteric relic of ancient times that was disappearing even in Europe, nor just a cheap way to get drunk, but an expression of the earth that held potential pleasure and fascination for everyone.

The changes that have taken place can broadly be divided into those associated with the practices, first of viticulture and then of vinification.

Since the 1950s, considerable research has been undertaken in order to develop the optimum vines for different climatic and environmental conditions. In particular a number of new crosses and hybrids (crosses from different species) have been introduced so that vines will not only be more disease resistant, but will also retain the taste characteristics of one variety while possessing different flowering or fruiting tendencies. Typical of the most successful recent crosses developed in Germany are the Kerner, a Trollinger × Riesling cross which produces Riesling type wines with light aroma, good sugar content and high acidity, and which by the late 1980s was the fourth most widely planted grape in the country, and the Ehrenfelser, a Riesling × Silvaner cross, which likewise produces good Riesling style wines but ripens earlier and has a higher yield than Riesling (Robinson, 1986; Stevenson, 1988). Much research has also been done on the identification of the best rootstocks for particular varieties of *Vitis vinifera* in different soil conditions. More recently, it has been recognised that different plants of the same variety of vine often have very different characteristics, and this has encouraged experiments with clonal reproduction, whereby numerous clones of the parent plant with the required characteristics are produced. The essential aim of all of these methods, though, has been to provide vine growers with plants that will provide a more regular and guaranteed supply of a standard grape product particularly suited to specific environmental conditions.

A second major change in viticultural practices in recent years has been the dramatic increase in its mechanisation. Compared with other crops, vines are labour intensive, requiring an annual cycle of ploughing, pruning, training, weeding, and numerous sprayings, let alone the harvesting of the grapes. Thus, faced with rapidly increasing labour costs, companies have been keen to introduce as much mechanisation as possible. In most instances this has involved the widening of rows between the vines when vineyards have been replanted, so that tractors can be used for ploughing, the application of sprays, and also the pruning of the vines. More recently, and particularly since the mid-1970s, grape harvesting machines have been introduced in order to reduce labour costs at the time of the vintage. In a census of the use of such machines undertaken in France in 1976 over 55 per cent of owners reckoned to have paid off the costs of their machines within four or five years (Lacombe, 1977), and with increased labour costs since then the amortisation time has decreased commensurately. The implications of mechanised harvesters, though, are much wider than merely a reduction in labour costs, since they also cut the amount of time taken to harvest a vineyard, thus enabling the grapes to be picked rapidly at the optimum moment, even at night. However, not all vineyards are suitable for such mechanisation, and this has led to a further differentiation in the costs of production between different regions. Thus, although mini-tractors can be used in parts of the Mosel and Douro valleys, for example, their steep slopes preclude the use of mechanical grape harvesters. Wine producers in these regions consequently suffer from an increasingly serious production cost disadvantage compared with other parts of the world.

Other recent changes in viticultural practice include the use of irrigation, in particular through drip irrigation systems, the introduction of new systems of vine training, and the application of foliar feeds. While the main emphasis of these changes in viticulture has been derived from a desire to reduce labour costs, the changes that have taken place in vinification have, in contrast, been designed to make the resultant wine more appealing to the consumer. The most important of these changes has been the introduction of temperature controlled cool fermentation for white wines, which has enabled producers in hot locations, such as the new vineyards of Australia and California, but also in long-established wine producing countries such as Italy, to make light dry white wines, rather than the often heavy and oxidised wines of old. Two of the traditional characteristics of wine were its variability from year to year, and its susceptibility to differences in storage condition. While this was relatively unimportant for the connoisseur, the development of mass-produced brands of wine for the lower end of the market required that the consumer always received wines that tasted the same, regardless of the year in which they were produced and the way in which they were stored. Consequently, most of the changes that have taken place have been designed to produce stable and uniform wines for the mass market. This has involved the introduction of numerous chemical additives to prevent bacterial spoilage and oxidation, as well as to control the levels of acidity in the wines. Moreover, the use of new types of filtering equipment, and the cooling of wines to precipitate tartrates prior to bottling, are now commonplace. Other recent changes in vinification practice include the pretreatment of grape juice for white wines prior to fermentation, the introduction of carbonic maceration, the prevention of oxidation through the use of inert gasses, the replacement of natural yeast by dry cultured yeasts for fermentation (Kunkee, 1984), and the use of centrifuges for separating musts from the remaining material after fermentation.

While the above comments apply most forcibly for producers of wine for the mass market, another important recent trend has been the rise in boutique wineries, and in the attention paid to quality wine production. As increasing numbers of wine consumers become more discerning in their tastes, and as wine buyers for the large retail chains travel ever more widely in their search for the best quality products at the most reasonable prices, there has been an increasing emphasis on the quality of wine production during the 1980s. This is also, in part, related to the challenge

346

that the technological developments in the New World laid down for wine producers in Europe. Until the 1970s, it was widely accepted that wines from the traditional producing areas of France and Germany were superior to those of the New World, and could therefore claim a substantial premium in price. However, in 1976, at a tasting of French and Californian wines organised by the English wine merchant Steven Spurrier, French authorities judged a red and a white Californian wine to be the best of each category (Blue, 1988: xiii). In Johnson's (1989: 457) words, 'The other wines jostled closely in a finishing line that proved quite simply that their qualities in expert French eyes, were approximately equal'. Thereafter, European wine producers have been forced to respond to the challenge, and considerable investments have now been made in technological improvements. Among the European producers to benefit most, have been those of Italy and Spain, who in the past suffered from climatic problems similar to those encountered by wine makers in California and Australia, and in particular from the difficulties associated with fermentation and wine storage in hot climates. The introduction of technologically advanced wine making equipment, able to control precisely every process from the arrival of the grapes to the bottling of the wine, has thus transformed wine production in many areas of the Old World as well as the New.

Symbols and the creation of fashion

The above examples of recent changes in the practice of viticulture and vinification have essentially been concerned with the economics of wine production. However, central to the success of the global alcohol corporations has been their ability to conjure up favourable images associated with their products in the minds of consumers. Indeed the early successes of companies such as Gilbey's and Victoria Wine owed much to the appeal of their advertising. Moreover, the successful attempts by wine and spirits producers to persuade consumers in countries such as Britain, Australia and the United States of America to consume more wine (Table 8) have been based in part on a social downgrading of the traditional image of wine in these countries. Furthermore, the laws and regulations affecting the advertising of wine, and indeed all types of alcohol, vary appreciably between different countries. In essence, three strategies have been adopted by companies keen to expand their markets: the creation of new types of wine, the capture of other companies' market shares, and the expansion of the overall market by the manipulation of fashion.

A classic example of the creation of new types of wine has been the rebirth of Moscadello di Montalcino in Tuscany. The American company Villa Banfi, which imported large amounts of Lambrusco to the

United States, were seeking to expand their sales in the mid-1970s, and recognising that an effervescent, sweet, low-alcohol white wine would fill a gap in the market, they purchased some 3,000 hectares of land in Tuscany, and built a new winery reputed to have cost some $100 million (Belfrage, 1985; Johnson, 1989). The vineyards were planted with a variety of vines, including Brunello, Cabernet Sauvignon and Chardonnay, but half of them were put down to Moscato in order to produce the Moscadello, which was a traditional, but little made, Montalcino wine. By the mid-1980s, a new wine, technologically transforming an almost forgotten local tradition, was therefore born, specifically designed to appeal to the American market.

Another good example of the manipulation of fashion, this time by Australian wine producers, was the development of bag-in-the-box wines. As Mayo (1986: 213) has commented:

> It was a brilliantly simple idea to make a plastic bag with a simple tap opened by deforming a handle, fill the bag with cheap wine, and enclose the whole in a cardboard carton with a carrying strap at the top. It was ideal for outdoor drinking, and has changed the face of picnics, and their detritus.

The importance of this development is reflected in Australian retail statistics which show that the purchase of white wines in plastic containers rose from 22.7 million litres in 1977/8 to 103.6 million litres in 1982/3, at a time when total bottle sales of red and white wine rose from 67.6 million litres to only 84.5 million litres (Mayo, 1986: 230). In 1984, bag-in-the-box wines had more than 60 per cent of the entire retail wine market in Australia (Mayo, 1986: 214).

The capture of market share, and efforts to increase overall wine consumption, have largely been achieved through advertising. In practice, though, the actual expenditure on wine advertising is suprisingly low, given the industry's retail value, which in 1987 in the United Kingdom amounted to some £3.5 billion (Key Note Reports, 1988). In the same year, Le Piat d'Or, which accounted for 2.5 per cent of the United Kingdom off-licence trade in still table wine, for example, had the highest expenditure on television and press advertising totalling some £1.95 million (Key Note Reports, 1988). This figure can be compared with the next highest advertising expenditure in the same year which was £451,000 for Black Tower, and with figures for sparkling wine, where the top two advertisers were Asti Martini spending £1,290,000 and Mateus Rosé spending £886,000. In addition to this advertising by individual companies, though, various national wine authorities have also undertaken overseas advertising campaigns in order to encourage export sales.

Most advertising is associated with the launching and reinforcing of brand names, and, as the example of the creation of Croft Original Sherry noted above illustrates, it is often used to encourage consumers to move into a previously unexploited sector of the drinks market. Thus, the attention paid to the advertising of dryer, lighter and whiter wines during the late 1970s and early 1980s, with Seagram for example launching an $18 million campaign in 1981 to promote its own brand of light wine (*Advertising Age*, 10 July 1981), can be interpreted in part as a move by wine producers towards this sector of the market at a time when sales of red wines were beginning to stabilise. Similarly, the deliberate targeting of women over this period, can be seen as an attempt by the alcohol corporations to tap the 'powerful social movements calling for greater female emancipation and participation' at a time when 'these social dynamics were joined by recessionary economic forces that depressed male consumption in several countries' (Cavanagh and Clairmonte, 1985: 133). Typical of these advertising campaigns aimed specifically at women were those for Jack Daniels whisky in the United States of America, and Cutty Sark's campaign in the French magazines *Biba* and *Elle* in 1989. The United Kingdom manager of Cutty Sark, for example, explained this shift to advertising in women's magazines as follows: 'Women have always been instrumental in the marketing of brands, but unfortunately they have been portrayed as a reward for the macho man drinking a certain product. Now we are recognising that many women are independent and career-minded' (*The Sunday Times*, 29th January 1989). Furthermore, during the late 1980s the shift towards the advertising of low alcohol and alcohol-free drinks, which have high profit margins, can also be seen as a direct attempt to create and exploit a new market area.

However, the increase in wine consumption during the 1960s in countries such as Britain and the United States of America was part of a much wider change in behaviour, associated with the *embourgeoisement* of society. Rising real incomes provided the context against which mass advertising could successfully propagate increasingly middle class aspirations throughout society, and as one of the symbols of this ideology wine consumption grew accordingly. In this context, it is interesting to note that in a survey of wine consumption in the United States in the mid-1970s, it was found that the households purchasing the largest volumes of wine had the highest incomes, a highly educated male head, smaller families, and were slightly further along in their life cycles as indicated by the age of the wife (Folwell and Baritelle, 1978). While advertisements for still table wine in the popular press were relatively uncommon, wine was increasingly brought to the attention of a wider public through its appearance, for example, in plays and films on television, which during the 1960s became increasingly available to a larger percentage of the population. Virtually all wine advertising since then has concentrated on the maintenance of this image, which projects wine as a symbol of material well-being and contentment. In the

European context, the tendency for people in wine producing countries to turn increasingly to beer during the 1970s and 1980s, and for beer drinking countries to turn increasingly to wine consumption, can likewise be explained in part by the emergence of a more pan-European cultural identity associated with the growing strength of the European Community.

Nevertheless, the regulations controlling alcohol advertising vary appreciably between different countries, reflecting the varying cultural symbolism that still pervades different societies. In France, for example, all advertising of alcohol on television is prohibited, and in a law passed in July 1987 all other alcohol advertising had to recommend moderation in drinking, and was prohibited from presenting alcohol as producing a beneficial physiological or psychological effect, from encouraging its consumption by minors, and from evoking any imagery associated with sexuality, sport, work or the driving of motor vehicles. Despite such regulations, many French wine advertisements, such as that used to promote the wines of the Côtes du Roussillon in 1988 and 1989 (Figure 39), with its deliberately seductive atmosphere, still reflect the continuing importance of the underlying symbolic linkage between wine and fertility. The British Code of Advertising Practice also reflects a degree of ambivalence in its attitude towards advertising for alcoholic drinks. Thus it specifies that 'Advertisements may emphasise the pleasures of companionship and social communication associated with the consumption of alcoholic drinks', but also that 'Advertisements should neither claim nor suggest that any drink can contribute towards sexual success, or make the drinker more attractive to the opposite sex' (Committee of Advertising Practice, 1988: 81). Regulations in other countries vary from relative tight control in Denmark and Germany, to much looser restrictions in Greece, Spain and Portugal. In general, therefore, it can be seen that southern European countries, with their long tradition of wine production, are much less strict in their alcohol advertising regulations than are the countries of northern Europe.

Turning to specific advertising campaigns, it is evident that most table wine advertisments, whether in the popular press or in more specialist wine magazines, tend simply to concentrate on the projection of a particular wine bottle or label, unlike the advertising campaigns for spirits and fortified or sparkling wines, which have much larger sales and advertising budgets, and which are concerned more specifically with the projection of an image associated with their products. Typical of the latter were the highly successful Smirnoff campaign of the mid-1970s under the heading 'I used to be . . . until I discovered Smirnoff', and the romantic imagery used in the Hennessy Cognac advertisements of the late 1980s projecting the image that it is 'fashionably old-fashioned'. Indeed, with spirits, the precise nature of the advertising varies greatly, with each

Figure 39 Advertisement for the wines of Côtes du Roussillon and Côtes du Roussillon Villages

brand of vodka, for example, having to create a different image in order to establish itself in the market place.

This concern with fashion reflects a further strategy adopted by the global corporations in their attempts to increase sales through the projection of certain drinks as being particularly fashionable. By changing the image of what is deemed to be in fashion, the market can be manipulated towards an expansion in the sales of a particular product. As Gladstone noted in the quotation cited at the beginning of this chapter, 'Taste is not an immutable, but a mutable thing'. Within the wine sector this trend can be seen most clearly in the attention given to wines made from Cabernet Sauvignon and Chardonnay grapes in the early 1980s, and more recently to Sauvignon and Chenin Blanc. Following the international controversies over the labelling of wines during the 1960s, producers in North America and Australia increasingly turned to the use of varietal names for their wines, and in particular to the high value Cabernet Sauvignon and Chardonnay varieties. The undoubted success of many of these wines, then led to increasing numbers of producers in other traditional wine making countries, such as Italy, to experiment with such 'noble' varieties of grapes, for which there was a ready market. Moreover, wines produced from these grapes could also be sold at a greater profit than those made from more traditional local varieties, and particularly in Italy the 1980s have therefore seen a gradual shift away from the cultivation of many of the indigenous varieties of vine, leading to a reduction in local diversity.

These trends have been enhanced by the increasing number of books, magazines, television programmes and columns in the popular press devoted to wine since the mid-1970s. Although numerous books aimed at the wine consumer had been published before 1970, it was Hugh Johnson's *World Atlas of Wine* published in 1971, which opened the floodgates of popular wine writing. In Britain, for example, it was followed in 1975 by the first edition of the magazine *Decanter*, and then by popular series of books on wine produced by a range of publishers such as Mitchell Beazley and Faber & Faber. Regularly weekly columns in the national press provided advice and guidance on the best wine buys, and a new breed of wine journalists was born, theoretically independent of the global alcohol corporations, but occasionally fuelled by them through the provision of wine tastings and vineyard visits. Moreover, the increased interest in wines, and the expansion in the number of people taking foreign holidays that took place during the 1970s, also opened up a whole new opportunity for the smaller producers particularly in France and Germany who began to sell a greater percentage of their wines from the cellar rather than through merchants. In California, the increased demand for wine that occurred in the 1970s and 1980s also provided the context for the emergence of numerous small wineries which were able to flourish through their attention to quality (Moulton, 1981; Stuller and Martin, 1989).

Wine in the realms of finance capital

While the global alcohol corporations have generally concentrated on the establishment of brand names for the mass market, they have also been keen to ensure that wine has retained its image as a prestigious symbol. Thus most corporations retain high quality wines such as vintage port and clarets within their portfolios. Typical of this tendency, has been the acquisition of well known châteaux in the Bordeaux region, not only by the global alcohol corporations but also by other large institutional groups, such as banks and insurance companies, who see them primarily as financial investments. Among the most recent of such acquisitions was the increased stake in Château Latour purchased by Allied Lyons for £56.2 million in 1989, which valued the vineyard at approximately £2 million per hectare (Parnell, 1989). With vineyards valued at such levels it has become more and more difficult for the traditional family owners of such estates to afford the 40 per cent inheritance taxes, and this has led to yet more estates being sold to outside concerns, often with little real interest in the wines other than for their prestige.

The value of such estates, though, has also been associated with a dramatic increase in the price of older vintage wines, and a growing tendency for them to be purchased specifically as investments. This trend has been enhanced by the emergence of regular wine auctions, with the auctioneers Christie's being the first of the major companies to establish a Wine Department in 1966, to be followed soon afterwards by Sotheby's in 1970 (Loftus, 1985; Spurrier and Ward, 1986). While wine had been sold at auction for centuries previously, it was only with the development of regular auctions in the 1970s that investors were able to realise their profits. However, this incipient financial market in wine suffered a serious setback in the early 1970s, initiating an extensive reorganisation of the Bordeaux wine trade. During the late 1960s, a poor quality harvest in 1968 and the small quantity of the 1969 vintage, had led to increasingly high prices for the top wines of Bordeaux. This coincided with a sharp increase in château-bottling, which 'was encouraged by the higher prices, since the price advantage to the consumer of bulk shipment in cask diminished as the cost of the wine itself rose' (Penning-Rowsell, 1985: 150). In 1971 and 1972 producers asked for, and were able to obtain very high prices for their wines. Opening prices for Château Palmer thus rose from Frs 10,000 per *tonneau* in 1969 to Frs 31,000 in 1971 and Frs 55,000 in 1972. However, in 1973 and 1974 the market collapsed, with prices tumbling from the levels of 1971 and 1972, elevated as they were by speculation (Pijassou, 1980; Peppercorn, 1982; Loftus, 1985). The coincidence of two large vintages in 1973 and 1974, with a rise in global oil prices and high international interest rates, turned a sellers' market into a buyers' one, and many Bordeaux *négociants*, who had fixed agreements to purchase wines at set prices, found that they had large quantities of wine which they simply could not sell. There were some bankruptcies, but on the whole the Bordeaux *négociants* closed ranks. With the recovery of the dollar by the late 1970s prices again rose, and despite fluctuations they have continued so to do ever since. Thus, for example, auction prices for Château-Lafite 1961 have risen from £720–840 a case in August 1979 to £2,090 a case in August 1988, while over the same period auction prices for Croft 1935 vintage port have risen from £240–70 a case to £850 a case (*Decanter*, August 1979, August 1988).

Recently, the main market for such wines has been in the United States of America and Japan, where investors have sufficient money at their disposal, and the health of the market owes much to the respective values of the US dollar and the yen against the French franc. One of the great keys to the success of the wine trade in the last twenty years, though, has been the transferance of this symbolism to wines for everyday drinking. The average consumer, by drinking a lesser wine of Bordeaux, or even a cheap Bulgarian Cabernet Sauvignon, is thus able to participate in the same act as the owner of a cellar full of expensive first growths from Bordeaux, when he or she she draws the cork and samples the contents of a bottle.

THE FUTURE OF CAPITAL

The wine industry, unlike other sectors of the drinks industry, remains relatively fragmented. Despite the changes in corporate industrial structure noted above, numerous small producers and retail outlets survive. In part this reflects the continued division in the market between mass-produced, low quality bulk wines, and the high quality, expensive wines purchased by the rich. However, it also reflects the very essence of wine itself. As Gladstone argued in his speech to the House of Commons as Chancellor of the Exchequer on 10th February 1860, 'Wine, I suppose, more than almost anything else that is produced from the earth by the labour of man, varies in quality and price' (*Hansard*, 10th February 1860: 844). It is this inherent diversity, and the options that it offers producers, that makes the subject of such complexity.

It is therefore particularly interesting to examine the relative advantages and disadvantages of the increased vertical integration of the industry for different types of wine producer, for wine retailers and for consumers. For the large global corporations that own vineyards, the financial returns on investment in wine production are not particularly high, and in some cases the ownership of vineyards and brand names is perhaps more for reasons of prestige than for strict financial return. However, such corporations are able to invest considerable sums of capital in the latest production technology, enabling them to supply from different vineyards both the low quality mass market and also the fine vintage wines demanded by those who can afford them. Moreover, they are also in a strong position to purchase either wine or grapes from smaller producers at beneficial rates, although at times of high demand and relatively low supply of quality grapes, as in parts of California in the late 1980s, such practices have led to a rapid escalation in grape prices which have also benefited the smaller grape growers. Furthermore, the comprehensive retail distribution networks of the global corporations permit them to maximise their sales in the international market. For the actual wine producing companies within such organisations, though, there can be great pressure from insensitive accounting measures, and with a commodity such as wine, which has considerable annual variability in both the quantity and quality of production, it is essential that a medium- to long-term view is taken on the return of investment.

Smaller vineyard owners and wine producers, providing they have sufficient capital to invest in the most advanced technology, and particularly if they are able to benefit from political interventionist measures and monopolistic demarcation practices, are also able to compete successfully in the quality wine market. Critical to the success of their enterprises, though, is access to the market, and relatively low levels of production mean that they depend heavily on the whims and judgement of journalists as well as the wine purchasers of major national and international retail companies. As Alison Green, the wine maker for the Firestone Vineyard in the Santa Ynez Valley in California has commented, 'Since critics have become such an important force in trying to sell wine, winemakers are following trends more than they used to . . . Unfortunately, it's often the critics' tastes rather than the winemaker's that's important' (Troutman, 1987). It is the small, traditional wine producers in unprestigious areas of the world, with low levels of capital investment, who are currently in the most difficult position. Unable to invest in the production technology to enable them to make the quality wines increasingly being demanded by consumers and retailers, they are subject to growing market pressure, particularly at a time when wine consumption in their own countries is decreasing (Table 8). One solution is for them to grow grapes or produce wine for one of the main international drinks corporations, from whose technology and advice they can benefit. However, this can be a risky undertaking subject to the whims of distant accountants, and one in which their financial bargaining position is weak. Another alternative is to join a wine cooperative, but many such cooperatives suffer from poor reputations concerning the qualities of their wines. This is very often the result of indiscriminate vinification, in which little attention is paid to the quality of grapes supplied by individual growers. However, cooperatives do have larger amounts of capital to invest in modern equipment than do individual producers, and can provide sufficient amounts of wine to satisfy the demands of the retailers. Both of these reasons give them considerable bargaining power, and with careful attention to production, as for example in the case of Les Caves des Hautes-Côtes in Burgundy, the qualities of the wines they produce can be high.

With respect to retailing, the large vertically integrated global corporations have distinct advantages in that they do not rely on only one national market or a limited range of drink products for their sales. Moreover, as in the example of the Dominic Group, they can maintain a number of different styles of retail outlet, each designed for a particular type of consumer. Nevertheless, because of their size, supermarkets and the large retail chains are unable to satisfy every niche in the market, and at a time of increasing affluence in many wine consuming countries it is still possible for quality conscious small independent wine retailers to flourish. The eventual outcome for the consumer has been a general increase in the quality and the range of wines available from all over the world. Thus in Britain, for example, there is little comparison between the off-licences of the 1960s, with their limited supplies of generally poor quality French, German and perhaps Spanish and Italian wines, and the selections of hundreds of wines from all over the world that are found at a range of prices in the retail outlets of the 1990s. While many retailers argue that these changes have been brought about by different patterns of consumer demand, it is clear that most are also active in influencing that demand. Economic interests alone do not determine the complexities and fortunes of the wine industry. The symbolic value of wine, and its wider role in society remain of utmost importance.

11

CONCLUSION

From the first discovery that the fermented juice of grapes produced a beverage which was not only pleasant to taste, but which also had the ability to create profound physiological effects on its drinker, wine has become imbued with'a range of different layers of meaning. Not only has it taken on economic significance as a product of the land from which profit can be extracted through the exploitation of labour, but it has also become a powerful symbol of the fundamental cycle of life, death and rebirth. Not only does it represent the essential being of life as a symbol of the essence of deity, but it is also the agency through which the drinker can enter the presence of deity.

The physiological effects of wine on those who consume it have changed little over the millenia, and although recent technological innovations have enabled new varieties of vine and methods of wine-making to be developed, the basic environmental controls on viticulture have also remained broadly the same. However, each generation has used wine and the vine as expressions of its own culture, building up layers of changing meaning upon these basic constants. In conclusion this chapter therefore seeks to elucidate some of the interactions between those elements in the culture of the vine that have remained relatively stable, and those that have experienced change, grouped under four headings: the ideological, concerned with the ways in which people have justified and legitimated their behaviour; the social, through which they have communicated with each other through personal relationships; the economic, concerned with systems of production, surplus expropriation and exchange; and the political, through which power and control have been implemented.

Beginning with ideological factors, it is apparent that the origins of viticulture and wine production were closely associated with the emergence of a particular religious experience, in which symbols in the environment were used as ways of explaining and understanding the complexities of human life. In particular, the vine came to be seen as a potent symbol of death and rebirth, and its product, wine, as a means of

357

coming into contact with the forces that were seen to control human destiny. However, as vines symbolically combined the opposition of life and death, so too could wine be seen as having both good and evil attributes. In limited quantities wine was beneficial and led to pleasurable experiences, but in excess it was dangerous, both for the individual and for society. The forces that wine unleashed therefore necessitated the codification of rules of behaviour which were deemed by society to be reasonable and that would not destroy its very fabric.

The development of Dionysiac rituals and symbolism can in part, therefore, be seen as a reflection of this initial integration of wine into the world view of the peoples of south-west Asia and the eastern Mediterranean. Once wine had become part of the religious experience of Greek culture, the cultivation of vineyards spread rapidly alongside the expansion of Greek cultural influence in the Mediterranean basin. By the time of the emergence of the Roman Empire, however, the religious role of wine and the vine had dwindled relatively in importance, and economic factors had emerged as being of most significance in influencing the organisation of viticulture and the wine trade. The incorporation of wine into Christian imagery and symbolism further served to ensure the maintenance of its cultural importance following the collapse of the western Roman Empire. Although the ritual use of wine in the Christian Eucharist has probably been exaggerated in the past as a factor leading to the survival of viticulture in the early medieval period, the position of wine in Christian culture was such that it was ensured a place in the cultural landscape of medieval Europe. A further ideological influence on the practice of viticulture was the expansion of Islamic teaching and power, which from the eighth century seriously curtailed the production of wine. Once again, though, it is apparent that vineyards continued to be cultivated in most Islamic countries, and that the main effect of Islamic prohibitions on alcohol consumption was probably to encourage the cultivation of sweet grapes for eating, rather than the complete destruction of vineyards.

Turning to social influences, it is evident that wine has served a number of different roles depending in part on the environmental context of its production. In prehistoric Mesopotamia and Egypt wine was mainly the preserve of the ruling elite of society, with beer being the usual drink of the poor. This largely reflected the scarcity of wine in these countries, and the difficulty of its production in such environments. With the spread of viticulture to the Mediterranean, where vines flourished more readily and where the successful fermentation of grapes was easy, wine became the preferred alcoholic beverage of all levels of society. This was in part influenced by the adoption of the cult of Dionysus and by the economic development of commercial viticulture, but it also reflected underlying environmental conditions. Subsequently, the shift of econ-

358

omic and political power northwards in the medieval period once again led to the adoption of wine as one of the symbols of the ruling classes, partly as a result of the difficulties of producing wine in northern Europe and the cost of its transport. At the same time, in southern Europe wine remained an important part both of the peasant agricultural system, and of peasant social life.

With the global expansion of European political power in the sixteenth and seventeenth centuries, viticulture was then taken as part of the cultural assemblage of the Spanish conquistadors into parts of the world from which the cultivation of the vine had previously been absent, and in these 'new' lands it came into conflict with fundamentally different cultures, with their own types of alcohol and their own related deities. Moreover, this was a time of increasing alcohol consumption within the economic heartland of Europe, and as trading conditions improved, enabling wine to become more readily available there, the ruling classes once again turned to new types of beverage as symbols of their social position. Rather than buying wines which lasted but a year, the socially privileged of northern Europe thus began to turn to the acquisition of new types of wine, such as champagne and the new French clarets, which were beyond the financial means of the poor. It has only been in the twentieth century that wine has become more readily available to most of the population of northern Europe, and the increased wine consumption of these countries in recent years can in part be seen as a reflection of people's aspirations to acquire a symbol of social well-being that lies deeply embedded within their culture.

As the above examples illustrate, the social role of wine has been closely influenced by economic considerations of demand and supply, of methods of production, and of the types of labour control used in that production. Central to any understanding of these economic processes, however, have been the physical characteristics of grape production and wine making. In particular, the annual variability in grape yield, and the time-lag between the planting of a vineyard and its first substantial production, have tended to create imbalances between levels of demand and supply, which have initiated crises in the viticultural economy. Good examples of such crises can be seen in the overplanting of vines in Campania following the eruption of Vesuvius in AD 79, in the fourteenth century agrarian crisis in Burgundy, and in the cyclical pattern of overproduction and crisis in Europe and North America at the end of the nineteenth and the beginning of the twentieth century. Often these crises have been initiated by external factors, which have provided significant disruptions to the traditional balance between the demand and supply of both grapes and wine. However, the reduction of supply at times of high demand has nearly always led to a period of high grape prices, thus encouraging the subsequent overplanting of vines. Because of the length of time taken before maximum yields are achieved there is always a delay of about four years in returning to previous levels of production, during which time prices tend to remain high thus encouraging yet more farmers to plant vines in the hope of windfall profits. Invariably this has led to an overproduction of grapes, a fall in prices, and consequently a crisis in which grape production no longer becomes profitable, and it has usually taken several such cycles before an approximate balance between demand and supply has been restored. During the twentieth century a number of technical and legislative changes have been introduced in order to dampen down such cycles, particularly in the European Community, but in California, for example, where such legislation is much less well developed, there continue to be marked cyclical variations in the prices of grapes.

Although viticulture and wine production have been characterised by very diverse systems of labour control, they have generally always been more labour intensive compared with other types of agriculture. In the Roman period peasant wine production existed as part of a largely subsistence based economy alongside the large villa estates designed for the production of a surplus through the exploitation of slave labour. Indeed, the survival of peasant wine production in various parts of Europe seems to have been one of the relatively constant features of viticulture and the wine trade ever since. In the medieval period a surplus of wine was also extracted by the landholding classes through various systems of rent in kind, sharecropping arrangements, and the direct exploitation of labour services. It is not possible to understand the subsequent introduction of wage labour systems, though, apart from the reorganisation of the wine trade as a whole which took place during the sixteenth century. Before about 1500 profit in the wine trade was achieved primarily as a result of demand in areas where wine was not produced, which merchants could supply to their financial advantage. Throughout most of the medieval period the wine trade therefore revolved around the production of mercantile profit. However, from the sixteenth century onwards, and particularly in the seventeenth and eighteenth centuries, money was increasingly being invested in the production of wine rather than simply in the trading process. Thus vineyards were established in lands from which they had previously been absent, as in the Atlantic islands and southern Africa, and capital became invested increasingly in the creation of new wines in traditional areas of viticulture. The essential characteristic of these new wines was that capital had been invested in their production in the direct expectation of increased profits. Consequently, rather than relying on traditional systems of land tenure and labour provision, it became increasingly important for producers of wines for the quality export market to turn to wage labour for their production.

CONCLUSION

This was also associated with a division of the wine market into two main sectors, which gave producers of wine a choice in the economic strategies which they adopted. For most of the medieval period there had been relatively little variation in the quality of wine available in the international market, with the main distinction being that between the sweet wines of the Mediterranean and the dryer and lighter ones produced in France and Germany. However, from the seventeenth century onwards, the growth of the urban population of northern Europe led to a large increase in demand for cheap, low quality wines, at about the same time as the ruling classes were seeking to consume high quality wines as symbols of their social prestige. A few wine producers, as in parts of the Médoc and Champagne, therefore concentrated on the upper end of the market, whereas the majority, most notably in regions close to the major cities, sought by every means possible, including the widespread adulteration of wines, to expand the quantity of their production more or less regardless of quality. This is but one example of the much more general observation that as demand for wine has grown in the past there has generally been a tendency for wine producers to concentrate on an increase in production with little regard being paid to the quality of their wines, whereas at times of decreasing demand greater emphasis has been placed on quality. This in part explains the changed emphasis in the first century AD from quality wine production to the supply of bulk wine for the rapidly growing urban market in Italy, and also the current increased attention being given in the late twentieth century to improvements in wine quality throughout the world at a time when wine consumption in most wine producing countries is in decline.

A further significant economic trend that has occurred over the last century has been the increasing vertical integration of the wine industry. While wine is vastly more variable than either beer or spirits, the production and sale of wine has become increasingly dominated by global corporations combining a number of production, wholesale and retailing companies. Even in the United States of America, where legislation prevents wine production companies from owning more than a very limited number of retail outlets for their products, a few large winery companies, most notably E. & J. Gallo and Heublein, the latter now part of Grand Metropolitan, have come to control the vast bulk of the wine industry.

Finally, wine has been subject to a range of political controls throughout its history. Taxes and customs duties have regularly been levied on wine production and exchange, forming part of the revenue of governments as diverse as those of imperial Rome, medieval England and contemporary Sweden. However, governments have also been concerned to regulate the wine trade in three other main ways: through the encouragement of domestic production by the restriction of imports, through attempts to control fraudulent manufacture and distribution of wine, and through the imposition of both direct and indirect restrictions on the consumption of wine. Among the clearest examples of political intervention were the closely similar attempts by Domitian in AD 79 and Philip II in 1595 to restrict wine production in peripheral parts of their empires. Similarly, the British government regularly used differential taxation on the wine trade during the medieval and early modern periods for the implementation of broader political aims, particularly with respect to its economic and political conflicts with France. Within their own polities, most governments have also reflected public concern over fraudulent practices by enacting legislation in an effort to ensure that unadulterated wine has been served in standard measures.

The most direct controls imposed on wine consumption, however, have resulted mainly from the interaction of a range of political, social and ideological interests associated with calls for the complete prohibition of alcohol. The adverse physiological influences of alcohol have led to widespread condemnation of its use, with the earliest concern over intemperance apparently dating from around 1300 BC in Egypt (Austin, 1985). While there is good evidence to indicate that high levels of alcohol consumption are directly associated at a national scale with high liver cirrhosis mortality and other medical problems (Davies and Walsh, 1983; Royal College of General Practitioners, 1986; Robinson, 1988), it has not mainly been on medical grounds that alcohol has traditionally been condemned by those advocating its prohibition. Instead, criticisms of alcohol have generally either focused on social issues, such as the breakdown of family life and an increase in criminal activity, or they have been couched in ideological language claiming that alcohol is inherently evil and should therefore be prohibited. Arrayed against these arguments have been the economic interests of the wine producers, who have usually been able to manipulate the balance of power in the political arena because of the amount of revenue that governments raise through their taxation of alcohol. However, with increasing attention being paid to the measurement of the real financial costs to society of alcohol consumption, in terms of health care provision, loss of labour, and criminal destruction (Grant and Ritson, 1983), the political strength of the anti-alcohol lobby would currently appear to be rising.

Most analyses of the social problems associated with alcohol abuse, however, fail satisfactorily to distinguish between the uses made of different kinds of alcohol (Davies and Walsh, 1983; Grant and Ritson, 1983), and in their defence supporters of wine consumption have frequently pointed to the positive advantages associated with the moderate consumption of wine with food (Ford, 1988). In one extreme case, for example, the Wine Advisory Board (1942) of California produced a book specifically on the therapeutic uses of wine, which argued that it could be used beneficially for the treatment of the digestive organs, the kidneys, the cardio-vascular system, the respiratory system, the nervous system, diabetes, acute infections, as well as in the treatment of the aged and convalescent. More recently, most advocates of wine have pointed to research which shows that spirits are absorbed much more rapidly into the blood than wine, and that the consumption of equivalent amounts of different kinds of alcohol can therefore lead to markedly different blood alcohol levels and physiological effects in the period immediately after their consumption (Leake and Silverman, 1966).

All societies have developed some form of escape from the harshness of material and social life, and in those which emerged around the shore of the Mediterranean this escape was sought in the solace of wine. Elsewhere, tobacco, different forms of fermented fruits and plants, and other narcotic substances have served the same purpose, and whether or not they have been prohibited has depended largely on the threat that they have posed to social order. The real debate over alcohol, therefore, concerns the relationship between the rights of individuals and those of society, determined and controlled in the political arena through state legislation. Indeed, increasing levels of alcohol consumption and alcoholism during the nineteenth century can in part be interpreted as a response to the dehumanising effects of squalid urban living conditions and the social requirements of a factory system which increasingly separated individuals from nature. One way in which people could return to nature in a very real sense, and escape from the bonds of social control, was through the consumption of alcohol.

The historical geography of viticulture reflects the complex web of social, political, economic and ideological interactions that have transformed the world over the last six thousand years. No one influence has completely dominated and determined the processes associated with vine cultivation and the wine trade at any particular instance, but there is some evidence that through time, and particularly with the rise to dominance of capitalist relations of production in the last two centuries, the role of ideology has become increasingly subordinated to that of the economy. This survey is, however, but a starting point, and the task that lies ahead is to examine in detail the very large areas of uncertainty that remain in our understanding of the historical geography of wine and the vine.

1

APPENDIX
Historic wine measures

Wine has been stored, transported and sold in a wide variety of vessels, and providing even approximate modern equivalents for some of these is extremely difficult. This is primarily because the same term could be used in different parts of the world to refer to vessels of different sizes. Thus Redding (1833) notes that, at the time he was writing, a *barrique* had a capacity of 120.00 litres in Limoux, 229.94 litres in Bordeaux and 240.00 litres in Nantes. As governments in the medieval period came to regulate the wine trade more closely, some of the more usual measures came to be standardised. Even then, though, the sizes of barrels usually varied from country to country, and as the above example illustrates, from region to region, depending on local custom. The following conversion table should therefore be used with considerable caution, but it is nevertheless presented as an approximate guide to some of the more common wine measures mentioned in the text.

AMPHORA

The sizes of classical amphorae varied greatly in capacity, with Dressel 1B (Peacock and Williams Class 4) amphorae being approximately 22.00 litres (4.84 gallons), and Dressel 2–4 (Peacock and Williams Class 10) averaging 25.2 litres (5.54 gallons). The medieval Venetian anfora was much larger, containing 518.5 litres (114.07 gallons).

BARREL

A general term used to describe wooden containers. When used as a specific measure, the medieval *barile* of Florence held 45.5 litres (10.01 gallons), and the fifteenth century English wine barrel held 31.5 gallons, (143.20 litres).

BOTTLES

Glass bottles initially varied very greatly in size. Younger (1966) notes that examples of fifteenth century bottles ranged in size from 26-52 fluid ounces (0.74-1.48 litres), while seventeenth century bottles in the Ashmolean museum have a capacity of 16-35 fluid ounces (0.45-0.99 litres). Modern wine bottles are usually either 0.70 or 0.75 litres (24.64-26.40 fluid ounces) in size.

BUTT

In England the butt was fixed by statute in the fifteenth century at 126 gallons (572.80 litres) (see also pipe). However, before then it varied from 108–140 gallons. In the fifteenth century in southern Italy the *botte* held about 454.5 litres (100 gallons) and in Bruges the butt was about 909.25 litres (200 gallons). The modern butt of sherry holds about 491 litres (108 gallons).

GALLON

In England, the medieval wine gallon held about 104 fluid ounces (2.95 litres), and in the eighteenth century the wine gallon contained 132.5 fluid ounces (3.76 litres). In 1824 the English wine and beer gallons were standardised at 160 fluid ounces (4.546 litres)

HOGSHEAD

The medieval hogshead varied greatly in size, but by the fifteenth century the English hogshead held 63 gallons (286.40 litres). The modern hogshead of brandy holds 60 gallons (272.75 litres), of port 57 gallons (259.12 litres), and of sherry 54 gallons (245.48 litres).

LEAGUER

A barrel of variable size, but in English use in India and Ceylon in the nineteenth century it contained about 150 gallons (681.90 litres).

PIPE

From the fifteenth century in England a pipe held 126 gallons (572.80 litres), but in Spain it varied from 100–105 gallons (454.60–477.33 litres) (see also butt). The modern port pipe holds 115 gallons (522.79 litres), the Tenerife pipe 100 gallons (454.60 litres) and the Marsala pipe 93 gallons (422.77 litres). In general usage in Portugal pipes for maturing port contain approximately 630 litres, and Douro pipes 550 litres.

TUN

In England from the fifteenth century the tun (or tunne) contained 252 gallons (1145.59 litres), and was equivalent to 2 pipes or butts, 3 punchions, 4 hogsheads, 6 tierces, 8 barrels, or 14 rundlets. The modern French *tonneau* holds 863.75 litres (190 gallons).

Sources: Redding (1833); Hall, H. and Nicholas, F.J. (eds.) (1929), 'Select tracts and table books relating to English weights and measures (1100-1742)', *Camden Miscellany* 15; Younger (1966); Peacock and Williams (1986); The Brewers' Society (1989), *Statistical Handbook* 1988, London: Brewing Publications; and the *Shorter Oxford English Dictionary*.

BIBLIOGRAPHY

Abraham-Thisse, S. (1984) 'The Hanse and France', in A. d'Haenens (ed.), *Europe of the North Sea and the Baltic: the World of the Hanse*, Antwerp: Fonds Mercator, 229-40.

Adam of Bremen (1917), *Adam von Bremen, Hamburgische Kirckengeschichte*, B. Schmeidler (ed.), Hannover: Hahnsche Buchhandlung.

Adams, L.D. (1984) *The Wines of America*, 3rd edn, London: Sidgwick & Jackson.

Addis, W.E. and Arnold, T (1951) *A Catholic Dictionary*, 15th edn, London: Routledge & Kegan Paul.

Ali, A.Y. (1983) *The Holy Qur'an: Text, Translation and Commentary*, Brentwood, Maryland: Amana Corp.

Allen, H. Warner (1950) *Number Three Saint James's Street: a History of Berry's the Wine Merchants*, London: Chatto & Windus.

Allen, H. Warner (1961) *A History of Wine: Great Vintage Wines from the Homeric Age to the Present Day*, London: Faber & Faber.

Allen, H. Warner (1963) *The Wines of Portugal*, London: George Rainbird in association with Michael Joseph.

Althusser, L. (1969) *For Marx*, Harmondsworth: Penguin.

Althusser, L. and Balibar, E. (1970) *Reading Capital*, London: New Left Books.

Amerine, M.A. (1965) 'The fermentation industries after Pasteur', *Food Technology* 19(5), 75-90.

Amerine, M.A. and Joslyn, M.A. (1970) *Table Wines: the Technology of their Production*, 2nd edn, Berkeley, California: University of California Press.

Amerine, M.A. and Wagner, R.M. (1984) 'The vine and its environment', in D. Muscatine, M.A. Amerine and B. Thompson (eds.) *The Book of California Wine*, Berkeley, California: University of California Press, 86-120.

Amerine, M.A. and Winkler, A.J. (1944) 'Composition and quality of musts and wines of Californian grapes', *Hilgardia* 15(6), 493-675.

Amerine, M.A., Berg, H.W., Kunkee, R.E., Ough, C.S., Singleton, V.L. and Webb, A.D. (1980) *The Technology of Wine Making*, 4th edn, Westport, Connecticut: AVI Publishing.

Anderson, B. (1982) *Vino: the Wines and Winemakers of Italy*, London: Papermac.

Andrews, K.R. (1978) *The Spanish Caribbean: Trade and Plunder 1530-1630*, New Haven: Yale University Press.

Antcliff, A.J. (1988) 'Taxonomy - the grapevine as a member of the plant kingdom', in B.G. Coombe and P.R. Dry (eds) *Viticulture, Volume 1: Resources in Australia*, Adelaide: Australian Industrial Publishers, 107-18.

Anon. (1989) 'For laying down and dying', *The Economist* 9 December 1989, 114–15.

Arberry, A.J. (ed.) (1954) *The Rubáiy'at of Omar Khayyám and other Persian Poems*, London: Dent.

Arlott, J. and Fielden, C. (1976) *Burgundy: Vines and Wines*, London: Davis-Poynter.

Arnaldus de Villanova (1943) *The Earliest Printed Book on Wine by Arnald of Villanova . . . . Now for the First Time Rendered into English with an Historical Essay by Henry E. Sigerist*, New York: Schuman's.

Arnoux, C. (1728) *Dissertation sur la Situation de la Bourgogne, sur les Vins qu'elle Produit, sur la Manière de Cultiver les Vignes, de Faire le Vin et de l'Eprouver*, London: P. Du Noyer.

Ash, H.B. (1941) 'Introduction' to *Lucius Junius Moderatus Columella on Agriculture*, London: William Heinemann, xiii–xxxi.

Aston, T.H. and Philpin, C.H.E. (eds) (1985) *The Brenner Debate: Agrarian Class Structure and Economic Development in Pre-Industrial Europe*, Cambridge: Cambridge University Press.

Attenborough, F.L. (1922) *The Laws of the Earliest English Kings*, Cambridge: Cambridge University Press.

Audibert, J. (1880) *L'Art de Faire le Vin avec les Raisins Secs*, 5th edn, Marseille: Millaud.

Ausonius (1919) trans. H.G.E. White, London: William Heinemann.

Austin, G.A. (1985) *Alcohol in Western Society from Antiquity to 1800: a Chronological History*, Santa Barbara, California: ABC-Clio Information Services.

Avineri, S. (ed.) (1969) *Karl Marx on Colonialism and Modernisation*, New York: Anchor Books.

Bacci, A. (1596) *De Naturali Vinorum Historia, de Vinis Italiæ, et de Conuiuiis Antiquorum, Libri Septem . . . .* , Rome: N. Mutii.

Baird, J.A. (ed.) (1979), *Wine and the Artist*, New York: Dover Publications.

Baisette, G. (1956) *Ces Grappes de ma Vigne*, Paris: Les Editeurs Français Réunis.

Baker, A.R.H. (1983) 'Devastation of a landscape, doctrination of a society: the politics of the phylloxera crisis in Loire-et-Cher (France) 1866–1914', *Wurzburger Geographische Arbeiten* 60, 205–17.

Barnett, R.D. (1980) 'A winged goddess of wine on an electrum plaque', *Anatolian Studies* 30, 169–78.

Barth, M. (1958) *Der Rebbau des Elsass und die Absatzgebiete seiner Weine*, Strasbourg: Editions F.-X. LeRoux.

Bartholomew, F.H. (1947) *The Count Found a Valley*, Sonoma, California: Rancho Buena Vista.

Barty-King, H. (1977) *A Tradition of English Wine*, Oxford: Oxford Illustrated Press.

Bassermann-Jordan, F. (1907) *Geschichte des Weinbaus unter besonderer Berücksichtigung der Bayerischen Rheinpfalz*, Frankfurt: Verlag von Heinrich Keller.

Baumann, C.M. and Michel, F.W. (1976) *German Wine Atlas and Vineyard Register*, Mainz: Stabilisierungsfonds für Wein.

Baumgart, W. (1982) *Imperialism: the Idea and Reality of British and French Colonial Expansion, 1880–1914*, Oxford: Oxford University Press.

Bautier, R.-H. (1970) 'The fairs of Champagne', in C. Cameron (ed.) *Essays in French Economic History*, Homewood, Illinois: Richard D. Irwin, 42–63.

Baxevanis, J.J. (1987) *The Wines of Champagne, Burgundy, Eastern and Southern France*, Totowa, New Jersey: Rowman & Littlefield.

Bazille, G., Planchon, J.-E. and Sahut (1868) 'Sur une maladie de la vigne actuellement régnante en Provence', *Comptes Rendus de l'Academie des Sciences* 67, 333-6.

Bechtel, G. (1976) *1907, La Grande Révolte du Midi*, Paris: Editions Robert Laffont.

Bede (1968) *A History of the English Church and People*, edited by L. Shirley-Price, revised by R.E. Latham, Harmondsworth: Penguin.

de Beer, E.S. (1955) *The Diary of John Evelyn, Volume III*, Oxford: Clarendon Press.

de Beer, G. (1969) *Hannibal: the Struggle for Power in the Mediterranean*, London: Thames & Hudson.

Belfrage, N. (1985) *Life Beyond Lambrusco: Understanding Italian Fine Wine*, London: Sidgwick and Jackson.

Bell, C. (1974) *Portugal and the Quest for the Indies*, London: Constable.

Bendinelli, G. (1931) 'La vite e il vino nei monumenti antichi in Italia', in A. Marescalchi and G. Dalmasso (eds), *Storia della Vite e del Vino in Italia: Volume I°*, Milano: Gualdoni, 23-274.

Berenyi, M.I. (1978) 'Les disparités régionales provoquées par la crise du Phylloxéra dans le vignoble de Tokaj-Hegyalja', in A. Huetz de Lamps (ed.), *Géographie Historique des Vignobles, Tome II*, Bordeaux: CNRS, 121-8.

Berger, A. (1978) *Le Vin d'Appellation d'Origine Contrôlée*, Paris: Institut National de la Recherche Agronomique et de Sociologie Rurales.

Berger, A., Arnaud, C. Badis, M.F. and Maamoun, M. (1980-811) *Le Négoce des Vins en France*, Paris: Institut National de la Recherche Agronomique d'Economie et de Sociologie Rurales.

Berlow, R.K. (1982) 'The "disloyal grape": the agrarian crisis of late fourteenth-century Burgundy', *Agricultural History* 56, 426-38.

Bibby, G. (1970) *Looking for Dilmun*, London: Collins.

Billiard, R. (1913) *La Vigne dans l'Antiquité*, Lyon: Librairie H. Lardanchet.

Billington, R.A. (1974) *Westward Expansion: a History of the American Frontier*, 4th edn, New York: Macmillan.

Birrell, A. (ed.) (1986) *New Songs from a Jade Terrace: an Anthology of Early Chinese Love Poetry Translated with Annotations and an Introduction*, Harmondsworth: Penguin.

Bishop, G.C. (1980) *Australian Winemaking: the Roseworthy Influence*, Hawthordene, South Australia: Investigator Press.

Blakeway, A. (1932-3) 'Prolegomena to the study of Greek commerce with Italy, Sicily and France, in the eighth and seventh centuries B.C.', *The Annual of the British School at Athens* 33, 170-208.

Blanc, G. (ed.) (1987) *Le Grand Livre de Bourgogne*, Paris: Chêne.

Blaut, J.M. (1975) 'Imperialism: the Marxist theory and its evolution', *Antipode* 7(1), 1-19.

Blegen, C.W. (1959) 'The Palace of Nestor excavations of 1958. Part 1', *American Journal of Archeology* 63, 121-7.

de Blij, H.J. (1983) *Wine: a Geographic Appreciation*, Totowa, New Jersey: Rowman and Allanheld.

de Blij, H.J. (1985a) 'Wine quality and climate: finding favourable environments capable of yielding great wines', *Focus* 35(2), 10-15.

de Blij, H.J. (1985b) *Wine Regions of the Southern Hemisphere*, Totowa, New Jersey: Rowman & Allenheld

Bloch, M. (1965) *Feudal Society*, trans. L.A. Manyon, London: Routledge & Kegan Paul.

Blue, A.D. (1988) *American Wine: a Comprehensive Guide*, rev. edn, New York: Harper & Row.

Bonamici, M. (1985) 'The Etruscan period', in S. Settis (ed.) *The Land of the Etruscans*, Firenze: Scala, 12-13.

Bondois, P.-M. (1929) 'Les bouteilles à Champagne et les verreries d'Argonne au XVIIIᵉ siécle', *Nouvelle Revue de Champagne et de Brie* 7, 82-99.

Bouloumié, B. (1981) 'Le vin etrusque et la première Hellenisation du midi de la Gaule', *Revue Archaeologique de l'Est et du Centre-Est* 75-81.

Bourdieu, P. (1977) 'Sur le pouvoir symbolique', *Annales, Economies, Sociétés Civilisations* 32, 405-11.

Bourne, A. (1980) 'A geologist's atlas of wine', *New Scientist* 88, 792-4.

Bowden, M.J., Kates, R.W., Kay, P.A., Riebsame, W.E., Warrick, R.E.A., Johnson, D.L., Gould, H.A. and Weiner, D. (1985) 'The effects of climate fluctuations on human populations: two hypotheses', in T.M.L. Wiglet, M.J. Ingram and G. Farmer (eds) *Climate and History: Studies in Past Climates and their Impact on Man*, Cambridge: Cambridge University Press, 479-513.

Boxer, C.R. (1969) *The Portuguese Seaborne Empire 1415-1825*, London: Hutchinson.

Bradford, S. (1983) *The Story of Port: the Englishman's Wine*, 2nd edn, London: Christie's Wine Publications.

Brady, R. (1984) 'Alta California's first vintage', in D. Muscatine, M.A. Amerine, B. Thompson. (eds) *The Book of California Wine*, Berkeley, California: University of California Press, 10-15.

Bragato, R. (1895) *Report on the Prospects of Viticulture in New Zealand*, New Zealand: Samuel Costall, NZ Government Printer.

Braudel, F. (1972-3) *The Mediterranean and the Mediterranean World in the Age of Philip II*, trans. S. Reynolds, London: Collins.

Braudel, F. (1981) *The Structures of Everyday Life: the Limits of the Possible*, trans. S. Reynolds, London: Collins (Civilization and Capitalism 15th-18th Century, volume I).

Braudel, F. (1982) *The Wheels of Commerce*, trans. S. Reynolds, London: Collins (Civilization and Capitalism 15th-18th Century, volume II).

Breasted, J.H. (1906-7) *Ancient Records of Egypt. Historical Documents from the Earliest Times to the Persian Conquest, Collected, Edited and Translated with Commentary*, Chicago: Chicago University Press.

Brehaut, E. (1933) 'Translator's notes' in Cato the Censor, *On Farming*, New York: Columbia Univeristy Press.

Brenner, R. (1976) 'Agrarian class structure and economic development in pre-industrial Europe', *Past and Present* 70, 30-75.

Brenner, R. (1977) 'The origins of capitalist development: a critique of Neo-Smithian Marxism', *New Left Review* 104, 25-93.

Brenner, R. (1982) 'Agrarian class structure and economic development in pre-industrial Europe: the agrarian roots of European capitalism', *Past and Present* 97, 16-113.

Briggs, A. (1985) *Wine for Sale: Victoria Wine and the Liquor Trade, 1860-1984*, London: B.T. Batsford.

Britnell, R.H. (1981) 'The proliferation of markets in England 1200-1349', *Economic History Review* 2nd series 34(2), 209-21.

Brogan, H. (1985) *Longman History of the United States of America*, London: Longman.

Bruegel, I. (1975) 'The Marxist theory of rent and the contemporary city: a critique of Harvey', in *Political Economy and the Housing Question*, London: Political Economy of Housing Workshop, 34–46.

Bruman, H.J. (1940) 'Aboriginal drink areas in New Spain', unpublished PhD thesis, University of California Berkeley.

Bruman, H.J. (1967) 'Man and nature in Mesoamerica: the ecologic base', in B. Bell (ed.) *Indian Mexico: Past and Present*, Los Angeles, California: UCLA Latin America Center, 13–23.

Burroughs, D. and Bezant, N. (1980) *The New Wine Companion*, London: Heinemann.

Busby, J. (1825) *A Treatise on the Culture of the Vine, and the Art of Making Wine. . . .*, Australia: printed by R. Howe, Government Printers.

Busby, J. (1830) *A Manual of Plain Directions for Planting and Cultivating Vineyards, and for Making Wine, in New South Wales*, Sydney: printed by R. Mansfield, for the executors of R. Howe.

Busby, J. (1833) *Journal of a Recent Visit to the Principal Vineyards of Spain and France. . . .*, London: Smith, Elder, & Co.

Busse, E. (1922) *Der Wein im Kult des Alten Testamentes. Religionsgeschichtliche Untersuchung zum Alten Testament*, Freiburg: Freiburger Theologische Studien, vol. 29.

Butlin, R.A. (1987) 'European rural transformations: some reflections on the context of agrarian capitalism', in H.-J. Nitz (ed.) *The Medieval and Early-Modern Rural Landscape of Europe under the Impact of the Commercial Economy*, Göttingen: Department of Geography, University of Göttingen, 87–104.

Cambiare, C.P. (1932) *The Black Horse of the Apocalypse (Wine, Alcohol and Civilization)*, Paris: Librairie Universitaire J. Gamber.

Camões, L. de (1981) *Os Lusíadas (The Lusiads)*, edited by F. Pierce, Oxford: Oxford University Press.

Capus, J. (1947) *L'Evolution de la Législation sur les Appellations d'Origine: Genèse des Appellations Contrôlées*, Paris: Louis Larmat.

Carandini, A. (1983) 'Columella's vineyard and the rationality of the Roman economy', *Opus* 2(1), 177–204.

Carosso, V.P. (1951) *The California Wine Industry: a Study of the Formative Years*, Berkeley, California: University of California Press.

Carpenter, T.H. (1986) *Dionysian Imagery in Archaic Greek Art: its Development in Black-Figure Vase Painting*, Oxford: Clarendon Press.

Carter, F.W. (1987) 'Cracow's wine trade (fourteenth to eighteenth centuries)', *The Slavonic and East European Review* 65(4), 537–78.

Carus-Wilson, E.M. (1954) *Medieval Merchant Venturers*, London: Methuen.

Carus-Wilson, E.M. (1962–3) 'The medieval trade of the ports of the Wash', *Medieval Archaeology* 6–7, 182–201.

Castan, P. (1927) 'Contribution à l'étude des levures de vin', *Annuaire Agricole de la Suisse* 28, 311–19.

Cato the Censor (1933) *On Farming*, trans. E. Brehaut, New York: Columbia University Press.

Cavanagh, J. and Clairmonte, F.F. (1985) *Alcoholic Beverages: Dimensions of Corporate Power*, London: Croom Helm.

Chandaman, C.D. (1975) *The English Public Revenue 1660–1688*, Oxford: Clarendon Press.

Charleston, R.J. (1984) *English Glass and the Glass used in England circa 400–1940*, London: George Allen & Unwin.

371

Charleston, R.J. and Angus-Butterworth, L.M. (1957) 'Glass', in C. Singer, E.J. Holmyard, A.R. Hall and T.I. Williams (eds) *A History of Technology, Volume III, from the Renaissance to the Industrial Revolution, c1500-c1750*, Oxford: Clarendon Press, 206-44.

Charleton, W. (1669) *The Mysterie of Vintners. . . .*, London: William Whitwood.

Chaucer, G. (1977) *The Canterbury Tales*, rev. edn, trans. N. Coghill, Harmondsworth: Penguin.

Chilcote, R.H. (1967) *Portuguese Africa*, Englewood Cliffs, New Jersey: Prentice Hall.

Childs, W.R. (1978) *Anglo-Castilian Trade in the Later Midde Ages*, Manchester: Manchester University Press.

Clark, P. (1983) *The English Alehouse: a Social History, 1200-1830*, London: Longman.

Clout, H. (1977a) 'Urban growth, 1500-1900', in H. Clout (ed.) *Themes in the Historical Geography of France*, London: Academic Press, 483-540.

Clout, H. (1977b) 'Industrial development in the eighteenth and nineteenth centuries', in H. Clout (ed.) *Themes in the Historical Geography of France*, London: Academic Press, 447-82.

Cohen, J.M. (ed.) (1963) *Bernal Diaz: the Conquest of New Spain*, Harmondsworth: Penguin.

Colgrave, B. and Mynors, R.A.B. (eds) (1969) *Bede's Ecclesiastical History of the English People*, Oxford: Clarendon Press.

Columella, Lucius Junius Moderatus (1941, 1954, 1955) *On Agriculture: vol.1, Res Rustica I-IV*, trans. H.B. Ash (1941), *vol.2, Res Rustica 5-9*, trans. E.S. Forster and E.H. Heffner (1954), *vol.3, Res Rustica, 10-12, and De Arboribus*, trans. E.S. Forster and E.H. Heffner (1955), London: Heinemann.

Comision Organizadora del Congreso Internacional Filoxérico de Zaragoza (1880) *Sesiones Celebrades desde el 1° al 11 de Octubre de 1880*, Zaragoza: Imprenta del Hospicio Provincial.

Commission Départementale de l'Hérault pour l'Etude de la Maladie de la Vigne (1877) *Phylloxéra, Expériences faits à Les Sorres. Resultats Pratiques de l'Application des Divers Procédés*, Montpellier: Grollier.

Committee of Advertising Practice (1988) *The British Code of Advertising Practice*, London: Committee of Advertising Practice.

Coombe, B.G. and Dry, P.R. (eds) (1988) *Viticulture, Volume 1: Resources in Australia*, Adelaide: Australian Industrial Publishers.

Cooper, A. (ed.) (1973) *Li Po and Tu Fu: Poems Selected and Translated with an Introduction and Notes*, Harmondsworth: Penguin.

Cooper, A.B. (1977/78) 'The family farm in Greece', *Classical Journal* 73, 162-75.

Corbridge, S. (1986) *Capitalist World Development: a Critique of Radical Development Geography*, Basingstoke: Macmillan.

Cornell, T. and Matthews, J. (1982) *Atlas of the Roman World*, Oxford: Phaidon.

Cosgrove, D. (1984) *Social Formation and Symbolic Landscape*, London: Croom Helm.

Cosgrove, D. and Daniels, S. (eds) (1988) *The Iconography of Landscape: Essays on the Symbolic Representation, Design and Use of Past Environments*, Cambridge: Cambridge University Press.

Crawford, A. (1977) *A History of the Vintners Company*, London: Constable.

Crescentiis, Petrus de (1471) *Petri de Crescentijs ciuis. Bononiensis Epistola in Libru Comodoru Ruralium. Enerabili in Xpo Patri . . . .*, Augsburg: Johannes Schussler.

Croft, J. (1787) *A Treatise on the Wines of Portugal; and what can be Gathered on the Subject and Nature of the Wines, &c.* . . ., York: printed by Crask & Lund.

Croft, P. (1973) *The Spanish Company*, London: London Record Society.

Croft-Cooke, R. (1955) *Sherry*, London: Putnam.

Croft-Cooke, R. (1957) *Port*, London: Putnam.

Croft-Cooke, R. (1961) *Madeira*, London: Putnam.

Cronquist, A. (1981) *An Integrated System of Classification of Flowering Plants*, New York: Columbia University Press.

Cunliffe, B. (1978) *Iron Age Communities in Britain*, 2nd edn, London: Routledge & Kegan Paul.

Dalmasso, L. (1933) 'La vite e il vino nella letteratura Romana', in A. Marescalchi and G. Dalmasso (eds), *Storia della Vite e del Vino in Italia, Volume II°*, Milano: Gualdoni, 3-124.

Dalmasso, L. (1937) 'Le vicende tecniche ed economiche della viticoltura dell'enologia in Italia', in A. Marescalchi and G. Dalmasso (eds), *Storia della Vite e del Vino in Italia, Volume III°*, Milano: Gualdoni, 165-612.

Darby, H.C. (1977) *Domesday England*, Cambridge: Cambridge University Press.

Darnton, R. (1986) 'Review article: the symbolic element in history', *Journal of Modern History* 58, 218-34.

Davenport, T.R.H. (1969) 'The consolidation of a new society: the Cape Colony', in M. Wilson and L. Thompson (eds) *The Oxford History of South Africa, Volume 1, South Africa to 1870*, Oxford; Clarendon Press, 272-332.

Davidson, W.M. and Nougaret, R.L. (1921) *The Grape Phylloxera in California*, Washington D.C.: Government Printing Office (U.S. Department of Agriculture Bulletin, No. 903).

Davies, P. and Walsh, D. (1983) *Alcohol Problems and Alcohol Control in Europe*, London: Croom Helm.

Davis, D. (1966) *A History of Shopping*, London: Adam and Charles Black.

Dawkins, R.M. (1902-3) 'Excavations at Palaikastro II', *Annual of the British School at Athens* 9, 290-328.

Dawood, N.J. (1974) *The Koran*, 4th rev. edn, Harmondsworth: Penguin.

Degrully, L. (1896) *Les Plants Américains et Sols Calcaires*, Montpellier: Coulet.

Delaforce, J. (1979) *The Factory House at Oporto*, London: Christie's Wine Publications.

Delaforce, J. (1982) *Anglicans Abroad: the History of the Chaplaincy and Church of St. James at Oporto*, London: S.P.C.K.

Delafosse, M. (1963) 'Les eaux-de-vie de Saintonge et d'Aunis et le fisc au debut du XVIII° siècle', *Revue du Bas-Poitou* 1, 16-22.

Dickenson, J.P. (1990) 'Viticultural geography: an introduction to the literature in English', *Journal of Wine Research* 1(1), 5-24.

Dickenson, J.P. and Salt, J. (1982) 'In vino veritas: an introduction to the geography of wine', *Progress in Human Geography* 6, 159-89.

Dietz, F.C. (1964) *English Public Finance 1558-1641*, 2nd edn, London: Frank Cass.

Diodorus Siculus (1939) *Bibliotheca Historica III*, trans. C.H. Oldfather, London: Heinemann.

Dion, R. (1952a) 'A propos des origines du vignoble bourguignon: l'archéologie et les textes', *Annales de Bourgogne* 34, 47-52.

Dion, R. (1952b) 'Métropoles et vignobles en Gaule Romaine: l'exemple bourguignon', *Annales, Economies, Sociétés Civilisations* 7, 1-12.

Dion, R. (1955) 'Le commerce des vins de Beaune au Moyen Age', *Revue Historique* 214(2), 209-21.

Dion, R. (1959) *Histoire de la Vigne et du Vin en France des Origines au XIX*ᵉ *siècle*, Paris: privately published (Republished by Flammarion, 1977).

Dioscorides of Anazarba (1934) *The Greek Herbal*, edited by R.T. Gunther, Oxford: Oxford University Press.

Dodd, C.H. (1965) *Historical Tradition in the Fourth Gospel*, Cambridge: Cambridge University Press.

Dodds, E.R. (1960) 'Introduction', in Euripides, *Bacchae*, Oxford: Clarendon Press, xi-lix.

Dodgshon, R.A. (1987) *The European Past: Social Evolution and Spatial Order*, Basingstoke: Macmillan.

Dollinger, P. (1970) *The German Hansa*, London: Macmillan.

Dollinger, P. (1984) 'The Hanse and the Rhine', in A. d'Haenens (ed.) *Europe of the North Sea and the Baltic: the World of the Hanse*, Antwerp: Fonds Mercator, 219-28.

Domercq, S. (1957) 'Etude et classification des levures de vin de la Gironde', *Annales de Technologie Agricole* 1, 6-58; 2, 59-183.

Donkin, R.A. (1957) 'The disposal of Cistercian wool in England and Wales during the twelfth and thirteenth centuries', *Cîteaux in de Nederlanden* 8, 109-31, 181-202.

Douglas, M. (1975) *Implicit Meanings: Essays in Anthropology*, London: Routledge & Kegan Paul.

Dressel, H. (1899) *Corpus Inscriptionum Latinorum, XV, Pars 1*, Berlin: Societas Regia Scientiarum.

Drinkwater, J.F. (1983) *Roman Gaul: the Three Provinces, 58 BC-AD 260*, London: Croom Helm.

Dry, P.R. and Gregory, G.R. (1988) 'Grapevine varieties', in B.G. Coombe and P.R. Dry (eds) *Viticulture, Volume 1: Resources in Australia*, Adelaide: Australian Industrial Publishers, 119-38.

Duncan, J.S. (1985) 'Individual action and political power: a structuration perspective', in R.J. Johnston (ed.) *The Future of Geography*, London: Methuen, 174-89.

Duncan-Jones, R. (1974) *The Economy of the Roman Empire: Quantitative Studies*, Cambridge: Cambridge University Press.

Dunford, M. and Perrons, D. (1983) *The Arena of Capital*, London: Macmillan.

Durliat, J. (1968) 'La vigne et le vin dans la région parisienne au début du IX*ᵉ siècle d'après le Polyptyque d'Irminon', *Le Moyen Age* 74, 387-419.

Enjalbert, H. (1953) 'Comment naissent les grands crus: Bordeaux, Porto, Cognac', *Annales, Economies, Sociétés, Civilisations* 8, 315-28; 457-74.

Enjalbert, H. and Enjalbert, B. (1987) *L'Histoire de la Vigne et du Vin, avec une Nouvelle Hiérarchie des Terroirs du Bordelais et une Sélection de «100 Grands Crus»*, Paris: Bordas.

Epic of Gilgamesh (1960) trans. N.K. Sanders, Harmondsworth: Penguin.

Euripides (1960) *Bacchae* edited by E.R. Dodds, Oxford: Clarendon Press.

Evans, A. (1935) *The Palace of Minos, Volume IV*, London: Macmillan.

Evans, A. (1988) *The God of Ecstasy: Sex-Roles and the Madness of Dionysos*, New York: St Martin's Press.

Faith, N. (1980) *Château Margaux*, London: Christie's Wine Publications.

Faith, N. (1983) *Victorian Vineyard: Château Loudenne and the Gilbeys*, London: Constable.

Faith, N. (1986) *Cognac*, London: Hamish Hamilton.

Faith, N. (1988) *The Story of Champagne*, London: Hamish Hamilton.

Fatio, V. (1879) *Le Phylloxéra: Instructions Sommaires à l'Usage des Experts Cantonaux et Fédéraux en Suisse*, Genève: Imprimerie Ramboz et Schuchardt.

Fatio, V. and Demole-Ador (1875) *Rapports sur le Traitement des Vignes de Pregny (République et Canton de Genève) en Vue de la Destruction du Phylloxéra*, Genève: Imprimerie Ramboz et Schuchardt.

Faucon, L. (1874a) *Guérison des Vignes Phylloxérées*, Montpellier: Coulet.

Faucon, L. (1874b) *Mémoire sur la Maladie de la Vigne et sur son Traitement par le Procédé de la Submersion*, Paris: Imprimerie Nationale.

Fel, A. (1977) 'Petite culture 1750-1850', in H. Clout (ed.) *Themes in the Historical Geography of France*, London: Academic Press, 215-46.

Ferrill, A. (1986) *The Fall of the Roman Empire: the Military Explanation*, London: Thames and Hudson.

Finberg, H.P.R. (1964) *The Early Charters of Wessex*, Leicester: Leicester University Press.

Finley, M.I. (1985) *The Ancient Economy*, 2nd edn, London: The Hogarth Press.

Fisher, H.S.E. (1971) *The Portugal Trade: a Study of Anglo-Portuguese Commerce 1700-1770*, London: Macmillan.

Fisher, W.B. (1978) 'Wine: the geographical elements. Climate, soil and geology are the crucial catalysts', *Geographical Magazine* 51(2), 86.

Fitz-James, M.A.M.L., La Duchesse de (1889) *La Viticulture Franco-Américaine (1869-1889). Les Congrés Viticoles, - Excursions Viticoles en France et en Algérie, - la Viticulture au point de vue Financier, - la Bouture a un Oeil*, Paris: G. Masson.

Fizelière, A. de la (1866) *Vins à la Mode et Cabarets au XVIIᵉ Siècle*, Paris: René Pincebourde.

Foëx, G. (1882) *Instruction sur l'Emploi des Vignes Américaines à la Reconstitution du Vignoble de l'Hérault*, Montpellier: Boehm.

Folwell, R.J. and Baritelle, J.L. (1978) *The U.S. Wine Market*, Washington D.C.: US Department of Agriculture (Agricultural Economic Report No. 417)

Forbes, R.J. (1948) *A Short History of the Art of Distillation from the Beginnings up to the Death of Cellier Blumenthal*, Leiden: E.J. Brill (Reprinted 1970).

Forbes, R.J. (1956) 'Food and drink' in C. Singer, E.J. Holmyard, A.R. Hall and T.I. Williams (eds) *A History of Technology, Volume II, the Mediterranean Civilizations and the Middle Ages, c.700 B.C. to c.A.D. 1500*, Oxford: Clarendon Press, 103-46.

Ford, G. (1988) *The Benefits of Moderate Drinking: Alcohol, Health and Society*, San Francisco: The Wine Appreciation Guild.

Fornachon, J.C.M. (1953) *Studies on the Sherry Flor*, Adelaide: Australian Wine Board.

Fornachon, J.C.M. (1957) 'The occurrence of malo-lactic fermentation in Australian wines', *Australian Journal of Applied Science* 8, 120-9.

Forrester, J.J. (1845) *Observations on the Attempts Lately Made to Reform the Abuses Practised in Portugal, in the Making and Treating of Port Wine . . .*, Edinburgh: J. Menzies.

Forster, R. (1961) 'The noble wine producers of the Bordelais in the eighteenth century', *Economic History Review* 2nd series 14(1), 18-33.

Fortunatus (1961) *Venanti Honori Clementiani Forunati Presbyteri Italici Opera Poetica*, edited by F. Leo, new edn, Berlin: Weidmann (1st edn 1881).

Francis, A.D. (1966) *The Methuens and Portugal 1691-1708*, London: Cambridge University Press.

Francis, A.D. (1972) *The Wine Trade*, London: Adam & Charles Black.

Franck, W. (1824) *Traité sur les Vins du Médoc et les Autres Vins Rouges et Blancs du Département de la Gironde*, Bordeaux: Imprimerie de Laguillotière.

Frank, T. (1936) 'On the export tax of Spanish harbors', *American Journal of Philology* 57, 87–90.

Frank, T. (ed.) (1940) *Economic Survey of Ancient Rome, Volume 5: Rome and Italy of the Empire*, Baltimore: Johns Hopkins University Press.

Gadille, R. (1967) *Le Vignoble de la Côte Bourguignonne. Fondements Physiques et Humains d'un Viticulture de Haute Qualité*, Paris: Publications de l'Université de Dijon.

Galet, P. (1979) *A Practical Ampelography: Grape Vine Identification*, Ithaca: Cornell University Press.

Galletier, E. (ed.) (1952) *Panégyriques Latins, Tome II: les Panégyriques Constantiniens (VI-X)*, Paris: Société d'Edition «Les Belles Lettres».

Galtier, G. (1960) *Le Vignoble du Languedoc Méditérranéen et du Roussillon: Etude Comparative d'un Vignoble de Masse*, Montpellier: Cause Graille Castelnau.

Galtier, G. (1968) 'La bataille des vins d'Henri d'Andeli. Un document sur le vignoble et le commerce des vins dans la France médiévale', *Bulletin de la Société Languedocienne de Géographie* série 3 2(3), 5–41.

Gandilhon, R. (1968) *Naissance du Champagne: Dom Pierre Pérignon*, Paris: Hachette.

Garlan, Y. (1983) 'Greek amphorae and trade', in P. Garnsey, K. Hopkins and C.R. Whittaker (eds) *Trade in the Ancient Economy*, London: Chatto & Windus, 27–35.

Garnsey, P. (1988) *Famine and Food Supply in the Graeco-Roman World*, Cambridge: Cambridge University Press.

Garnsey, P. and Saller, R. (1987) *The Roman Empire: Economy, Society and Culture*, London: Duckworth.

Garnsey, P., Hopkins, K. and Whittaker, C.R. (eds) (1983) *Trade in the Ancient Economy*, London: Chatto and Windus.

de Genouillac, H. (1909) *Tablettes Sumériennes Archaïques: Matériaux pour Servir à l'Histoire de la Société Sumérienne, Publiés avec Introduction, Transcription et Tables*, Paris: H. de Genouillac.

George, R. (1984) *The Wines of Chablis and the Yonne*, London: Sotheby's Publications.

Gerhard, P. (1972) *A Guide to the Historical Geography of the New World*, Cambridge: Cambridge University Press.

Gervais, P. (1904) 'La crise phylloxérique et la viticulture européenne, *Revue de Viticulture* 22(2) 36–42.

Geuss, R. (1981) *The Idea of a Critical Theory: Habermas and the Frankfurt School*, Cambridge: Cambridge University Press.

Gibbon, E. (1776–88) *History of the Decline and Fall of the Roman Empire*, London: printed for W. Strahan and T. Cadell in the Strand.

Giddens, A. (1981) *A Contemporary Critique of Historical Materialism, Volume 1: Power, Property and the State*, London: Macmillan.

Giraldus Cambrensis (1951) *The First Version of the Topography of Ireland*, trans. J.J. O'Meara, Dundalk: Dundalgan Press.

Girouard, M. (1984) *Victorian Pubs*, New Haven: Yale University Press.

Glennie, P. (1987) 'The transition from feudalism to capitalism as a problem for historical geography', *Journal of Historical Geography* 13(3), 296–302.

Glubb, J.B. (1963) *The Great Arab Conquests*, London: Hodder and Stoughton.

Goldthwaite, R.A. (1987) 'The Medici bank and the world of Florentine capitalism', *Past and Present* 114, 3-31.

Gonçalves, F.E. (1984) *Portugal: a Wine Country*, Lisbon: Editora Portuguesa de Livros Técnicos e Científicos.

Goodenough, E.R. (1956) *Jewish Symbols in the Greco-Roman Period: Volumes 5 and 6*, New York: Pantheon.

Goor, A. (1966) 'The history of the grape vine in the Holy Land', *Economic Botany* 20, 46-64.

Gordon, C.H. (1971) *Forgotten Scripts: the Story of their Decipherment*, Harmondsworth: Penguin.

Gordon, M.M. González (1972) *Sherry: the Noble Wine*, London: Cassell.

Gracia, J.J.E. (1976) 'Rules and regulations for drinking wine in Francesc Eiximenis' "Terç del Crestià" (1384)', *Traditio* 32, 369-85.

Graells, M. de la P. (1881) *La Phylloxera Vastatrix. Memoria que Sobre la Historia Natural de este Insecto . . . .*, Madrid: Colegio Nacional de Sordo-Mudos y de Ciegos.

Granett, J., Bisabri-Ershadi, B. and Carey, J. (1983) 'Life tables of phylloxera on resistant and susceptible grape rootstocks', *Entomologicia Experimentalis et Applicata* 34, 13-19.

Grant, M. (1976) *The Fall of the Roman Empire*, Radnor: Annenburg.

Grant, M. and Ritson, B. (eds) (1983), *Alcohol: the Prevention Debate*, London: Croom Helm.

Gras, N.S.B. (1918) *The Early English Customs System: a Documentary Study of the Institutional and Economic History of the Customs from the Thirteenth to the Sixteenth Century*, Cambridge, Massachusetts: Harvard University Press.

Gray, H.L. (1933) 'English foreign trade from 1446 to 1482', in E. Power and M.M. Postan (eds) *Studies in English Trade in the Fifteenth Century*, London: George Routledge & Sons, 1-38.

Gregory of Tours (1927) *The History of the Franks*, trans. O.M. Dalton, Oxford: Clarendon Press.

Gregory, D. (1978) *Ideology, Science and Human Geography*, London: Hutchinson.

Gregory, G.R. (1988) 'Development and status of Australian viticulture', in B.G. Coombe and P.R. Dry (eds) *Viticulture, Volume 1: Resources in Australia*, Adelaide: Australian Industrial Publishers.

Griffin, J. (1985) *Latin Poets and Roman Life*, London: Duckworth.

Grove, J.M. (1988) *The Little Ice Age*, London: Methuen.

Guerrero, R. (1978) 'Notes sur un vignoble vieux de quatre siècles: le Chili Méditerranéen', in A. Huetz de Lemps (ed.) *Géographie Historique des Vignobles, Tome II*, Bordeaux: CNRS, 143-55.

Guirand, F. (1968a) 'Assyro-Babylonian mythology', in *New Larousse Encyclopedia of Mythology*, 2nd edn, London: Paul Hamlyn, 49-72.

Guirand, F. (1968b) 'Greek mythology', in *New Larousse Encyclopedia of Mythology*, 2nd edn, London: Paul Hamlyn, 85-198.

Gurney, O.R. (1964) *The Hittites*, rev. edn., Harmondsworth: Penguin.

Guyot, J. (1860) *Culture de la Vigne et Vinification*, Paris: Librairie Agricole de la Maison Rustique.

Haas, V. (1977) *Magie und Mythen im Reich der Hethiter: 1. Vegetationskulte und Pflanzenmagie*, Hamburg: Merlin Verlag.

Habermas, J. (1976) *Legitimation Crisis*, trans. T. McCarthy, London: Heinemann.

377

Habermas, J. (1978) *Knowledge and Human Interests*, 2nd edn, trans. J.J. Shapiro, London: Heinemann.

d'Haenens, A. (ed.) (1984) *Europe of the North Sea and the Baltic: the World of the Hanse*, Antwerp: Fonds Mercator.

Hahn, E. (1896) *Die Haustiere und ihre Beziehungen zur Wirtschaft des Menschen*, Leipzig: Duncker & Humblot.

Halász, Z. (1962) *Hungarian Wine through the Ages*, Budapest: Corvina Press.

Halkin, J. (1895) *Etude Historique sue la Culture de la Vigne en Belgique*, Liège: L. Grandmont-Donders.

Hall, H. (ed.) (1896) *The Red Book of the Exchequer III*, London: Stationery Office.

Hallgarten, S.F. (1951) *Rhineland Wineland: a Journey through the Wine Districts of Western Germany*, London: Paul Elek.

Hallgarten, F. (1987) *Wine Scandal*, London: Sphere.

Halliday, J. (1985) *The Australian Wine Compendium*, London: Angus and Robertson.

Haraszthy, A. (1862) *Grape Culture, Wines and Wine-Making. With Notes upon Agriculture and Horticulture*, New York: Harper & Brothers.

Hardie, W.J. and Cirami, R.M. (1988) 'Grapevine rootstocks', in B.G. Coombe and P.R. Dry (eds) *Viticulture, Volume 1: Resources in Australia*, Adelaide: Australian Industrial Publishers, 154-76.

Harding, D.W. (1974) *The Iron Age in Lowland Britain*, London: Routledge & Kegan Paul.

Harrison, B. (1971) *Drink and the Victorians: the Temperance Question in England, 1815-1872*, London: Faber & Faber.

Hautefontaine, P. (1979-80) '1729-1979, Ruinart Père et fils, la plus ancienne maison de Champagne', *La Revue Française de Généalogie* 5, 8-12.

Henry, T. (1986) *Harveys of Bristol*, London: Good Books.

Hermansen, G. (1981) *Ostia, Aspects of Roman City Life*, Edmonton, Alberta: University of Alberta Press.

Herodotus (1954) *The Histories*, trans. A. de Sélincourt, Harmondsworth: Penguin.

Hervey, J. (1894) *The Diary of John Hervey, First Earl of Bristol. With Extracts from his Book of Expenses 1688-1742*, Wells: Ernest Jackson.

Hesiod (1973) *Theogony, Works and Days. Theognis: Elegies*, trans. D. Wender, Harmondsworth: Penguin.

Hesnard, A. (1980) 'Un dépôt Augustéen d'amphores à la Longarina, Ostie', *Memoirs of the American Academy in Rome* 36, 141-56.

Heuzé, G. (1852) 'Maladie de la vigne: procédé Grison', *Revue Horticole* 168-70.

Hewitt, W.B. (1958) 'The probable home of Pierce's disease virus', *American Journal of Enology and Viticulture* 9, 94-8.

Higounet, C. (ed.) (1962-72) *Histoire de Bordeaux*, Bordeaux: Fédération Historique du Sud-Ouest.

Higounet, C. (1968) 'Cologne et Bordeaux: marchés du vin au Moyen Age', *Revue Historique de Bordeaux et du Département de la Gironde* nouvelle série 17(2), 65-80.

Higounet, C. (ed.) (1974) *La Seigneurie et le Vignoble de Château Latour: Histoire d'un Grand Cru du Médoc (XIV' - XX'; Siècle)*, Bordeaux: Fédération Historique du Sud-Ouest.

Hilton, R.H. (1965) 'Freedom and villeinage in England', *Past and Present* 31, 174-91.

Hilton, R.H. (1978) 'A crisis of feudalism', *Past and Present* 80, 3-19.

378

Hilton, R.H. (1982) 'Lords, burgesses and hucksters', *Past and Present* 97, 3-15.

Hobsbawm, E.J. (1975) *The Age of Capital 1848-1875*, London: Weidenfeld & Nicolson.

Hobsbawm, E.J. (1987) *The Age of Empire 1875-1914*, London: Weidenfeld & Nicolson.

Hoon, E.E. (1986) *The organisation of the English Customs System, 1696-1786*, new introduction by R.C. Jarvis, Newton Abbot: David & Charles.

Hopkins, K. (1978) *Conquerors and Slaves*, Cambridge: Cambridge University Press.

Horace (1987) *Horace: Satires and Epistles. Persius: Satires*, trans. N. Rudd, rev. edn, Harmondsworth: Penguin.

Horkheimer, M. (1976) 'Traditional and critical theory', in P. Connerton (ed.) *Critical Sociology*, Harmondsworth: Penguin, 206-24.

Howard, M.C. and King, J.E. (1985) *The Political Economy of Marx*, 2nd edn, London: Longman.

Howkins, B. (1982) *Rich, Rare and Red: the International Wine & Food Society's Guide to Port*, London: Heinemann.

Huetz de Lemps, A. (1968) 'Apogeo y decadencia de un viñedo de calidad: el de Ribadavia', *Anuario de Historia Economica y Social* 1, 207-25.

Huetz de Lemps, A. (ed.) (1978) *Géographie Historique des Vignobles*, Bordeaux: Centre National de la Recherche Scientifique.

Humoristes Associés (1980) *Le Vin*, France: Humoristes Associés.

Hutchinson, R.B. (1969) The California Wine Industry, unpublished PhD dissertation, University of California Los Angeles.

Hutchison, J.N. (1984) 'Northern California: from Haraszthy to the beginnings of Prohibition', in D. Muscatine, M.A. Amerine and B. Thompson (eds) *The Book of California Wine*, Berkeley, California: University of California Press, 30-49.

Hyams, E. (1987) *Dionysus: a Social History of the Wine Vine*, 2nd edn, London: Sidgwick & Jackson.

al-Idrisi (1836-40) *La Géographie d'Edrisi*, trans. P.-A. Jaubert, Paris: Société de Géographie.

I.N.R.A. (Institut National de la Recherche Agronomique) (1978) *Génétique et Amélioration de la Vigne*, Paris: INRA.

Instituicão da Companhia Geral da Agricultura dos Vinhos do Alto Douro (1756), Lisbon: Miguel Rodrigues.

Isaac, E. (1970) *Geography of Domestication*, Englewood Cliffs, New Jersey: Prentice-Hall.

Isnard, H. (1975) 'La viticulture algérienne, colonisation et décolonisation', *Méditerranée* 23(4), 3-10.

Jackson, D. and Schuster, D. (1981) *Grape Growing and Wine Making: a Handbook for Cool Climates*, Martinborough, New Zealand: Alister Taylor.

Jacob, F.C., Archer, T.E. and Castor, J.G.B. (1964) 'Thermal death time of yeast', *American Journal of Enology and Viticulture* 15, 69-74.

James, M.K. (1957) 'The medieval wine dealer', *Explorations in Entrepreneurial History* 10(2), 45-53.

James, M.K. (1971) *Studies in the Medieval Wine Trade*, edited by E.M. Veale, Oxford: Clarendon Press.

James, W. Bosville (1855) *Wine Duties Considered Financially and Socially. Being a Reply to Sir. James Emerson Tennent, on 'Wine: Its Taxation and Uses'*, London: Longman, Brown, Green & Longmans.

379

Jameson, M.H. (1977-8) 'Agriculture and slavery in Classical Athens', *Classical Journal* 73, 122-45.

Jeffs, J. (1982) *Sherry*, 3rd. edn, London: Faber & Faber.

Jekel, B. (1983a) 'Soil and the taste of wine', *Decanter* 8(5), 59.

Jekel, B. (1983b) 'Soil and wine - the debate continues', *Decanter* 8(10), 15.

Johnson, H. (1971) *The World Atlas of Wine*, London: Mitchell Beazley.

Johnson, H. (1989) *The Story of Wine*, London: Mitchell Beazley.

Johnson, H. (1994) *World Atlas of Wine*, London: Mitchell Beazley (4th edition).

Johnston, P.A. (1980) *Vergil's Agricultural Golden Age: a Study of the Georgics*, Leiden: E.J. Brill.

Jones, G. (1964) *The Norse Atlantic Saga, Being the Norse Voyages of Discovery and Settlement to Iceland, Greenland, America*, London: Oxford University Press.

Jones, P. (1966) 'Italy', in M.M. Postan (ed.) *The Cambridge Economic History of Europe, Volume 1: the Agrarian Life of the Middle Ages*, 2nd edn, Cambridge: Cambridge University Press, 340-431.

Jullien, A. (1816) *Topographie de tous les Vignobles Connus, Contenant: leur Position Topographique . . . . Suivie d'une Classification Générale des Vins*, Paris: the author; Madame Huzard; L. Colas.

Jungmann, J.A. (1959) *The Mass of the Roman Rite: its Origins and Development*, new and abridged edn, London: Burns & Oates.

Katzen, M.F. (1969) 'White settlers and the origins of a new society, 1652-1778', in M. Wilson and L. Thompson (eds) *The Oxford History of South Africa, Volume I, South Africa to 1870*, Oxford: Clarendon Press, 187-232.

Kelly, F. (1988) *A Guide to Early Irish Law*, Dublin: Dublin Institute for Advanced Studies.

Kennedy, W. (1913) *English Taxation, 1640-1799: an Essay on Policy and Opinion*, London: G. Bell & Sons.

Kerényi, C. (1962) *The Religion of the Greeks and Romans*, London: Thames & Hudson.

Kerényi, C. (1976) *Dionysos: Archetypical Image of Indestructible Life*, London: Routledge & Kegan Paul.

Kershaw, I. (1973) 'The great famine and agrarian crisis in England 1315-22', *Past and Present* 59, 3-50.

Key Note Reports (1988) *Wine: an Industry Sector Overview*, 6th edn, London: ICC Information Group.

Khayyam, O. (1981) *The Ruba'iyat of Omar Khayyam*, trans. P. Avery and J. Heath-Stubbs, Harmondsworth: Penguin.

Klauser, T. (1969) *A Short History of the Western Liturgy: an Account and Some Reflections*, London: Oxford University Press.

Kleberg, T. (1957) *Hôtels, Restaurants et Cabarets dans l'Antiquité Romaine: Etudes Historiques et Philologiques*, Uppsala: Almqvist & Wiksells Boktryckeri.

Kliewer, W.M. (1964) 'Influence of environment on metabolism of organic acids and carbohydrates in Vitis vinifera. I. Temperature', *Plant Physiology* 39(6), 869-80.

Kliewer, W.M. and Schultz, H.B. (1964) 'Influence of environment on metabolism of organic acids and carbohydrates in *Vitis vinifera*. II. Light', *American Journal of Enology and Viticulture* 15, 119-29.

Knowles, D. and Obolensky, D. (1969) *The Christian Centuries, Volume 2: the Middle Ages*, London: Darton, Longman & Todd.

Kosminsky, E.A. (1956) *Studies in the Agrarian History of England in the Thirteenth Century*, New York: Kelley & Millman.

Kozma, P. (ed.) (1975) *Le Contrôle de l'Alimentation des Plantes Cultivées*, Budapest: Akadémiai Kiadó.

Kramer, F.L. (1967) 'Eduard Hahn and the end of the "Three Stages of Man"', *The Geographical Review* 57(1) 73–89.

Krell, D. (ed.) *The California Missions: a Pictorial History*, Menlo Park, California: Lane Publishing Co.

Kuhnholtz-Lordat, G. (1963) *La Genèse des Appellations d'Origine des Vins*, Mâcon: Buguet-Comptour.

Kunkee, R.E. (1984) 'Selection and modification of yeasts and lactic acid bacteria for wine fermentation', *Food Microbiology* 1, 315-32.

Kunkee, R.E. and Goswell, R.W. (1977) 'Table wines', in A.H. Rose (ed.) *Alcoholic Beverages*, London: Academic Press, 315-86 (Economic Microbiology, volume 1).

Kyvig, D.E. (1979) *Repealing National Prohibition*, Chicago: Chicago University Press.

Lachiver, M. (1988) *Vins, Vignes et Vignerons: Histoire des Vignobles Français*, Paris: Fayard.

Laclau, E. (1979) *Politics and Ideology in Marxist Thought*, London: Verso.

Lacombe, R. (1977) 'Enquête sur l'utilisation des machines à vendanger en France de 1972 à 1977', *Le Progrès Agricole et Viticole* 22, 2-21.

Laffer, H.E. (1949) *The Wine Industry of Australia*, Adelaide: Australian Wine Board.

Laliman, L. (1860) *Coup d'Oeil Agricole et Social, Réformes Viticoles, Cépages Indigènes de l'Amérique*, Paris: Lacroix et Baudry.

Laliman, L. (1879) *Etudes sur les Divers Travaux Phylloxériques et les Vignes Americaines . . .*, Paris: Librairie Agricole.

Laliman, L. (1889) *Notice Chronologique sur l'Origine des Vignes Americaines Résistant au Phylloxéra . . .*, Paris: Masson.

Lamb, H. (1965) 'Britain's changing climate', in C.G. Johnson (ed.) *The Biological Significance of Climatic Change in Britain*, London: Academic Press, 3-31.

Lamb, H.H. (1982) *Climate, History and the Modern World*, London: Methuen.

Lang, M. (1959) 'The Palace of Nestor excavations of 1958, part II', *American Journal of Archaeology* 63, 128-37.

Langland, W. (1966) *Piers the Ploughman*, trans. with Introduction by J.F. Goodridge, Harmondsworth: Penguin.

Lattimore, R. (1951) 'Introduction', in *The Iliad of Homer*, Chicago: University of Chicago Press, 11-54.

Laube, J. (1988) 'Who owns the Napa Valley', *The Wine Spectator* 13(14), 36-45.

Laufer, B. (1919) *Sino-Iranica. Chinese Contributions to the History of Civilization in Ancient Iran*, Chicago: Field Museum of Natural History (Anthropological Series, Vol. 15, No. 3).

Lavalle, M.J. (1855) *Histoire et Statistique de la Vigne et des Grands Vins de la Côte-d'Or*, Paris: Dusacq.

Lawrence, R. de Treville, III (ed.) (1989) *Jefferson and Wine: Model of Moderation*, 2nd rev. edn, The Plains, Virginia: The Vinifera Wine Growers Association.

Leake, C.D. and Silverman, M. (1966) *Alcoholic Beverages in Clinical Medicine*, Chicago: Year Book Medical Publishers.

Lee, H.G. and Lee, A.E. (1987) *Virginia Wine Country*, White Hall, Virginia: Betterway Publications.

Leggett, H.B. (1941) *The Early History of Wine Production in California*, San Francisco: Wine Institute.

Leighly, J. (ed.) (1963) *Land and Life: a Selection from the Writings of Carl Ortwin Sauer*, Berkeley, California: California University Press.

Leipoldt, C.L. (1952) *300 Years of Cape Wine*, Cape Town: Stewart.

Le Roy Ladurie, E. (1972) *Times of Feast, Times of Famine: a History of Climate since the Year 1000*, trans. B. Bray, London: Allen & Unwin.

Le Roy Ladurie, E. and Baulant, M. (1980) 'Grape harvests from the 11th to the 19th century', *Journal of Interdisciplinary History* 10, 839–49.

Lerner, F. (1984) 'Hanseatic merchandise', in A. d'Haenens (ed.) *Europe of the North Sea and the Baltic: the World of the Hanse*, Antwerp: Fonds Mercator, 131–46.

Levadoux, L. (1956) 'Les populations sauvages et cultivés de *Vitis vinifera* L.', *Annales de l'Amélioration des Plantes* 1, 59–118.

Levick, B. (1982) 'Domitian and the provinces', *Latomus* 41, 50–73.

Levi-Strauss, C. (1966) *The Savage Mind*, London: Weidenfeld & Nicolson.

Lewis, A. (1984) 'The Hanse and England', in A. d'Haenens (ed.) *Europe of the North Sea and the Baltic: the World of the Hanse*, Antwerp: Fonds Mercator, 241–48.

Lewis, N. and Reinhold, D.M. (1951) *Roman Civilization. Sourcebook I: the Republic*, New York: Harper & Row.

Ley, C.D. (ed.) (1947) *Portuguese Voyages 1498–1663*, London: J.M. Dent & Sons.

Lichine, A. (1981) *Alexis Lichine's New Encyclopedia of Wines and Spirits*, London: Cassell.

Lillywhite, B. (1963) *London Coffee Houses: a Reference Book of Coffee Houses of the Seventeenth, Eighteenth and Nineteenth Centuries*, London: George Allen & Unwin.

Lindquist, S.-O. (ed.) (1985) *Society and Trade in the Baltic during the Viking Age*, Visby: Gotlands Fornsal.

Livermore, H.V. (1958) *A History of Spain*, London: George Allen & Unwin.

Livermore, H.V. (1971) *The Origins of Spain and Portugal*, London: George Allen & Unwin.

Livermore, H.V. (1973) *Portugal: a Short History*, Edinburgh: Edinburgh University Press.

Livingstone-Learmonth, J. and Master, M.C.H. (1983) *The Wines of the Rhône*, 2nd edn, London: Faber & Faber.

Lloyd, S. (1978) *The Archaeology of Mesopotamia: from the Old Stone Age to the Persian Conquest*, London: Thames & Hudson.

Lloyd, S. and Mellaart, J. (1958) 'Beycesultan excavations: fourth preliminary report', *Anatolian Studies* 8, 93–125.

Lloyd, S. and Mellaart, J. (1962–72) *Beycesultan*, London: British Institute of Archaeology at Ankora.

Lloyd, T.H. (1982) *Alien Merchants in England in the High Middle Ages*, Brighton: Harvester Press.

Loftus, S. (1985) *Anatomy of the Wine Trade: Abe's Sardines and Other Stories*, London: Sidgwick & Jackson.

Lopez, R.S. (1987) 'The trade of medieval Europe: the south', in M.M. Postan and E. Miller (eds) *The Cambridge Economic History of Europe, Volume II: Trade and Industry in the Middle Ages*, Cambridge: Cambridge University Press, 306–401.

Lorimer, J.G. (1908–15) *Gazetteer of the Persian Gulf, 'Oman and Central Arabia*, Calcutta: Government Printing House (reprinted Farnborough: Gregg, 1970).

Loubère, L.A. (1978a) 'Géographie du commerce du vin: la France et l'Italie pendant la crise du phylloxéra, in A. Huetz de Lemps (ed.) *Géographie Historique des Vignobles, Tome II*, Bordeaux: CNRS, 91–4.

Loubère, L.A. (1978b) *The Red and the White: a History of Wine in France and Italy in the Nineteenth Century*, Albany: State University of New York Press.

Lowe, J.J. and Walker, M.J.C. (1984) *Reconstructing Quaternary Environments*, London: Longman.

Lucia, S.P. (1963) *A History of Wine as Therapy*, Philadelphia: J.B. Lippincott.

Lutz, H.F. (1922) *Viticulture and Brewing in the Ancient Orient*, Leipzig: J.C. Hinrichs'sche Buchhandlung.

Luxembourg, R. (1972) *The Accumulation of Capital*, London: Allen Lane.

MAFF (Ministry of Agriculture, Fisheries and Food) (1980) *Grapes for Wine*, London: HMSO.

McCarthy, R.G. (ed.) (1959) *Drinking and Intoxication: Selected Readings in Social Attitudes and Controls*, New Haven, Connecticut: Yale Centre of Alcohol Studies.

McCarthy, R.G. and Douglas, E.M. (1949) *Alcohol and Social Responsiblity: a New Educational Approach*, New York: Thomas Y. Crowell.

McGhee, R. (1985) 'Archaeological evidence for climatic change during the last 5000 years', in T.M.L. Wigley, M.J. Ingram and G. Farmer (eds) *Climate and History: Studies in Past Climates and their Impact on Man*, Cambridge: Cambridge University Press, 162–79.

Macintyre, S. and Tribe, K. (1975) *Althusser and Marxist Theory*, 2nd edn, Cambridge: the authors.

McKee, I. (1947) 'The beginnings of California winegrowing', *Historical Society of Southern California Quarterly* 29, 59–71.

Macqueen, J.G. (1986) *The Hittites and their Contemporaries in Asia Minor*, rev. edn, London: Thames & Hudson.

Maddicott, J.R. (1975) *The English Peasantry and the Demands of the Crown 1294–1341*, Oxford; Past and Present Society.

Magnusson, M. and Pálsson, H. (eds.) (1965) *The Vinland Sagas: the Norse Discoveries of America*, Harmondsworth: Penguin.

Mahaffy, J.P. (1890) 'The work of Mago on agriculture', *Hermathena* 7, 29–35.

Mai, D.H. (1987) 'Neue Früchte und Samen aus paläozänen Ablagerungen Mitteleuropas', *Feddes Repertorium* 98(3-4), 197–229.

Mandel, E. (1976) 'Introduction', in K. Marx, *Capital Volume 1*, trans. B. Fowkes, Harmondsworth: Penguin, 11–86.

Manniche, L. (1987) *Sexual Life in Ancient Egypt*, London: KPI.

Marescalchi, A. and Dalmasso, G. (eds.) (1931-7) *Storia della Vite e del Vino in Italia*, Milano: Gualdoni.

Marques, A.H. de Oliveira (1972), *History of Portugal*, New York: Columbia Univeristy Press.

Marquette, J. (1978) 'La vinification dans les domaines de l'archevêque de Bordeaux à la fin du Moyen Age', in A. Huetz de Lemps (ed.) *Géographie Historique des Vignobles, Tome I*, Bordeaux: CNRS, 123–47.

Marx, K. (1971) *Critique of Political Economy*, London: Lawrence & Wishart (First published 1859).

Marx, K. (1976) *Capital, Volume 1*, trans. B. Fowkes, Harmondsworth: Penguin (First published 1867).

Marx, K. (1981) *Capital: a Critique of Political Economy, Volume Three*, Harmondsworth: Penguin (First published 1894).

Masson-Oursel, P. and Morin, L. (1968) 'Indian mythology', in *New Larousse Encyclopedia of Mythology*, 2nd edn, London: Paul Hamlyn, 325-78.

Maxwell, H. (1907) *Half-a-Century of Successful Trade: Being a Sketch of the Rise and Development of the Business of W. & A. Gilbey 1857-1907*, London: printed at the Pantheon Press for W. & A. Gilbey.

Mayerson, P. (1985) 'The wine and vineyards of Gaza in the Byzantine period', *Bulletin, American Schools of Oriental Research*, 257, 75-80.

Mayo, O. (1986) *The Wines of Australia*, London: Faber & Faber.

Melelli, A. and Perari, R. (1978) 'Le développement du vignoble spécialisé dans la région ombrienne', *Geographia Polonica* 38, 193-205.

Minst, K.J. (1966-70) *Lorscher Codex. Urkundenbuch der ehemaligen Fürstabtei Lorsch*, Lorsch: Verlag Laurissa.

Mócsy, A. (1974) *Pannonia and Upper Moesia: a History of the Middle Danube Provinces of the Roman Empire*, London: Routledge & Kegan Paul.

Morris, J. (1988) *The White Wines of Burgundy*, London: Octopus.

Motolinía, Fray Toribio (1950) *Motolinía's History of the Indians of New Spain*, trans. and ed. E.A. Foster, Berkeley, California: Cortés Society.

Mouillefert, P. (1876) *Le Phylloxéra, Moyens Proposées pour le Combattre: Etat Actuel de la Question*, Paris: G. Masson.

Moulton, K.S. (1975) 'The California Wine Industry and its Economic Relationships', unpublished statement prepared for the hearing on amendments to the California Marketing Order for Wine, San Francisco, California, May 23, 1975.

Moulton, K.S. (ed.) (1981) *The Economics of Small Wineries*, Berkeley, California: Cooperative Extension, University of California.

Moulton, K.S. (1984) 'The economics of wine in California', in D. Muscatine, M.A. Amerine and B. Thompson (eds) *The Book of California Wine*, Berkeley, California: University of California Press, 380-405.

Murray, M.A. (1963) *The Splendour that was Egypt*, London: Sidgwick & Jackson.

Naranjo, Z. (1978) 'Observations sur la crise du phylloxéra et ses conséquences dans le vignoble de Xérés', in A. Huetz de Lemps (ed.) *Géographie Historique des Vignobles, Tome II*, Bordeaux: CNRS, 63-76.

Negrul, A.M. (1938) 'Evolution of cultivated forms of grapes', *Comtes Rendus (Doklady) de l'Académie des Sciences de l'U.R.S.S.* 18(8), 585-8.

Negrul, A.M. (1960) 'Arkheologicheskie nakhodi semian vinograd', *Sovetskaya Arkheologiya* 1, 111-19.

Neitzert, D. von (1987) 'Göttingens Wirtschaft, an Beispielen des 15. und 16. Jahrhunderts', in D. Denecke and H.-M. Kühn (eds) *Göttingen: Geschichte einer Universitätsstadt, Band 1, von den Anfängen bis zum Ende des Dreißigjährigen Krieges*, Göttingen: Vandenhoeck & Ruprecht, 298-345.

New Catholic Encyclopedia (1967), New York: McGraw Hill.

Newman, P.C. (1978) *Bronfman Dynasty: the Rothschilds of the New World*, Toronto: McClelland & Stewart.

Niederbacher, A. (1988) *Wine in the European Community*, 2nd edn, Luxembourg: Office for Official Publications of the European Communities.

Nilsson, M.P. (1940) *Greek Popular Religion*, New York: Columbia University Press.

Nilsson, M.P. (1975) *The Dionysiac Mysteries of the Hellenistic and Roman Age*, New York: Arno Press, 1975.

North, D.C. and Thomas, R.P. (1973) *The Rise of the Western World: a New Economic History*, Cambridge: Cambridge University Press.

Núñez, D.R. and Walker, M.J. (1989) 'A review of palaeobotanical findings of early *Vitis* in the Mediterranean and of the origins of cultivated grape-vines, with special reference to prehistoric exploitation in the western Mediterranean', *Review of Palaeobotany and Palynology* 61, 205–37.

Olken, C.E., Singer, E.G. and Roby, N.S. (1982) *The Connoisseur's Book of California Wines*, New York: Alfred A. Knopf.

Ordenanzas con que se ha de Gobernar y Guardar la Entrada de Vino y Venta del en la Cuidad de Valladolid (1975), Valladolid: Editorial Sever Cuesta (Reproduction of manuscript in Universidad de Valladolid, Sección de Santa Cruz, 12462).

Ordish, F.G. (1951) 'A hundred years of lime-sulphur', *Agriculture*, 111–15.

Ordish, G. (1953) *Wine Growing in England*, London: Rupert Hart-Davies.

Ordish, G. (1987) *The Great Wine Blight*, 2nd edn, London: Sidgwick & Jackson.

Orffer, C.J. (ed.) (1979) *Wine Grape Cultivars in South Africa*, Cape Town: Human & Rousseau.

Ortiz, A.D. (1971) *The Golden Age of Spain*, trans. J. Casey, London: Weidenfeld & Nicolson.

Osborne, R. (1987) *Classical Landscape with Figures: the Ancient Greek City and its Countryside*, London: George Philip.

Palmer, R.E.A. (1980) 'Customs on market goods imported into the city of Rome', *Memoirs of the American Academy in Rome* 36, 217–33.

Panella, C. (1981) 'La distribuzione e i mercati', in A. Giardina and A. Schiavone (eds.) *Società Romana e Produzione Schiavistica, Vol. II: Merci, Mercati e Scambi nel Mediterraneo*, Roma-Bari: La Terza, 55–80.

Parker, T. (1979) *Inglenook Vineyards: 100 Years of Fine Winemaking*, Rutherford, California: Inglenook Vineyards.

Parnell, C. (1989) 'Takeover trail', in C. Parnell (ed.) *The Fine Wines of Bordeaux*, London: Decanter, 3–6.

Pasteur, L. (1866) *Etudes sur le Vin, ses Maladies, Causes qui les Provoquent, Procédés Nouveaux pour le Conserver et pour le Vieiller*, Paris: Imprimerie Impériale (2nd edn, Paris: F. Savy, 1875).

Patrick, C.H. (1952) *Alcohol, Culture and Society*, Durham, North Carolina: Duke University Press.

Peacock, D.P.S. and Williams, D.F. (1986) *Amphorae and the Roman Economy: an Introductory Guide*, London: Longman.

Penanrun, De (1868) 'Sur la nouvelle maladie de la vigne', *Bulletin de la Société d'Agriculture de Vaucluse*, 7 July 1868, 258–62.

Penning-Rowsell, E. (1985) *The Wines of Bordeaux*, 5th edn, Harmondsworth: Penguin.

Peppercorn, D. (1982) *Bordeaux*, London: Faber & Faber.

Pepys, S. (1971) *The Diary of Samuel Pepys, Volume IV, 1663*, edited by R. Latham and W. Matthews, London: G. Bell & Sons.

Pepys, S. (1972) *The Diary of Samuel Pepys, Volume VI, 1665*, edited by R. Latham and W. Matthews, London: G. Bell & Sons.

Periplus of the Erythraean Sea (1912) *The Periplus of the Erythræn Sea: Travel and Trade in the Indian Ocean by a Merchant of the First Century*, trans. W.H. Schoff, London: Longmans, Green & Co.

Peters, G.L. (1989) *Wines and Vines of California*, Belmont, California: Star Publishing.

Petronius (1986) *Petronius: the Satyricon. Seneca: the Apolocyntosis*, trans. J.P. Sullivan, rev. edn, Harmondsworth: Penguin.

Peynaud, E. (1987) *The Taste of Wine: the Art and Science of Wine Appreciation*, London: Macdonald.

Pijassou, R. (1974) 'Le marché de Londres et la naissance des grands crus médocains (fin XVIIe siècle-début XVIIIe siècle)', *Revue Historique de Bordeaux* 139-50.

Pijassou, R. (1980) *Un Grand Vignoble de Qualité: le Médoc*, Paris: Librairie Jules Tallandier.

Pinard, J. (1987) 'Les nouvelles activités manufacturières dans les campagnes du centre-ouest de la France du XVIe au XVIIIe siècle' in H.-J. Nitz (ed.) *The Medieval and Early-Modern Rural Landscape of Europe under the Impact of the Commercial Economy*, Göttingen: Department of Geography, University of Göttingen, 295-306.

Pinney, T. (1984a) 'The early days in southern California', in D. Muscatine, M.A. Amerine and B. Thompson (eds) *The Book of California Wine*, Berkeley, California: University of California Press, 2-9.

Pinney, T. (1984b) 'The Gold Rush era', in D. Muscatine, M.A. Amerine and B. Thompson (eds) *The Book of California Wine*, Berkeley, California: University of California Press, 16-21.

Pinney, T. (1989) *A History of Wine in America: from the Beginnings to Prohibition*, Berkeley, California: University of California Press.

Pirenne, H. (1933) 'Un grand commerce d'exportation au Moyen Age: les vins de France', *Annales d'Histoire Economique et Sociale* 5, 225-43.

Planchon, J.-E. (1877) 'La question du phylloxéra en 1876', *Revue des Deux Mondes* 28, 241-77.

de Planhol, X. (1977) 'Le vin de l'Afghanistan et de l'Himalaya Occidental', *Revue Géographie de l'Est* 17, 3-26.

Plato (1951) *The Symposium*, trans. W. Hamilton, Harmondsworth: Penguin.

Pliny, The Elder (1945-1950) *Natural History*, trans. H. Rackham, London: Heinemann.

Pomerol, C. (ed.) (1989) *The Wines and Winelands of France: Geological Journeys*, London: Robertson McCarta.

Porton, G.G. (1976) 'The grape-cluster in Jewish literature and art of late antiquity', *Journal of Jewish Studies* 27, 159-76.

Postan, M.M. (ed.) (1966) *The Cambridge Economic History of Europe, Volume I: the Agrarian Life of the Middle Ages*, 2nd edn, Cambridge: Cambridge University Press.

Postan, M.M. (1987) 'The trade of medieval Europe: the north' in M.M. Postan and E. Miller (eds.) *The Cambridge Economic History of Europe, Volume II: Trade and Industry in the Middle Ages*, 2nd edn, Cambridge: Cambridge University Press, 168-305.

Postan, M.M. and Hatcher, J. (1978) 'Population and class relations in feudal society', *Past and Present* 78, 24-37.

Postan, M.M. and Miller, E. (eds.) (1987) *The Cambridge Economic History of Europe, Volume II: Trade and Industry in the Middle Ages*, 2nd edn, Cambridge: Cambridge University Press.

Pounds, N.J.G. (1973) *An Historical Geography of Europe 450 B.C.-A.D. 1330*, Cambridge: Cambridge University Press.

Pounds, N.J.G. (1979) *An Historical Geography of Europe, 1500-1840*, Cambridge: Cambridge University Press.

Prats, B. (1983) 'The terroir is important', *Decanter* 8(7), 16.

Prescott, W.H. (1936) *History of the Conquest of Mexico and History of the Conquest of Peru*, New York: The Modern Library.

Prisnea, C. (1964) *Bacchus in Rumania*, Bucharest: Meridiane Publishing Ho.

Purcell, N. (1985) 'Wine and wealth in ancient Italy', *Journal of Roman Studies* 75, 1-19.

Putnam, M.C.J. (1979) *Virgil's Poem of the Earth: Studies on the Georgics*, Princeton: Princeton University Press.

Ramishvili, R. (1983) 'New archaeological evidence on the history of viniculture in Georgia', *Matsne* (Tbilisi) 2, 127-40.

Ramsden, E. (1940) 'James Busby: the prophet of Australian viticulture', *Royal Australian Historical Society Journal and Proceedings* 26(5), 361-86.

Randolph, T.J. (ed.) (1829) *Memoir, Correspondence and Miscellanies from the papers of Thomas Jefferson, Volume II*, Charlottesville: F. Carr, & Co.

Rankine, B.C., Fornachon, J.C.M., Boehm, E.W. and Cellier, K.M. (1971) 'Influence of grape variety, climate and soil on grape composition and on the composition and quality of table wines', *Vitis* 10, 33-50.

Rathbone, D.W. (1983) 'The slave mode of production in Italy: review article', *Journal of Roman Studies* 73, 160-8.

Ray, C. (1984) *Robert Mondavi of the Napa Valley*, Novato, California: Presidio Press.

Read, J. (1982) *The Wines of Portugal*, London: Faber & Faber.

Read, J. (1986) *The Wines of Spain*, London: Faber & Faber.

Redding, C. (1833), *A History and Description of Modern Wines*, London: Whittaker, Teacher and Arnot.

Renouard, Y. (1970) 'The wine trade of Gascony in the Middle Ages', in R. Cameron (ed.), *Essays in French Economic History*, Homewood, Illinois: Richard Irwin Inc., 64-90.

Renoy, G. (1985) *Les Mémoires du Bourgogne*, Paris: BAV Edition.

Ribéreau-Gayon, P. and Sudraud, P. (1957), 'Les anthocyannes de la baie dans le genre *Vitis*', *Comptes Rendus* 244, 233-5.

Richard, J. (1978) 'Aspects historiques de l'évolution du vignoble bourguignon', in A. Huetz de Lemps (ed.), *Géographie Historique des Vignobles, Tome I*, Bordeaux: CNRS, 187-96.

Richter, G. (1980) 'Three years of plot measurements in vineyards of the Moselle-Region - some preliminary results', *Zeitschrift für Geomorphologie N.F. Supplementband* 35, 81-91.

Rickard, J. (1988) *Australia: a Cultural History*, London: Longman.

Robertson, G. (1982) *Port*, rev. edn, London: Faber & Faber.

Robinson, J. (1986) *Vines, Grapes and Wines*, London: Mitchell Beazley.

Robinson, J. (1988) *On the Demon Drink*, London: Mitchell Beazley.

Rodier, C. (1920) *Le Vin de Bourgogne (la Côte d'Or)*, Dijon: Louis Damidot.

Rohr, J.B. von (1730) *Viticultura Germaniae Oeconomica, oder Haußwirthliche auf Teutfchland gerichtete Nachricht von dem Wein=bau, . . .*, Leipzig: Bey Joh. Friedrich Brauns fel, Erben.

Rose, A.H. (ed.) (1977) *Alcoholic Beverages*, London: Academic Press (Economic Microbiology, Volume 1).

Roseira, M.J.Q. (1973) 'A região do Vinho do Porto', *Finisterra* 81, 116-29.

Rossiter, J.J. (1981) 'Wine and oil processing at Roman farms in Italy', *Phoenix* 35, 345-61.

Rossiter J.J. and Haldenby, E. (1989) 'A wine-making plant in Pompeii Insula II.5', *Echos du Monde Classique/Classical Views* 33(8), 229-39.

Roudié, P. (1985) 'Long-distance emigration from the port of Bordeaux', *Journal of Historical Geography* 11(3), 268-79.

Roudié, P. (1988) *Vignobles et Vignerons du Bordelais (1850-1950)*, Paris: CNRS.

Round, J.H. (1900) 'Essex vineyards', *Transactions Essex Archaeological Society* new series 7, 249-51.

Royal College of General Practitioners (1986) *Alcohol - a Balanced View*, London: Royal College of General Practitioners.

Rubio, M.L. (1978) 'Note sur la crise de phylloxéra dans la Province de Cordoue', in A. Huetz de Lemps (ed.) *Géographie Historique des Vignobles, Tome II*, Bordeaux: CNRS, 55-62.

Ruck, C.A.P. (1982) 'The wild and the cultivated: wine in Euripides' *Bacchae*', *Journal of Ethnopharmacology* 5, 231-70.

Ruddock, A.A. (1951) *Italian Merchants and Shipping in Southampton 1270-1600*, Southampton: University College.

Sabellico, A. (1986) *Note Pratiche di Legislazione Vinicola*, 3rd rev. edn, Milan: Associazone Enotecnici Italiani.

Sanceau, E. (1970) *The British Factory Oporto*, Oporto: British Association.

Sawyer, P.H. (ed.) (1968) *Anglo-Saxon Charters: an Annotated List and Bibliography*, London: Royal Historical Society.

Schaff, P. and Wace, H. (eds) (1954) *A Select Library of Nicene and Post-Nicene Fathers of the Christian Church, Second Series, Volume VI, St. Jerome: Letters and Select Works*, Grand Rapids, Michigan: Wm. B. Eerdmans Publishing.

Schildhauer, J. (1985) *The Hansa: History and Culture*, Leipzig: Edition Leipzig.

Schoenman, T. (ed.) (1979) *The Father of California Wine*, Santa Barbara, California: Capra Press.

Schreiber, G. (1980) *Deutsche Weingeschichte. Der Wein in Volksleben, Kult und Wirtschaft*, Köln: Rheinland-Verlag.

Schröder, K.H. (1978) 'L'ancienne extension de la viticulture dans le nord-est de l'Europe centrale', in A. Huetz de Lemps (ed.) *Géographie Historique des Vignobles, Tome II*, Bordeaux: CNRS, 15-21.

Seguin, G. (1965) 'Etude de quelques profils de sol du vignoble Bordelais', unpublished thesis, Thèse de Doctorat de $3°$ cycle, University of Bordeaux.

Seguin, G. (1970) 'Les sols de vignobles du Haut-Médoc: influence sur l'alimentation en eau de la vigne et sur la maturation du raisin', unpublished thesis, Thèse de Docteur Es Sciences Naturelles, University of Bordeaux.

Sellers, C. (1899) *Oporto, Old and New. Being a Historical Record of the Port Wine Trade, and a Tribute to British Commercial Enterprise in the North of Portugal*, London: Herbert E. Harper.

Seltman, C. (1957) *Wine in the Ancient World*, London: Routledge & Kegan Paul.

Settis, S. (ed.) (1985) *The Land of the Etruscans*, Firenze: Scala.

Seward, D. (1979) *Monks and Wine*, London: Mitchell Beazley.

Sharpe, R.R. (1894-1912) *Calendar of Letter Books Preserved among the Archives of the Corporation of the City of London, Letter Book I*, London: J.E. Francis.

Shaw, T.G. (1854) *Wine, in Relation to Temperance, Trade and Revenue*, London: Gilbert Bros.

Shrewsbury, J.F.D. (1971) *A History of Bubonic Plague in the British Isles*, Cambridge: Cambridge University Press.

Sichel, P.M.F. (1983) *The Wines of Germany*, London: Faber & Faber (rev. edn of F. Schoonmaker (1956) *The Wines of Germany*)

Sidonius (1915) *The Letters of Sidonius*, trans. O.M. Dalton, Oxford; Clarendon Press.

Simon, A.L. (1906-9) *History of the Wine Trade*, London: Wyman & Sons.

Simon, A.L. (1926) *Bottlescrew Days: Wine Drinking in England during the Eighteenth Century*, London: Duckworth.

Simon, A.L. (1934) *Port*, London: Constable.

Simon, A.L. (1962) *The History of Champagne*, London: Ebury Press.

Sittler, L. (1949) 'Le commerce du vin de Colmar, jusqu'en 1789', *Revue d'Alsace* 89, 37-56.

Sivéry, G. (1969) *Les Comtes de Hainauit et le Commerce du Vin du XIVe siècle et au début du XVe siècle*, Lille: Centre Régional d'Etudes Historiques de l'Université de Lille.

Smith, C. Delano (1979) *Western Mediterranean Europe: a Historical Geography of Italy, Spain and Southern France since the Neolithic*, London: Academic Press.

Smith, R.S. (1966) 'Spain', in M.M. Postan (ed.) *The Cambridge Economic History of Europe, Volume I: the Agrarian Life of the Middle Ages*, 2nd ed, Cambridge: Cambridge University Press, 432-8.

Sorenson, L. and Berger, H. (no date) *Beringer: a Napa Valley Legend*, St. Helena, California: Silverado Publishing.

Sòriga, R. (1933) 'La vite e il vino nella letteratura e nelle figurazioni Italiane del medioevo' in A. Marescalchi and G. Dalmasso (eds) *Storia della Vite e del Vino in Italia*, Milano: Gualdoni, 127-89.

Spufford, P. (1988) *Money and its Uses in Medieval Europe*, Cambridge: Cambridge University Press.

Spurr, M.S. (1986) *Arable Cultivation in Roman Italy, c.200 B.C.-c A.D. 100*, London: The Society for the Promotion of Roman Studies (Journal of Roman Studies Monograph, No.3).

Spurrier, S. and Dovaz, M. (1983) *Academie du Vin Wine Course*, London: Century.

Spurrier, S. and Ward, J. (1986) *How to Buy Fine Wines: Practical Advice for the Investor and Connoisseur*, Lexington, Massachusetts: Stephen Greene Press.

Stanislawski, D. (1970) *Landscapes of Bacchus: the Vine in Portugal*, Austin, Texas: University of Texas Press.

Stanislawski, D. (1973) 'Dark Age contributions to the Mediterranean way of life', *Annals, Association American Geographers* 63, 397-410.

Stanislawski, D. (1975) 'Dionysus westward: early religion and the economic geography of wine', *The Geographical Review* 65, 427-44.

Starr, C.G. (1982) *The Roman Empire, 27 B.C.-A.D. 476: a Study in Survival*, Oxford: Oxford University Press.

Stead, M.J. (1984) 'Winemaking in Ancient Egypt', *Decanter* 9(11), 82-5.

Steckley, G.F. (1980) 'The wine economy of Tenerife in the seventeenth century: Anglo-Spanish partnership in a luxury trade', *Economic History Review* 2nd series 33, 335-50.

Stevenson, A.C. (1985) 'Studies in the vegetational history of S.W. Spain. II. Palynological investigations at Laguna de las Madres, S.W. Spain', *Journal of Biogeography* 12, 293-314.

Stevenson, T. (1988) *Sotheby's World Wine Encyclopedia*, London: Dorling Kindersley.

Stevenson, W.I. (1978) 'La vigne américaine, son rayonnement et importance dans la viticulture héraultaise au XIXe siècle', in *Economie et Société en Languedoc-Roussillon de 1789 à nos Jours*, Montpellier: Centre d'Histoire Contemporaine du Languedoc Méditerranéen et du Roussillon, 69-83.

Stevenson, W.I. (1980) 'The diffusion of disaster: the phylloxera outbreak in the *département* of the Hérault, 1862-80', *Journal of Historical Geography* 6(1), 47-63.

Stevenson, W.I. (1981) 'Viticulture and society in the Hérault (France) during the phylloxera crisis, 1862-1907', unpublished PhD thesis, University of London.

Strabo (1949–54) *The Geography of Strabo*, trans. H.L. Jones, London: William Heinemann.

Stuller, J. and Martin, G. (1989) *Through the Grapevine*, New York: Wynwood Press.

Suetonius (1930) *C. Suetoni Tranquilli: de Vita Caesarum, Libri VII-VIII*, trans. G.W. Mooney, London: Longmans, Green & Co.

Sutcliffe, D. (1934) 'The vineyards of Northfleet and Teynham in the thirteenth century', *Archaeologia Cantiana* 46, 140–49.

Sutcliffe, S. (1981) 'Viticulture, vinification and the care of wine', in S. Sutcliffe (ed.) *André Simon's Wines of the World*, 2nd edn, London: Macdonald Futura, 12–21.

Szakály, F. (1979) 'A közép-Duna menti bortermelés fénykora (a XVI. század derekán)', *Dunataj* 2, 12–24.

Tacitus, Cornelius (1970) *The Agricola and the Germania*, trans. H. Mattingly, rev. S.A. Handford, Harmondsworth: Penguin.

Taylor, M. and Thrift, N. (1982) *The Geography of Multinationals*, London: Croom Helm.

Tchernia, A. (1980) 'Quelques remarques sur le commerce du vin et les amphores', *Memoirs of the American Academy in Rome* 36, 305–12.

Tchernia, A. (1983) 'Italian wine in Gaul at the end of the Republic', in P. Garnsey, K. Hopkins and C.R. Whittaker (eds), *Trade in the Ancient Economy*, London: Chatto & Windus, 87–104.

Tchernia, A. (1986) *Le Vin de l'Italie Romaine: Essai d'Histoire Economique d'après les Amphores*, Rome: Ecole Française de Rome.

Teiser, R. and Harroun, C. (1983) *Winemaking in California*, New York: McGraw Hill.

Teiser, R. and Harroun, C. (1984) 'The Volstead Act, rebirth and boom', in D. Muscatine, M.A. Amerine, and B. Thompson (eds) *The Book of California Wine*, Berkeley, California: University of California Press, 50–81.

Teisseyre, M.C. (1978) 'Le rôle des sociétés savantes bordelaises au moment de la crise du phylloxéra', in A. Huetz de Lemps (ed.) *Géographie Historique des Vignobles, Tome I*, Bordeaux: CNRS, 201–9.

Tennent, J.E. (1855) *Wine, its Use and Taxation. An Inquiry into the Operation of the Wine Duties on Consumption and Revenue*, London: James Madden.

Theophrastus (1916) *Enquiry into Plants*, trans. A.F. Hort, London: William Heinemann.

Thevenot, E. (1952) 'Les origines du vignoble bourguignon: les conditions de son établissement', *Annales de Bourgogne* 34, 245–57.

Thirsk, J. and Cooper, J.P. (eds) (1972) *Seventeenth Century Economic Documents*, Oxford: Clarendon Press.

Thorpe, B. (ed.) (1840) *Ancient Laws and Institutions of England*, London: Record Commission.

Thorpy, F. (1983) *Wine in New Zealand*, Auckland: Penguin.

de Thureau-Dangin, F., Genouillac, H. and Delaporte, L. (1910–14) *Inventaire des Tablettes de Tello Conservées au Musée Impérial Ottoman*, Paris: Mission Française de Chaldée.

Tiffney, B. (1979) 'Nomenclature revision', *Review of Palaeobotany and Palynology* 27, 91–2.

Tiffney, B. and Barghoorn, E.S. (1976) 'Fruits and seeds of the Brandon Lignite. 1. Vitaceae', *Review of Palaeobotany and Palynology* 22, 169–91.

Tigay, J.H. (1982) *The Evolution of the Gilgamesh Epic*, Philadelphia: University of Pennsylvania Press.

Troutman, A. (1987) 'Winemaking more than romance and excitement', reprint from *Los Angeles Times* 10 May 1987.

Turner, E.G. (1958) 'A Ptolemaic vineyard lease', *Bulletin of the John Rylands Library* 31(1), 148-61.

Turrill, W.B. (1959) *The Royal Botanic Gardens, Kew: Past and Present*, London: Herbert Jenkins.

Unwin, P.T.H. (1981) 'Rural marketing in medieval Nottinghamshire', *Journal of Historical Geography* 7(3), 231-51.

Unwin, P.T.H. (1990) 'Towns and trade: 1066-1500', in R.A. Dodgshon and R.A. Butlin (eds), *An Historical Geography of England and Wales*, 2nd edn, London: Academic Press, 123-49.

Vaizey, J.E. (1960) *The Brewing Industry, 1886-1951: an Economic Study*, London: Pitman.

Van Rank tot Drank (1990), Brussel: Didier Hatier.

Varro (1912) *On Farming: M. Terenti Varronis Rerum Rusticarum*, trans. L. Storr-Best, London: G. Bell & Sons.

Viala, P. and Vermorel, V. (1901-10) *Traité Général de Viticulture: Ampélographie*, Paris: Masson.

Viaud, J. (1968) 'Egyptian mythology', in *New Larousse Encyclopedia of Mythology*, 2nd edn, London: Paul Hamlyn, 9-48.

Virgil (1974) *Eclogues, Georgics, Aeneid I-VI*, trans. H.R. Fairclough, rev. edn, London: Heinemann.

Vives, J.V. (1970) *Approaches to the History of Spain*, Berkeley, California: University of California Press.

Vizetelly, H. (1880) *Facts about Port and Madeira, with Notices of the Wines Vintaged around Lisbon and the Wines of Tenerife . . .*, London: Ward Lock & Co.

Voss, R. (1985) *The European Wine Industry: Production, Exports, Consumption and the EC Regime*, London: Economist Intelligence Unit.

Walford, A.R. (1940) *The British Factory in Lisbon, and its Closing Stages Ensuing upon the Treaty of 1810*, Lisbon: Instituto Británico em Portugal.

Wallace, P. (1972) 'Geology of wine', in *International Geological Congress, 24th Session, Canada 1972. Section 6: Stratigraphy and Sedimentology*, 359-65.

Wallerstein, I. (1974) *The Modern World-System: Capitalist Agriculture and the Origins of the European World-Economy in the Sixteenth Century*, London: Academic Press.

Wallerstein, I. (1979) *The Capitalist World-Economy*, Cambridge: Cambridge University Press.

Wallerstein, I. (1980) *The Modern World-System II: Mercantilism and the Consolidation of the European World-Economy, 1600-1750*, London: Academic Press.

Wallerstein, I. (1983) *Historical Capitalism*, London: Verso.

Warner, C.K. (1960) *The Winegrowers of France and the Government since 1875*, New York: Columbia University Press.

Watts, D. (1987) *The West Indies: Patterns of Development, Culture and Environmental Change since 1492*, Cambridge: Cambridge University Press.

Waugh, A. (1957) *Merchants of Wine: Being a Centenary Account of the Fortunes of the House of Gilbey*, London: Cassell and Company.

Weaver, R.J. (1976) *Grape Growing*, New York: John Wiley & Son.

Webster, D., Webster, H. and Petch, D.F. (1967) 'A possible vineyard of the Romano-British period at North Thoresby, Lincolnshire', *Lincolnshire History and Archaeology* 2, 55-61.

Weidemann, T. (1981) *Greek and Roman Slavery*, London: Croom Helm.

Weinhold, R. (1978) *Vivat Bacchus: a History of the Vine and its Wines*, Watford: Argus Books.

Weiter-Matysiak, B. (1985) *Weinbau im Mittelalter*, Köln: Rheinland-Verlag (Geschichtlicher Atlas der Rheinlande, Beiheft VII/2).

Wellmann, I. (1974) 'Communautés de viticulteurs dans la Hongrie des XVIIe-XVIIIe siècles', *La Pensée* 55-70.

Wender, D. (1973) 'Introduction', in *Hesiod: Theogony, Works and Days. Theognis: Elegies*, Harmondsworth: Penguin, 11-22.

West, M.L. (ed.) (1978) *Hesiod: Works and Days*, Oxford; Clarendon Press.

West, R.G. (1977) *Pleistocene Geology and Biology, with Especial Reference to the British Isles*, 2nd edn, London: Longman.

Westermeyer, J. (1986) *A Clinicial Guide to Alcohol and Drug Problems*, New York: Praeger.

Westwood, I.O. (1869) 'New vine diseases', *The Gardeners' Chronicle and Agricultural Gazette*, 30 January 1869, 109.

White, K.D. (1975) *Farm Equipment of the Roman World*, Cambridge: Cambridge University Press.

Whitmore, W.W. (1853) *The Wine Duties*, London: Longman, Brown, Green & Longmans.

Wilkinson, J.G. (1878) *The Manners and Customs of the Ancient Egyptians, Volume 1*, new edn rev. by S. Birch, London: John Murray.

Williams, D. (1977) 'A consideration of the sub-fossil remains of *Vitis vinifera* L. as evidence for viticulture in Roman Britain', *Britannia* 8, 327-34.

Williams, R.N. and Shambaugh, G.F. (1988) 'Grape phylloxera (Homoptera: Phylloxeridae) biotopes confirmed by electrophoresis and host susceptibility', *Annals of the Entomological Society of America* 81(1) 1-5.

Willis, J.C. (1973) *A Dictionary of the Flowering Plants and Ferns*, rev. H.K. Airy Shaw, 8th edn, Cambridge: Cambridge University Press.

Wilson, J.V.K. (1972) *The Nimrud Wine Lists: a Study of Men and Administration at the Assyrian Capital in the Eighth Century B.C.*, London: British School of Archaeology in Iraq.

Wine Advisory Board (1942) *The Therapeutic Use of Wine*, San Francisco, California: Wine Advisory Board.

Winkler, A.J. (1938) 'The effect of climatic regions', *Wine Review* 6, 14-16, 32.

Winkler, A.J. (1962) *General Viticulture*, Berkeley, California: University of California Press.

Winkler, A.J. (1973) *Viticultural Research at University of California, Davis, 1921-1971*, Davis, California: University of California, Davis (University of California, Davis History Series).

Winkler, A.J., Cook, J.A., Kliewer, M.M. and Lider, L.A. (1974) *General Viticulture*, 2nd edn, Berkeley, California: University of California Press.

Wolf, E. (1982) *Europe and the People Without History*, London: Faber & Faber.

Woolley, C.L. (1934) *Ur Excavations, Volume II*, Oxford: Oxford University Press.

Wright, O. (1955) *Voyage of the Astrolabe - rendered in English*, Wellington: A.H. and A.W. Reed.

Xenophon (1922) *Works*, trans. C.L. Brownson, London: Heinemann.

Young, A. (1792) *Travels During the Years 1787, 1788, and 1789. Undertaken more Particularly with a View of Ascertaining the Cultivation, Wealth, Resources, and Natural Prosperity of the Kingdom of France*, Bury St

Edmunds: Printed by J. Rackham; for W. Richardson, Royal-Exchange, London.

Younger, W. (1966) *Gods, Men and Wine*, London: The Wine and Food Society.

Zapletal, V. (1920) *Der Wein in der Bibel*, Freiburg: Biblische Studien.

Zimányi, V. (1987) *Economy and Society in Sixteenth and Seventeenth Century Hungary (1526-1650)*, Budapest: Akadémiai Kiadó.

Zimányi, V., Makkai, L., Kirilly, Z. and Kiss, I.N. (1968) 'Mezögazdasági termelés és termelékenység Magyarországon a késöi feudalizmus korában (1550-1850)', *Agrartorteneti Szemle* 10, 39-93.

INDEX

'Abbasids 153
abuse, alcohol *see* alcohol
acetaldehyde 55
acetic acid 46, 48, 55; bacteria 47
acetification 314
Acetobacter 47-8, 253-4
acids 34, 46, 48, 55, 292; acetic 46, 48, 55; lactic 46-8; malic 46-8
acqua vitae 236
Acquitaine 162, 184
Adam of Bremen 160-1
additives 311, 313
adulteration 196
advertising 334, 341, 348-52
Aegean 120
aesthetics 25
Afghanistan 76-7
Africa 10, 214; north 344; southern 13, 43, 50, 245, 308, 311, 320, 335
afterlife 91
Age of Discovery 205-32
ageing: bottles 57; wood 56
agriculture: expertise 66; peasant 123, 164, 167, 359; subsistence 123, 132
Agrionia 89
Aix-la-Chapelle 147
Akbar 155
al-Idrisi 153, 175
Alamanni 117, 136, 138, 146
Alba Helvia 111
alcohol: abuse 338-9, 362; consumption 335; ethyl 46; low 342, 349; methyl 312; seventeenth century consumption 239; social concern 328; fermentation 46, 49-50, 53-4, 264, 358
alcoholism 234, 338-9, 362
ale 177, 240

alehouses 327
Algarve 263
Algeria 283, 295, 308, 344
Almagro 218
Almeirim 111
Alps 112, 181
Alsace 185, 230
Althusser, L. 6-7
altitude 45
Ambrose 137
America 10; Latin 210-11, 216-20, 231, 296; North 13, 58, 270, 308, 352; South 215; United States of 343, 347, 354, 361; vines 284, 290-1, 293
American War of Independence 248
Ammonites 85
ampelography 29-32
amphorae 55, 62, 94-5, 97, 103, 113, 115-16, 119, 121-4, 126, 129; Egyptian 69; Spanish 119
Amsterdam 213, 259; banks 212
Anatolia 73-7
Anglo-Gascon wine trade 3, 12-13, 18, 166, 187, 193-202
Anjou 229, 275
Anthesteria 89
anthocyanins 36, 57
anti-alcohol lobby 304, 338, 342
Antwerp 221, 227
Appellations d'Origine Contrôlées 313-16
Ararat 82
archaeology 61-3, 94, 107
Argentina 342
Armagnac 238, 314
Armenia 67
Arnaldus de Villanova 179
Arras 176

art, Minoan 77
Ashur 66
Ashurbanipal 66-7
Asia, eastern 58
aspect 45
assize courts 196
Assyria 64, 71
Atlantic 204, 286
Attica 89, 96
attitudes, seventeenth century wine 266-9
auctions 353
Augustine 137
Aurelian 132
Ausonius 118
Australasia 10
Australia 13, 50, 270, 292-3, 296, 308, 311-12, 335, 344, 346-8, 354; origins of viticulture 296-301
Austria 221, 285, 312
autovinification 53
Auxerre 118
Avignon 178
Azores 214
Aztecs 215-16

Babylon 64, 67-8, 71, 83, 87, 110
Bacchae 58, 86-7, 89
Bacchus 61, 77, 85-91, 127, 141
Bacci, Andrea 169
bacteria 28, 46-9; acetic acid 47; lactic acid 47
Baetica 123
bag-in-the-box 348
bagaçeira 53
Bahrain 80
balché 59, 217
Baltic 180, 184-5, 190-1, 198, 227, 229, 232, 238, 241
banks 13, 212; Amsterdam 212; Florentine 212; Genoese 212
banquets 73, 93
Banyuls 314
Bar-sur-Aube 181
Barbaresco 324
Barcelona 176
Bardi 212
Barolo 56, 324
barrels 28, 62, 129, 165, 252-3, 258; fermentation 171
bars, Roman 125, 128
Basque 187, 235
Bassareus 87

bastard 187
Bataille des Vins 182
batellage 188
Bau 78
Bayonne 194
Beaujolais 15, 33, 55
Beaulieu 24-5, 339
Beaune 15-16, 52, 117, 309
Beccero 168
Bede, The Venerable 157
beer 71, 87, 131, 177, 235-6, 240-1, 294, 326-7, 329, 332, 335, 341, 361; Egyptian 68; Mesopotamian 68
Benedictine 241
Benedictines 179
Bergerac 198
Bergstrasse 148
Beringer 25
Berlin 227
Berry's 327-30
Beycesultan 75
Bible 63, 75, 81-5, 139, 157
Bilbao 194
bills of exchange 247, 313
Blachon 1-2
Black Sea 101
Blaxland, G. 297
blood, human 81
Blue Mountains 297
Boeotia 89, 96-7
Bolivia 34, 42, 218
Bordeaux 44, 56, 184, 194, 198, 227, 229, 237, 253, 258; New French Clarets 256-9; wine classification 278-82
Bordeaux Mixture 38, 292
Bordeaux Parlement 258
Bordelais 197, 314
Boston 249
Botrytis cinerea 44, 257
bottles 266; glass 13, 55, 254-6
bottling 45
boutique wineries 346
Brabant 191, 230
brand names 334, 341, 353
brands 334, 341, 346
brandy 234, 236-40, 251, 254, 262, 331, 334
Braudel, F. 205, 212-14, 234-5, 236, 238
Brazil 214, 246
Bremen 175
brewers 330, 332, 341

Bristol 190, 195, 330-1
Britain 252, 308, 347, 356; origins of viticulture 118
Brittany 235
brothels, Assyrian 66
Bruges 175-6, 227-8
Brunello 348
Brunello di Montalcino 324
Brussels 176
Bual 248
Bulgaria 323, 354
burgesses 177
Burgos 223
Burgundians 138, 144
Burgundy 14-18, 33, 44, 116, 178, 203, 221, 230, 275, 290, 307, 311, 359; Duchy of 17; Dukes of 230; origins of viticulture 116-17; wine classification 27, 282-3
Busby, J. 13, 298-300

Cabernet Franc 229-30
Cabernet Sauvignon 22, 39, 43, 55, 348, 352, 354
Cádiz 105, 111, 222, 253
Cadmus 87
Cahors 198, 257
Calabria 98
Calah 66
California 22-6, 36, 39, 50, 270, 292-3, 296, 307-9, 311-12, 344, 346-7, 355, 360; origins of viticulture 301-6
calvados 234
Campania 98, 111, 113, 115, 118, 120, 127, 129, 359
Canaan 82, 93
canals 272
Canaries 245, 268; origins of viticulture 245-8
cap 48
Cape Colony 245, 268, 296-7; origins of viticulture 250-2
capital 13, 57, 237, 242, 306, 354-6, 360; finance 309, 353-4; circulation 212; investment 255, 268, 296
capitalism 206-10, 225, 342
Caracalla 117
carbon disulphide 290
carbonic maceration 55, 346
Carta Mercatoria (1303) 192-4
Carthage 75, 102, 107-8, 152
Castile 185, 214-16, 223; armies 184
Castillon 201

castles 176
Catalaunian Plains, Battle of 138
Catholic Church, rituals 218
Catholicism 247, 250
Cato 12, 63, 102-4, 107, 109-10, 112, 168
Caucasus 59
Cave des Hautes-Côtes 52, 356
Centaurs 88
ceremonies, Bacchic 90
Chablis 15
chalk 292
Champagne 43, 54, 184, 229, 268, 275, 314, 361; demarcation 314-15; fairs 181, 185
champagne 54, 253, 255, 262, 311, 315, 319-20, 334, 359; origins of 259-62
Chaptal 50, 295
Chardonnay 15, 43-4, 348, 352
Charente 237-9
Charlemagne 146
Charles II 239, 259, 262
Charles V 221
Charles VII 201
Charleton, W. 276
Chartreuse 241
Château Latour 353
Château Loudenne 334
Châteauneuf-du-Pape 178, 315
Chaucer 186-7
chemical additives 311
Chenin Blanc 229, 250, 352
Chianti 53
Chile 43, 218
China 76, 185
Chinese Empire 211
Chios 89, 97, 110-11
chlorophyll 36
Christianity 2, 133, 136, 215; influence on wine 165; sacramental requirements 162; symbolism 81-5, 139-44, 358
church 175, 225, 282
cider 235-6, 241, 294
Cistercians 15, 178-9
Citeaux 15, 178
city state, Greek 96
Civil War, English 239, 247
claret 229-30, 258, 262, 277, 353, 359
classification: German 318; wine 276-83, 311
climate 28, 34-5, 42, 236, 293, 344; change 68, 162, 164, 175, 229, 274-6;

Egypt 71
Clitumnus, Temple of 142
clones 293
Clos de Vougeot 178
cloth 180-1
Clovis 146
Cluny 15, 178
Code of Hammurabi 64, 66
coffee houses 240, 327
Cognac 39, 238, 314
Colares 293
Colmar 230
Colombard 238
Colombia 218
colonialism 13, 359
colonies 242, 245, 296-7, 308; penal
    296; Greek 98, 101
Columella 12, 63, 94, 102, 105-10, 112,
    116, 129, 168
commerce 124, 205; economy 102;
    medieval organisations 189-92
commodities 208
commodity production 207
Common Agricultural Policy 321
communications 272, 283, 289
Communion 2, 139, 142, 218, 301
Companhia Geral da Agricultura das
    Vinhas do Alto Douro 265-6
Compiègne 187
conquistadors 218, 250, 359
Constantine 117, 133, 136-7
Constantinople 137, 149
consumption 348; alcohol 335
convivium 127
convoy 200; Venetian 191
cooperatives 355
copper 237
copper sulphate 38
Córdoba 152-3
cordon 33
cork 261
corkscrew 261
corporate structure 25, 347, 352, 354,
    356
corruption 313
Cortés 215, 216
Cos 97, 110
costs, labour 345-6; transport 188
Côte Challonais 15
Côte de Nuits 15
Côte d'Or 14-18, 117, 230, 279
coulure 42
Council of Constance (1414-18) 144

Council of Trent 144
courts: assize 196; English 239; French
    184, 239; papal 178
courtiers 280, 325
credit 13, 211-13
Crescentiis, Petrus de 168
Crete 77, 186, 233, 293
criminality 362
Criolla 217, 301
crisis 306-10
crop failure 43
crosses, vines 30, 345
crusades 186
crustulum 128
culture 231, 357-8
cultural landscape 3-4, 8, 14-26
cultural traits 10
Cumae 98
cups, drinking 62
customs duties 189, 242, 244, 328-9,
    361
Cyclops 86
Cyprus 182, 186, 293, 320

Damascus 153
dancing girls 73, 99
Danube 101, 117, 137
Danzig 227
Dão 19
dates 68, 71
de Agri Cultura 102-4
de Arboribus 106
de Re Rustica 105-9
death 60, 78-9, 357-8
debt 325
Defoe, D. 156
demand 361; Greek wine 99-101;
    polarisation 271
demarcation of vineyards 265, 278,
    311-25, 355; Portugal 265
demography 202, 205; Roman 124
demonstrations, political 314
Denominazione di Origine
    Controllata 322, 324
Denominazione di Origine
    Controllata e Garantita 324
Diáz, Bernal 216
diethylene glycol 312
Dijon 147
Dijon Parlement 282
Dilmun 80
Diocletian 131, 136; reforms 136
Diodorus Siculus 75, 124

Dion, R. 4
Dionysus 61, 77, 85-91, 131, 140-1, 358; ritual 10, 141, 358
Dioscorides 97
discoveries, fifteenth century 214-16
distillation 234, 240-2, 322-3
distillery 238, 240-1
distinction, social 93
dolia 120
Domesday Book 157
Dominic Group 337, 340, 356
Domitian 114, 220, 276
Dordogne 197
Douai 176
Douro valley 18-22, 262-6, 268, 278, 294, 307, 346
downy mildew 37
drainage 45
Dutch East India Company 250-2

Eanes, Gil 214
Earth-mother 78
Eau Grison 284
eau-de-vie 238, 257
Ebro 120
economy 359-60; depression 304; expansion 205; global 13, 205, 209-11; Graeco-Roman 94-133; Roman agrarian 108; villa 108
Edelfäule 44
Edicts: Diocletian (AD 301) 131; Domitian (AD 92) 114, 276; of Milan (AD 313) 137
Edward I 192
Edward III 212
Egypt 68-73, 78-9, 358, 362
Ehrenfelser 30, 342
Eleusinian Mysteries 89
élevage 27, 54-7
Elizabeth I 327
Ely 165
emigration 123
empires 135
England 12, 19, 163-5, 177, 181, 182, 184-6, 190, 198-9, 221-2, 225, 229, 233, 235, 241, 246, 261, 274; early medieval viticulture 157; medieval vineyards 172
Enkidu 81, 84
environment, physical 1, 24, 42-6, 293, 302
Epernay 184, 230, 259, 261-2, 314
Epic of Gilgamesh 62, 79-81, 84

ergot 90
eroticism 10
espalier 33
estufas 248
ethanol 46
Etheridge, Sir G. 259
Ethiopia 215
ethyl acetate 55
ethyl alcohol 46
Etruria 98, 120
Etruscans 98
Eucharist 2, 139, 142, 218, 301
Euphrates 63-4, 77, 110
Euripides 86
Europe 78, 270; eastern 210
European Community 21, 317, 321-3, 331, 360
Eutypiose 39
Euvitis 29-30, 37
evapotranspiration 43
exchange 10-14, 94
excise duty 252
exploration: Portuguese 205, 214-16; Spanish 205, 214-16

Factories, English 262
fairs: Champagne 181, 185; Rouen 184
Falernum 104, 128
Falstaff 222
famine 96, 205
fan leaf 37, 39
fashion 269, 342-4, 347-54
feasts 73, 93
fermentation: alcoholic 46, 48-50, 53-4, 264, 358; barrels 171; cool 24, 45, 50, 276; early modern period 229-30; hot, 53; malo-lactic 48, 53-4; short 53, 229-30; temperature control 54, 344, 346
fertility 9, 78, 87, 93, 347, 350; agrarian 92; gods 78-81; human 61, 81
fertilization 33
feudalism 203, 206-9, 225; crisis 202
filtration 346
finance: capital 309, 353-4; internationalisation 326
fining 55
fiscal policy 131, 137, 201, 242-5, 322, 361
flagellation 91
Flanders 175-6, 180-2, 184-6, 191, 193, 198, 227
fleets: Italian 188; salt 191; wine 166,200

flor 254
Florence 168, 176, 212; banks 212
Florida 218
fluid, divine, 78-81
fortification 262
fortified wine 54
fossil 29
France 58, 167, 176-7, 182, 184, 198-9,
    209, 226-32, 235, 242, 253, 261, 263,
    269, 274, 278, 294-5, 298-9, 308,
    313-16, 321-2; revolution (1789) 18,
    213, 274, 282
Francis I 236
Franciscans 22
Franco-Prussian War 18
Frankfurt 23, 181
Franks 117, 136, 138, 146, 152
Franschhoek 251
fraud 276, 312
freight charges 188
frost 34, 42-3, 229, 274
fructose 34

Ga-Tum-Dug 78
Gaea 78
Gaillac 198, 257
Galen 179
Galicia 187
galley, Mediterranean 186
Gama, Vasca da 214
Gamay 17, 33, 55, 203, 275
Gard 285, 287, 289
Garonne 116, 120
Gascon Charter 193
Gascony 162, 184, 188-9, 198-200, 221,
    242
Gaul 2, 127, 133; origins of viticulture
    113-19; wine demand 124
Gay-Lussac equation 49
Gdansk 227
Genoa 168, 176, 212; banks 212
geology 44
Georgics 105
German Mass *see* Mass
Germanic tribes 133; wine
    consumption 145
Germany 167, 177, 184, 188-91,
    226-31, 235, 253, 269, 285, 303, 308,
    311-12, 316-19, 323, 341
Geshtin 78
Gewürztraminer 44
Ghent 175-6
Gilbey's 332-40, 347

Gilgamesh 180-1
gin 234, 241, 255, 328
gin-shops 327
Giraldus Cambrensis 157
Gironde 287
glacials 29
Gladstone, W. 310, 328-9, 332-3, 352,
    354
glass 255, 268; bottles 13, 55, 354-6;
    furnaces 261
Glastonbury Abbey 157
global corporations 25, 347, 352, 354,
    356
glucose 34
gobelet 33
goddesses: wine 78-81
gods: fertility, 78-81; wine 61, 77-81,
    85-91, 131, 140-1, 358
Gold Rush 302, 308
Goths 138
Göttingen 178
Graecinus 106, 108
grafts 292, 294
grain: prices 203; supply 115
Granada 215
Grand Metropolitan 326, 332-40, 361
grapes 54; composition 34, 36;
    physiology 32-7; pips 36; prices 25,
    355, 359; ripening 34; seeds 36, 61;
    selection, early modern 233; skins
    36; Roman varieties 102-3, 116
grappa 53
Graves 257
Great Custom (Bordeaux) 197
Great Discoveries 214-16
Great Silk Road 159
Greater Dionysia 89
Greece 9, 57-8, 93-4, 149, 283, 358;
    ancient colonies 101; ancient wine
    trade 101; wine demand 99-101
Gregory of Tours 118, 147
Grœnlendinga Saga 160-1
Groot Constantia 251
Gros Plant 238
Guadalquivir 222
Guadiana 120
guingettes 272
Guyenne 184

Habermas, J. 7
Habsburg Empire 221, 224
hail 43
Hainault, Counts of 231, 259

Haloa 89
Han Dynasty 159
Hansa 185, 190, 192
Haraszthy, A. 303-4
harvest dates, grapes 187, 237
harvesting machines 345
Harveys of Bristol 327, 330-1
Haut Pays 197-8, 200, 257
Haut-Brion 256, 258
Hautes-Côtes 15
Hautvillers 184, 261; Abbey of 259-60
health care 362
heat 43; energy 49; summation 43
Heidelberg 148
Heidelberg Tun 253
Hérault 28-9, 291
Herculaneum 125, 129
Hermitage 279
Herodotus 63, 66-7, 71, 98
Hervey, J. 258
Hesiod 95-8
Heublein 25, 361
hides 129
hieroglyphs 69, 73
Hilarion 149
Hippocrates of Cos 179
historical geography of viticulture 4-5
historical materialism 6
History of the Franks 118
Hittites 75
Holland 209, 221, 225, 251
Homer 62, 86, 97
honey 97
hops 235
Horace 127
Huguenots 218, 248, 251
Hull 195
human action 5, 7
humidity 43, 45
Hundred Years War 197-202
Hungary 157, 211, 221, 224-6, 283, 285, 293, 323
hunger 96
Huns 138
Hunter Valley 298-9
Huss, J. 142
hybrids 345

Iberia 111, 113, 210, 231, 250, 253
ideology 8, 91, 357-8, 362; early modern 211-14
Ile-de-France 184, 229-30, 272
Iliad 86, 97

Illyrians 111
Incas 215
India 159, 165, 205, 214
industrial structure 354
Inglenook 24-5, 339
Innini 78
inns 327
Inquisition 215
integration: horizontal 339; sectoral 325; vertical 342, 354, 361; wine industry 326
interglacials 29
International Distillers and Vintners 335-7, 342
International Phylloxera Congress, Bordeaux 291
intoxication 71
investment 353
Iran 60, 63, 155
Iraq 60, 63
Ireland 229
irrigation 22, 43, 346
Isis 79
Islam 157, 184, 186, 215; conquests 152-3; effects on viticulture 150-5; trade 205
Israel 83
Italy 50, 53, 57-8, 93, 102, 127, 168, 176-7, 180, 189, 283, 285, 295, 312, 321-3, 343, 347, 352, 361
Ivriz 75-6, 79

James I 242, 249
Jamestown 249
Japan 58, 312, 354
Jefferson, T. 273, 279, 308
Jerez de la Frontera 222, 234, 254, 257, 293-4
Jericho 73
Jerome 137, 149
Jesuits 218, 301
Jews 81-5, 216
journalism 352
Judaism 81-5, 216
Julius Caesar 126
Jura 101
Justinian 143, 145, 149

Kenya 42
Kerner 30, 32, 345
Kew, Royal Botanic Gardens 283
Khaemwese, tomb 69-70
Khamseh of Nizami 155-6

Khayyam, Omar 154-5
King Alfred, Laws of 157
Kish 78
Kloeckera apiculata 47
Kloster Eberbach 179
Knights Hospitallers 179
Knights Templars 179
Köln 147, 164, 176-7, 181, 230
Königsberg 227
Koran 150-1
kottabos 99
Kraków 175, 185
Krug, C. 24-5, 303-4
Kurdistan 91
kvass 241
kylix 99

La Rochelle 184, 191, 227, 237-8
labour 12, 21, 23, 207, 225-6, 231-2;
    control 210, 360; costs 345-6; day
    258, 294; power 207; relations 11, 93;
    services 360; slave 96, 108-9, 132;
    wage 168, 360
lactic acid 46-8; bacteria 47
Lagash 64, 78
Lagny 181
Lake Mareotis 111
Lamego 263
land: prices 25; tenure 167-8, 210,
    345-6, 360
Langland, W. 177
Languedoc 295, 299, 308
Laon 230
Las Sorres 289
latifundia 108
Latin America *see* America; origins of
    viticulture 216-20
Latium 105, 111, 113, 127
lead oxide 314
leases 232
Lebanon 75, 80
legislation 308, 329, 332-3, 342, 361,
    363; European 321-3; French 313-6;
    German 316-9; international
    319-23; Italian 322, 324; phylloxera
    290-4; Roman 90
Lenaea 89
Lepe 186-7
*Les Très Riches Heures de Jean de
    France, Duc de Berry* 169-70
Lesbos 89, 97, 110
Li Po 160
libation 77, 81

Liber Commodorum Ruralium . . .
    168
Liber Pater 90
Libya 110
life 93, 357
Liguria 111, 169
liknon 91
limestone 44
liqueur de tirage 54
liqueurs 239, 241
Lisbon 247, 262-3
Little Ice Age 162, 229
Livy 90
lobby, anti-alcohol 304, 338, 342
lodges, port 266-7
Loir-et-Cher 291
Loire 117-18, 144, 147, 157, 229, 237,
    272, 275
Lombardy 176
London 176, 190-2, 195-6, 239-40,
    257, 269, 299, 327, 330
Lorsch Abbey 148
Los Angeles 302, 309
Lot 85
Louis XI 202
Louis XII 236
Louis XIV 251
Louis XV 261
love 93, 99, 160
low alcohol beverages 342, 349
Low Countries 187, 221, 274
Lübeck 190, 227
Lucca 176
Lull, R. 236
Lydia 87, 98
Lyon 112, 116, 123, 271

Macarthur, J. 297
macération carbonique 55, 346
Mâconnais 15
Madeira 214, 245, 263, 268; origins of
    viticulture 245-8
Madrid 247
Maecenas 171-2
maenads 89
Maghfeld, G. 194-5
Mago 101
Mainz 178-9
Malaga 257
malic acid 36, 48
malmsey 185, 187, 191
malo-lactic fermentation 48, 53-4
Malvasia 246, 248

mandrake 73
Maras 76
marc 53, 313
Marche 169
Margaux 258
market: economy 11, 176; mass 330, 346, 353; share 348
markets 190; rural 181; urban 230
Marne 162
Marseille 98, 113, 271, 295
Marx, K. 206, 211, 312–13
Mass 2, 139, 142, 218, 301; German 144; Tridentine 144
mass market 330, 346, 353
Massif Central 272
Massilia 98, 113
maturation 27, 54–7
Maxentius 136
Maximian 136
measurements, wine 364–6
mechanisation 33, 345
medical problems 362
medicine, wine 102, 179
Medina del Campo 223–4
Mediterranean 28, 98, 101, 169, 179, 180, 182, 232–3, 273, 292, 307, 358
Médoc 258, 268–9, 273, 278–82, 314, 333, 361, 364
Melon de Bourgogne 275
Memphis 68
mercatores 120, 123
merchants 24, 177, 231, 253, 269, 277, 316, 341; British 263–5; cloth 180; Dutch 227, 234, 236–40, 253, 257; English 189, 193–202, 234, 246, 248, 253, 261; Flemish 189, 238; French 253; Gascon 190, 192–202; Genoese 191, 221; German 181, 253; Hanseatic 185, 190–2; Italian 181; north European 232; Scottish 248; Spanish 253; urban 307
Merlot 55
Mesopotamia 59, 63–8, 78, 87, 358
metaphor 8, 60
metaphorical relations 9
Methen 69
méthode champenoise 54
Methuen Treaties (1703) 3, 12–13, 21, 264, 278, 328
methyl alcohol 312
Meursault 15
Mexico 22, 215–16, 220, 231, 296, 301
microclimate 45

Midi 283, 294, 307, 314
Milan 168, 176
mildew 307; downy 37
military repression 315
milk 78
millerandage 42
Minho 19, 21, 48, 263, 278
Minoan civilization 77
Mission vines 22, 217, 301–2
missions 220, 301
Moabites 85
modern world system 13, 209–11
modiato 189
Moghuls 155
Moisac 198
monasteries 171, 175–6, 282; viticultural influence 134
monasticism 147
Monbazillac 257
Mondavi, C. 24
Mondavi, R. 23, 25
money 237, 354
monopolies 309, 312, 355
monopoly rent 313
Monte Testaccio 131
Montpellier 184, 289, 299
Moors 186, 215
Moscadello di Montalcino 347–8
Mosel(le) 116–18, 144, 147, 157, 161, 163, 185, 346
Moses 83
Motolinía 217, 220
Mount Sinai 83
Mouton-Rothschild 280
Müller-Thurgau 30, 43
mulsum 128, 132
Muscadinia 29–30
Muscat (vine) 149, 230, 250
Muses 87
musicians 73
must 36; refrigeration 50; temperature 49
Mycenae 86; civilization 77
mysticism 85; Orphic 89
mythology 60, 127

Nantes 118, 164, 229, 275
Napa Valley 22–6, 339
Naples 176, 221
Napoleon Bonaparte 252, 298
Napoleon III 46
Narbonensis 111–13, 116, 126
Narbonne 116, 184, 285

narcotics 363
Navigation Act (1660) 247-8
navigation 188
Nazirites 84, 141
Nebamun 73-4; tomb of 69
Nebbiolo 43, 55
Nebuchadnezzar 67
Negev 149
négociants 18, 257, 307, 309, 315, 354, 363
negotiatores 123
Nestlé 25
Netherlands 221, 230, 232, 235, 242, 250-2, 257, 296
New Custom 193
New French Clarets *see* Bordeaux
New Songs from a Jade Terrace 159
New Testament 84, 139-41
New York 304
New Zealand 293, 300, 312; origins of viticulture 296-301
Newfoundland 161
Niebaum, G. 24, 339
Nile 68, 77, 79
Nimrud 66, 93
Nineveh 66
Nippur 62
nobility 148, 171, 175, 232
noble rot 44
noblesse de robe 258
non-alcoholic beverages 332
Normandy 157, 193, 235
North Africa *see* Africa
North America *see* America
North Sea 180, 241
Nuits St Georges 117
Nürnberg 176
Nymphs 88
Nysa 87

Odoacer 137
Odyssey 62, 86
oenologists 27
off-licences 329, 340, 356
oïdium 37, 283-4, 287, 307
Old Testament 81-5, 140
oligopoly 326
olives 96, 102, 108
Oman 152
Oporto 234, 262, 266
origins of viticulture 58-93, 296; Australia 296-301; Britain 118; Burgundy 116-17; California 301-6;

Canaries 245-8; Cape Colony 250-2; champagne 259-62; Gaul 113-19; Latin America 216-20; Madeira 245-8; New Zealand 296-301; port 262-6; Virginia 248-9
Orléans 227, 229, 272
Orpheus 89
Osiris 79, 89
Ostia 128
Ottoman Empire 137, 186, 211, 233
Ousir 79
overproduction 308, 310

Pa-Geshtin-Dug 78
Paarl 43, 251
Palaikastro 77
Palestine 67, 73-4
Pan 88
Pannonia 117-18
Papacy, Avignon 178
papyri 68, 71
Paramatta 297
Paris 118, 144, 148, 164, 176, 178, 180, 184, 229-30, 271-2, 283, 289, 299
Paris Basin 162, 164, 175
passes 181
Pasteur, L. 46
Pavia 176
peasant agriculture 123, 164, 167, 359
peasantry 132, 167-8, 203, 207, 209, 258, 273, 359-60; Mediterranean 167
pectins 34
Pennsylvania 249
Pentheus 89
Pepin 146
Pepys, S. 233, 256
Perelada 319
Pérignon, Dom Pierre 259, 261
Periplus of the Erythraean Sea 151
peronospera 37
Persia 77, 83, 137, 155, 159; Empire 152
Peru 42, 231
Peruzzi 212
Petronius 128
pH 36
phallus 91
Philadelphia (Egypt) 62
Philip II 220, 223, 362
Philippe Auguste 182
Philippe le Hardi (Philip the Bold) 17, 169
Phocaeans 113

Phoenicians 75
photosynthesis 33
Phrygia 87
Phylloxera vitifoliae 18, 21, 37, 39-42, 307; arrival in Europe 284-7; carbon disulphide treatment 290; chemical treatment 291; commissions 290; flooding treatment 290; global spread 292-6; legislation 290-4; social and economic implications 294-6; spread within France 287-92
Picardy 193
Pierce's disease 37, 39
Piers the Ploughman 177
pilgrimage 150
pilotage 188
Pinhão 20, 51, 263
Pinot Noir 15, 17, 43, 169, 203, 230, 259
piquette 53, 99, 230
Pisa 176
pitch 17
pithoi 101
Pizzaro 215, 218
plague 17, 145, 200
Plato 99
Pliny 110, 111, 114-16
ploughing 105
poetry 85, 99; Chinese 159; Islamic 151, 154-5; Persian 154-5; Roman 127-8
Poitiers 184
Poitou 184
Poland 175
polis 96
political control 314, 321, 361
pollination 33
polyculture 11, 273
Polyphemus 86
Pombal, Marquis de 265, 278
Pomerania 227
Pommard 15
Pompadour, Madame de 262
Pompeii 91, 128-9
Pontac, Arnaud de 256-7
population 205, 271; decline 202; urban 272-3, 361
port 21, 54, 253, 255, 269, 294, 311-13, 334, 353; companies 264; lodges 266-7; origins of 262-6
Portugal 12, 18-22, 43-4, 48, 57, 111, 186, 214-16, 245, 257, 262-6, 278, 285, 293, 321, 323, 328

potlatch 124
pourriture noble 44, 257
poverty 232, 274, 359; urban 232, 234
power, symbolic 9
Prague 176
Pramnian wine 97
Priapus 88, 91-2
prices 308, 310, 322, 324; grape 355, 359; retail 196, 198; Roman wine 131; vineyard 306; wine 196, 264, 307, 321, 353, 360; wine retail 243; wine wholesale 243
pride, class 166
priests 83; Christian 140
primitive accumulation 208-9
prisa 189
privilege 311-25
Probus 117-18
production 306, 311, 361; agrarian 94; commodities 207; costs 311; Greek agrarian 95-9; relations of 205; technology 355; wine 202, 231-2
profit 14, 102, 108, 166, 231, 311, 313, 338, 360; appropriation 213; early modern 211-14; margins 349; mercantile 360; motive 133, 250; Roman viticulture 107-10
prohibition 22, 304, 306; Islamic 150
prostitutes 129; temple 81
protest, political 314, 324
Provence 285
Provins 181
pruning 33
Puligny-Montrachet 15
pulque 59, 217
Puritanism 239
Pylos 86

Quaternary 29, 58
Queen Mary's Psalter 171, 173
Queseda 218
Quinta da Foz 51

racking 56, 253, 261
railways 283, 289
rainfall 43, 78
Ramses III 71
ratafias 241
rationality 207
Ravenna 142, 149
rebirth 78-9, 357
recipes 102
reconquest, Iberia 184, 186

recta prisa 190, 193
reek wines 188, 200
refrigeration 50, 56
Reims 230, 259, 261-2, 314
relations of production 205
religion 90, 134, 357-8; medieval 178-9
remontage 53
rent 360; monopoly 313
Res Rusticae 104-5
resin 97
Restoration (1660) 239, 243, 259, 327
retailing 337, 340, 355; Britain 325-43; medieval 180; wine prices 195-6
retailers 307, 311
Revocation of the Edict of Nantes (1685) 251
revolution, French (1789) 18, 213, 274, 282
Rheinhessen 148
Rhine 44, 116-17, 120, 137-8, 148, 157, 161, 163, 178, 180, 182, 185, 227, 230, 233, 243, 271, 319
Rhodes 186, 293
Rhône 120, 178, 285, 287
Riebeeck, Jan van 250
Riesling 30, 32, 44
ripening 43
Riquewihr 230
ritual 60, 94; Christian 218; Dionysiac 10, 141, 358; fertility 81, 89
robot 225-6
Roll of Abbot Irminon 147
Romania 323
Rome 9, 55, 123, 171; ancient vinification 101-13; ancient viticulture 101-19; bars 128; Christian Empire 135-6, 139; collapse of western Empire 133-9, 175; Empire 358; legislation 90; profitability of viticulture 105-10; provinces 107; spread of viticulture 110-19; taverns 128; villa estates 108, 123-4, 126, 164; wall 132; wine adulteration 276; wine classification 111-12; wine demand 123-31; wine merchants 116, 120; wine prices 131; wine supply 123-31; wine trade 116, 119-31; wine transport 129-30
romney 185, 187, 191, 196
roots 33
rootstock 290
Roseworthy College, South Australia 3, 344

Rouen, fair 184
Rousillon 299
Rubidage 187, 221
Rueda 223
ruling classes 359
run 234, 241
Rum Rebellion 297
Russia 191, 235

Sabazius 87
Saccharomyces cerevisiae 47-8
sack 243, 254
sacrifice 77, 81, 83, 141
Saint-Denis, Abbey of 275
Saint Emilion 201
Saint-Evremond 259
Saint-Germain-des-Prés 147, 161, 164
Saint-Omer 176
Salamanca 223
Samos 110
San Francisco 22, 24
San Joaquin Valley 24, 339
Sanlúcar 222
Santa Costanza, Church of 142-3
Santiago de Compostela 153
Sâone 101
Satyrs 88, 91
Sauvignon Blanc 352
scandal 311-25
Scandinavia 227
Scheurebe 30
Schiava 169
schist 44
Schloss Johannisberg 179
Scotland 229
Scuppernong 218, 249
sea-water 107
Seagram 326, 337, 340, 343, 349
Second Vatican Council 144
seduction 10, 350
Segovia 223
Seine 117, 144, 148, 157, 162, 180
Semele 86-7
Sémillon 44, 257
Sercial 248
serfs 225
Sète 295, 322
Setúbal 257
Sevilla 176, 235
sexuality 350
Shakespeare, W. 221-3
share-cropping 103, 168
shelter 45

sherry 311-12, 331, 334; legislation 320-1; shippers 320
ships 166, 188, 191, 200, 286; Genoese 180; Italian 188; Mediterranean 186; sailing 285-6; Venetian 180
Sicily 221
Sidonius Appolinaris 145
Siduri 80
Silenus 88
Silvaner 30
Skrælings 161
slavery 210, 231
slaves 12, 96, 108-9, 119, 123-4, 126, 132, 168, 242; wine consumption 103
slope 45
smallpox 218
society: reorganisation 126; differentiation 291; urban 232
soils 28, 33, 44, 45, 106, 293, 345; acid 292; impervious 33; structure 44; texture 289
Solon 96
Soma 78
Song of Songs 85
Sonoma 303
sources, prehistoric 61-3
South America see America
Southern Africa see Africa
Southampton 191, 195
Spain 59, 75, 113, 126, 169, 182, 184, 186, 191, 209, 213-16, 221-4, 229, 235, 242, 245, 248, 257, 283, 285, 294-5, 299, 301, 321, 331, 347, 359
Spanish Company 253
spices 97
spirits 236, 240-2, 326-7, 329, 332, 335, 341, 361
Steelyard 191
Steen 250
Stel, Simon van der 251
Stellenbosch 251
Stettin 227
stowage 188
Strabo 63, 110-11
Strasburg Tun 253
structural constraint 5
structuration 5
structures 7
Subsidy of Tonnage and Poundage 243
subsistence economy 11, 102, 123, 132, 166

Suebi 138, 144
Suetonius 114
sugar 34, 50
sulphur 258
sulphur dioxide 48-9, 53-4
Sumeria 64, 76, 87
sunshine 42
supermarkets 341
Sweden 237
Swift, J. 256
Switzerland 230, 283, 285, 293
Sydney 299-300
symbolism 2, 6, 8-10, 57, 60-1, 77-8, 81-2, 84, 91, 93-4, 131-2, 134, 144, 175, 177-8, 240, 255, 350, 356, 358; Christian 9, 63, 81-5, 135-6, 139-44, 154, 358; death 60, 78-9, 357-8; Dionysiac 154; erotic 10, 73; fertility 91-3; grape cluster 82; Jewish 9, 63, 81-5; love 160; prehistoric 77-91; rebirth 78-9, 357; sacrificial 66; sexual 91
symposia 99, 127
symposium 127
syndicates, anti-phylloxera 291
Syracuse 96
Syria 60, 63, 67, 73-7, 110

Tacitus 145
Tagus 111
Talmud 75, 149
tannin 36, 53, 55-7
Tanzania 34, 42
tariffs 342
Tarraconensis 119, 123
tartrate crystals 56
taverns 269, 327; Roman 128
taxation 131, 137, 201, 242-5, 322, 361; English wine 242-5; wine 189, 242
Tchelistchef, A. 24, 339
technology 24, 166, 355; change 271, 302-3, 311, 326, 342-7, 357
temperature 42; fermentation 54, 344, 346
Ten Commandments 81-2
tenants 168
Tenerife 245-8
tension, erotic 73
tenure 167
terraces 21
terroir 45
Tertiary 29, 58
tesgüino 59, 217

Thasos 97, 101, 104, 111, 131-2
Thebes 69-70, 73-4; tomb paintings 69-70
Theophrastus 96-7
theory 7
theriacs 179
Thirty Years War 233, 242, 271
Thompson Seedless 24
Thrace 87, 97, 118
Tierra del vino 223
Tigris 63-4, 77
Tokaj 226, 293
Tokay aszu 226
Toledo 152
tomb paintings 71, 74
tonnage 243
Tordesillas, Treaty of 214
Toulouse 124
Tourraine 275
trace elements 44
trade 94; Anglo-Gascon 3, 12-13, 18, 166, 187, 193-202; Islamic 205; early modern 211-14; long-distance 180-1, 213; luxuries 205; retail 180; Roman 269; wine 202-3, 229, 245-8
training systems 33
trans-national corporations 311
Transcaucasus 60
transport 113, 272; charges 188; railways 283, 289; ratios 113; river 120, 178; wine 129, 162; wine costs 162
transubstantiation 142
Treaty of Rome (1958) 321
Treaty of Windsor (1386) 19
Trebbiano 169
trenches 106
Tridentine Mass *see* Mass
Trier 117
tropics 34
Troy 86
Troyes 181
Tu Fu 160
Turin 279
Turkey 59-60, 63, 75, 82, 91, 110, 149, 224, 283, 285
Tuscany 98, 169, 179, 347-8
Tyre 75

Ummayads 152
United States of America *see* America
University of California, Davis 3, 24, 344

Upper Douro valley 18-22, 44, 262-6, 268, 278, 294, 307, 346
Ur 62, 64, 68
urban: economies 176; expansion 175; market 230; poverty 232, 234; population 272-3, 361; revival 175-6; society 177-8; wine demand 277, 307
urbanisation 176
Uruk 80
USSR 60, 63
Utrecht 175

Valdivia 218
Valladolid 223, 254; wine ordinances 223-4
Vandals 138, 144
Varro 12, 102, 104-5, 107, 110
vases 99
Vaucluse 285
Venantius Fortunatus 147
Venice 168, 176, 255
véraison 34
Verdelho 248
Vernaccia 169
vermouth 295
Versailles, Palace of 283-4
Vesuvius 91, 115, 359
Viana do Castelo 247, 263
Victoria Wine Company 329-30, 334, 347
Vienne 112, 116
Vikings 157
Vila Nova de Gaia 19, 266-7
Villa Banfi 347-8
Villa dei Misteri 91
villa estates 108, 123-4, 126, 164
Villanova, Arnaldus de 236
vin de goutte 53, 71
vin de presse 53, 71
vine varieties: Brunello 348; Bual 248; Cabernet Franc 229-30; Cabernet Sauvignon 22, 39, 43, 55, 348, 352, 354; Chardonnay 15, 43-4, 348, 352; Chenin Blanc 229, 250, 352; Colombard 238; Criolla 22, 217, 301-2; Ehrenfelser 30, 342; Gewürztraminer 40; Gros Plant 238; Kerner 30, 32, 345; Malvasia 246, 248; Melon de Bourgogne 275; Merlot 55; Mission 22, 217, 301-2; Müller-Thurgau 30, 43; Muscat 149, 230, 250; Nebbiolo 43, 55; Pinot

Noir 15, 17, 43, 169, 203, 230, 259;
Riesling 30, 32, 44; Sauvignon
Blanc 352; Scheurebe 30; Schiava
169; Scuppernong 218, 249;
Sémillon 44, 257; Sercial 248;
Silvaner 30; Thompson Seedless 24;
Trebbiano 169; Verdelho 248;
Vernaccia 169; Zinfandel 303
vinegar 48
vines 96, 102, 108: American 284,
290-1, 293; crosses 30, 345; diseases
37-42; fertilization 27; grafting 103,
292; hybrids 345; irrigation 27;
layering 103, 106; leaves 29;
monoculture 175; morphology 58;
nutrients 44; origins 29; parasites
270; pests 37-42; physiology 28, 32-7;
planting 105; pruning 27, 103,
106; roots 33, 44; supports 104;
training 27, 33, medieval 172;
varieties 29-32, 169, 357; selection,
early modern 229-30; fossils 29;
influence of physical environment
42-6; Roman varieties 106; seeds 29;
wild 58-9
vineyards 102, 359: abandonment 292-4;
classification 278-83; commercial
165; cultivation 232; demarcation
265, 278, 311-25, 355; episcopal 147;
maintenance 105; mechanisation
345; monastic 14; nobility 148;
planting 202; prices 306; princely
146; replanting 292; restructuring
274-6; riverside location 161
vinho verde 48, 263
vinification 27, 45-54, 276, 302, 311,
345-6; eighteenth century
improvements 258; medieval 172;
origins 63-77; Roman 101-13
Vínland 160-1
vintage 107, 187, 237, 305, 313; tomb
paintings 69-70; traditional 19;
wines 13-14, 129, 165, 200, 254-6,
353, 355
Vintners' Company 195
Vintry 190
Virgil 105-6, 171-2
Virginia 268; origins of viticulture
248-9
Virginia Company 249
Visby 190
Visigoths 138, 144, 146, 152
viticulture 27, 311, 345; early medieval

extent 155-64; effects of Islam 150-55;
French fifteenth and sixteenth
century 226-31; German fifteenth
and sixteenth century 226-31; global
distribution 28, 35; Greek extent
97-9; hearth 63-5; introduction to
Gaul 113-19; Latin America 216-20;
medieval 166-204; origins 63-77;
peasant 147, 165; Roman 101-19;
Roman spread 110-19; specialised
182
Vitis 29-30, 58
Vitis vinifera 29-30, 33, 37, 42, 214,
285-6, 294, 345
vodka 234, 241, 350, 352
Volstead Act 304, 306
Voltaire 262

wage labour 168
Wales 181
Wallerstein, I. 13, 209-11, 221, 225-6,
232
War of the Spanish Succession 247,
257
war 197-202, 221, 233, 237, 242, 244,
247, 252, 257, 261, 263, 271, 289, 304,
315, 331
water table 33
whisky 234, 241, 334
wine: additives 97; adulteration 196;
ageing 224, 253, 266, 268; Anglo-
Gascon trade 3, 12-13, 18, 166, 187,
193-202; auctions 353; British
merchants 319; cheap 124, 271, 330;
classifications 276-83, 311;
commercialisation 283;
consumption 329; date 68, 71;
demand 275; English taxation 242-5;
exchange currency 2; Falernian
127-8; fleets 166; fortification 254-62;
fraud 276-83; Gallic imports
124; goddesses 78-81; gods 78-81;
Greek demand 99-101; high alcohol
231-2; high quality 310, 371;
imports 132; industrial integration
326; investment 353; labelling 311-12;
legislation 313-23; light 185; low
quality 269, 271-2, 277, 310, 361;
luxury 271; maturation 54-7;
measurements 364-6; medicinal 107,
179; medieval market 175-80;
medieval merchants 189-92;
medieval prices 188; mixing 73;

Monarite 110; *passito* 95; physiological effects 93; political use 132; Portuguese regulations 265-6; presses 61-2, 103, 171; price 264, 275, 310, 321, 353; price fixing 27; production 287; quality 361; racking 188; reek 200; retail prices 198-9, 243; rice 159; Roman classification 111-12; Roman demand 123-31; Roman export 124; Roman merchants 116, 120; Roman prices 131; Roman supply 123-31; seventeenth century 266-9; seventeenth and eighteenth century production 233-69; shops 66, 75; social role 99, 359; specialisation 273; specialist merchants 341; stabilising 45; storage 306; supply 285; sweet 229, 248, 254, 257; sweet Mediterranean 185-6, 221; taxation 202, 242; transport 113, 277, 283; urban demand 124, 168, 269, 277, 307; vintage 129, 200, 254-6; white 172; wholesale prices 243; writers 352 wineries 346; boutique 346; cleanliness 107
Wolxheim 230
women 349
wool 180-1
Wyclif, J. 142

Xenophon 99

yeast 46-9, 54, 254, 346; cultured 24, 45, 53; fermentation capacity 47; natural 53; strains 54
yield 275, 287, 294, 308, 359-60
Yonne 118, 162
Young, A. 274

Zagreus 87
Zagros mountains 59, 64, 76, 80
Zeus 86-7
Zinfandel 303

Breinigsville, PA USA
13 December 2009
229154BV00002B/9/A

9 780415 144162